CLIMATE CHANGE ARCHAEOLOGY

Climate Change Archaeology

Building Resilience from Research in the World's Coastal Wetlands

ROBERT VAN DE NOORT

OXFORD
UNIVERSITY PRESS

OXFORD
UNIVERSITY PRESS

Great Clarendon Street, Oxford, OX2 6DP,
United Kingdom

Oxford University Press is a department of the University of Oxford.
It furthers the University's objective of excellence in research, scholarship,
and education by publishing worldwide. Oxford is a registered trade mark of
Oxford University Press in the UK and in certain other countries

First Edition published in 2013

Impression: 1

Published in the United States of America by Oxford University Press
198 Madison Avenue, New York, NY 10016, United States of America

British Library Cataloguing in Publication Data
Data available

Library of Congress Control Number: 2013941955

ISBN 978–0–19–969955–1

As printed and bound by
CPI Group (UK) Ltd, Croydon, CR0 4YY

Contents

Preface

That the Earth's climate is changing as a direct consequence of the burning of fossil fuels is beyond doubt, and this will present humanity with one of its biggest challenges in the twenty-first century. The study of the climate in the past forms the scientific basis of our understanding of climate change in the future. However, when it comes to understanding the impact of climate change on the environment, and how this will affect human communities, the past is mostly ignored by the climate change science community.

The aim of this book is to address this paradox and to reposition archaeological research within current climate change debates. This is realized through the application of an approach I have called 'climate change archaeology', which views the past as a repository of ideas and concepts that can help build the resilience of communities in a time of rapid climate change. The decision to focus the study on coastal wetlands was based, primarily, on the realization that these landscapes, and the communities that live here, will be among the first to be affected significantly by a climate change-induced rise in sea levels. Four case studies were selected—the North Sea, the Sundarbans, Florida's Gulf Coast, and the Iraqi Marshlands—for this comparative study. These case studies were chosen to provide diversity, not only in the type of coastal wetland but also in terms of past and present sociopolitical and economic circumstances.

To a significant extent, this book is the consequence of my work across different disciplines over the last decade. My various positions at the University of Exeter have required me to work with academics from many different disciplines, and this has made me much more aware of the benefits of cross-disciplinary research. Whilst my research focus remains firmly on archaeology and archaeological methods, I have sought to learn and incorporate many of the significant advances that have been made in other disciplines, and in particular those pertaining to the study of the Earth's climate.

I thank the Arts and Humanities Research Council for its support in funding the work undertaken for three of the case studies—the Sundarbans, Florida's Gulf Coast, and the Iraqi Marshlands (AH/J002984/1)—and the universities of South Florida and Exeter for additional financial support for the visit to Florida. A large number of individuals have helped me in different ways, and I want to acknowledge in particular Neil Adger, Bob Austin, Katherine Brown, Christopher Callow, Dilip Chakrabarti, Jess Collins, John Curtis, Victoria Donnolly, Gary Ellis, Rich Estabrook, Sean Goddard, Mavis Gutu, Sarah Hoyle, Jeff Moates, Jason Moser, Andrea Schoefbeck, Brent

Weisman, Nancy White, and the knowledgeable crew of *M.V. Paramhamsa* for their help and hospitality over the last two years.

Aspects of this research have been publicly presented and I want to acknowledge the comments, suggestions, and support received during the discussions following research seminars at the universities of Bristol, Durham, Edinburgh, Exeter, Groningen, South Florida, and York. The research was also presented at a number of conferences, and I have sought to recognize and address the constructive comments from the delegates of the 2011 Theoretical Archaeology Group conference at the University of Birmingham, and the European Association of Archaeologists annual conferences in 2010 in The Hague and 2012 in Helsinki. I would also like to thank AVIARE at the Weedon Island Preserve, St Petersburg, Florida, for their magnificent hospitality. Finally, I am grateful to OUP's Hilary O'Shea, Taryn Das Neves, and Kizzy Taylor-Richelieu, and to my editor Richard Mason, for their role in bringing this project to fruition.

Robert Van de Noort

List of Figures

List of Abbreviations

AOGCM	Atmosphere-Ocean General Circulation Model
ESL	Eustatic Sea Level
GCMs	General Circulation Models
GHGs	greenhouse gases
ICZM	Integrated Coastal Zone Management
IHOPE	Integrated History of People on Earth
IPCC	Intergovernmental Panel on Climate Change
LGM	Last Glacial Maximum (c. 21,000 years ago)
MIS	Marine (oxygen) Isotope Stage
NGO	non-governmental organization
ppb	parts per billion
ppm	parts per million
RSL	Relative Sea Level
SRES	Special Report on Emission Scenarios
UN	United Nations
UNEP	United Nations Environment Programme
UNFCCC	United Nations Framework Convention on Climate Change
UNWCED	United Nations World Commission on Environment and Development

Note to the Reader

Radiocarbon convention

All radiocarbon dates used in this book were calibrated according to the maximum intercept method (Stuiver and Reimer 1986), using OxCal 4.0 (Bronk Ramsey 2006), and with end points rounded outwards to 10 years (Mook 1986). Date ranges or mid-range dates obtained through radiocarbon dating include the term 'cal', for calibrated. Historical dates, and dates obtained through dendrochronology, are provided as calendar years.

Confidence and likelihood

The IPCC distinguishes confidence in scientific understanding and the likelihood of specific outcomes. Statements in scientific understanding have only been included in this text where the IPCC considers the confidence high (8 out of 10 change) or very high (at least a 9 out of 10 change), unless explicitly stated otherwise. Statements on the likelihood of specific outcomes have only been included in this text where this likelihood is considered by the IPCC to be very likely (>90% probability), extremely likely (>95% probability), or virtually certain (>99% probability), unless explicitly stated otherwise.

Place names and geographical names

For geographical names, the modern English version has been used where this is available and in common usage, notably for the names of countries, major rivers, and larger places. However, new names have been adopted where national governments have changed place names in recent decades, for example Kolkata instead of Calcutta. Names of places, structures, and objects derived from transliteration from Bengali (for the Sundarbans) or Arabic (for the Iraqi Marshlands) have followed their prevalent use in academic literature, although consistency is often lacking.

1

Introduction

> ... a radical change in the physical geography of the world must have powerful implications for the human geography—where people live, and how they live their lives *(Nicholas Stern 2006: iv).*

Climate change is one of the greatest challenges, if not the greatest, facing humanity in the twenty-first century. In the past, climate change was a natural process. But current global warming is, to a large degree, linked to the production of greenhouse gases (GHGs) from the burning of fossil fuels and its impact on radiative forcing, namely, the solar energy received by the Earth minus the energy reflected and radiated back into space. The scenarios for the rate of future climate change are dependent on our ability to reduce GHGs in the atmosphere, and much current research is focused on this. Nevertheless, even under the most optimistic of these scenarios, more heat from the sun will be trapped in the Earth's atmosphere in the twenty-first century than was the case in previous centuries, inevitably leading to higher average global temperatures. A direct consequence of this will be that the Eustatic Sea Level (ESL) will rise, because landlocked ice will melt and add water to the seas, and a warmer sea takes up more space due to thermal expansion. Current scenario planning for this suggests that by the end of the twenty-first century, Eustatic Sea Levels could be up to 0.59 m higher than they are today (Bernstein et al. 2007: 45).

The estimated 400 million people around the world living on land elevated less than 10 m above current sea levels will be among the first to experience the environmental impact of climate change. Coastal communities will need to adapt to this. Some may be able to fortify and heighten the embankments that have kept their possessions protected from the sea up till now and extend their ways of life, at least for the immediate future. Others may not be so fortunate, and will be left with no other option than to abandon their homes and migrate to higher and dryer land. The poorest coastal communities are likely to have the least resilience to sea-level change, as adaptation is often a costly matter (Stern 2006).

Climate change is not a phenomenon unique to the twenty-first century. Archaeologists have been studying past climate change and its effects on human societies for at least 150 years (e.g. Trigger 1996: 130–8; see chapter 2).

It might be thought that this research would have produced some significant insights into the ways past societies coped with climate change, which could in turn be utilized to inform modern communities. Unfortunately, studies of how societies adapted to climate change and its environmental impacts in the past have made no contribution to the debates on the ways in which humanity will need to adapt to climate change in the future. This is certainly the case in the reports of the Intergovernmental Panel on Climate Change (IPCC). But it can also be observed in many of the national assessments of climate change, such as the Stern Report produced for the UK (Stern 2006), and in much of the academic literature on adaption to climate change. This disregard exists not because archaeologists are unwilling to take part in the debates: indeed, there are many scholars who have produced long-term perspectives on past climate change and its impact on human societies (e.g. McIntosh, Tainter, and McIntosh 2000; Burroughs 2005; Fagan 2008). The main reason why this body of work has been largely ignored in current climate change debates is undoubtedly because the past appears to provide few, if any, 'lessons' for the future. There is a broad realization that 'the present situation is dire precisely because there is no clear precedent for *global* environmental mismanagement' (Yoffee 2010: 178), and it is undeniably true that we live now with a globalized 'fossil fuel energy system' that has no parallel in the past (McNeil 2010: 364). Both global environmental mismanagement and the globalized energy system have emerged as the direct consequence of a rapid increase in the world's population, especially in the developing world, and significantly increased consumption levels, concentrated in the developed world.

The crux of this book is that it presents archaeological research as a repository of adaptive pathways on the part of past human communities. From this understanding, ideas and concepts that help build the resilience of communities in the face of rapid climate change can be gathered. Adaptive pathways include changes in social, political, economic, technological, and ritual aspects of past societies; it is fully recognized that such changes also occur for reasons wholly unconnected to climate and environment, but any adaptation may contain important messages that can strengthen the resilience of communities in a period of rapid climate change and its environmental impacts.

The concept of resilience has evolved from its origins within ecology to encompass resilience of social institutions (see chapter 2). Socioecological resilience has been defined as the ability of communities to cope with the impacts of external environmental change, including climate-driven environmental and sea-level change (e.g. Berkes and Folke 1998; Adger et al. 2005). However, the contribution that the study of the past, including the distant past, has made towards strengthening socioecological resilience has not been recognized. As well as providing insights into long-term geological, geographical, and environmental processes, and occasionally technical solutions that have been forgotten, studying the past is of particular importance as the knowledge gained connects in a very direct way to people's sense of place and their self-reflected place in a world that needs to adapt to climate change. In other words, the past and long-term perspectives on adaptation to change can act as a source of inspiration for modern communities. Future adaptive pathways built on strong histories, which reflect communities' sense of place, are likely to be more acceptable to local communities and more environmentally sustainable in the long term.

Theoretically, this book is positioned not within the realm of debates that continue to consider the environmental determinist versus environmental relativist dichotomy, echoing the debates between functionalist, modernist/processual, and post-modernist/post-processual archaeologists, but rather in one where the natural and human aspects of the world are understood to be closely and dynamically interlinked, producing hybrid and interconnected narratives. The role of studies into resilience, historical ecology, and sustainability in bringing the interwoven nature of these relationships to the fore is acknowledged (see chapter 2). The recognition of the role of human agency in global warming has brought together environmental and social scientists in the now broad and interdisciplinary field of climate change studies. Interconnectedness is acknowledged too through the use of integrated systems theory, which aims to understand the interactions and feedback mechanisms between the Earth systems and the human systems; however, the more fully integrated concept of a 'peopled Earth', where individuals and communities can influence their adaptive pathways, is preferred (Fig. 1.1).

By looking at the past for inspiration, it may well be possible to build the socioecological resilience of local communities in a context that recognizes their sense of place, their heritage, and their past. In writing this book, I aspire to contribute to current climate change debates, and as such this book could be described as a form of 'action archaeology'. In *Archaeology Matters: Action Archaeology in the Modern World*, Jeremy Sabloff (2008: 48) points out that the contribution of archaeology comes in two forms: as 'general models of successful and unsuccessful trajectories of sustainable growth over long periods of time' and 'through specific on-the-ground research projects'. I have my reservations about the contribution of archaeology in the form of 'general models' or 'grand narratives', such as recent studies on the collapse of civilizations (notably Diamond 2005)—these effectively consider societies as indivisible units, and tend to favour single-causal explanations such as environmental change, leading to environmental deterministic narratives. Furthermore, reflecting on the very rapid growth of the world population in the last three centuries (Fig. 1.2), such generalizations from the past are simply not relevant to the modern overpopulated world.

However, I am in full agreement with Sabloff that on-the-ground projects have considerable potential to contribute to communities' socioecological resilience in the modern world. I recognize that the past is a repository of information that contains elements helpful to the building of resilience within

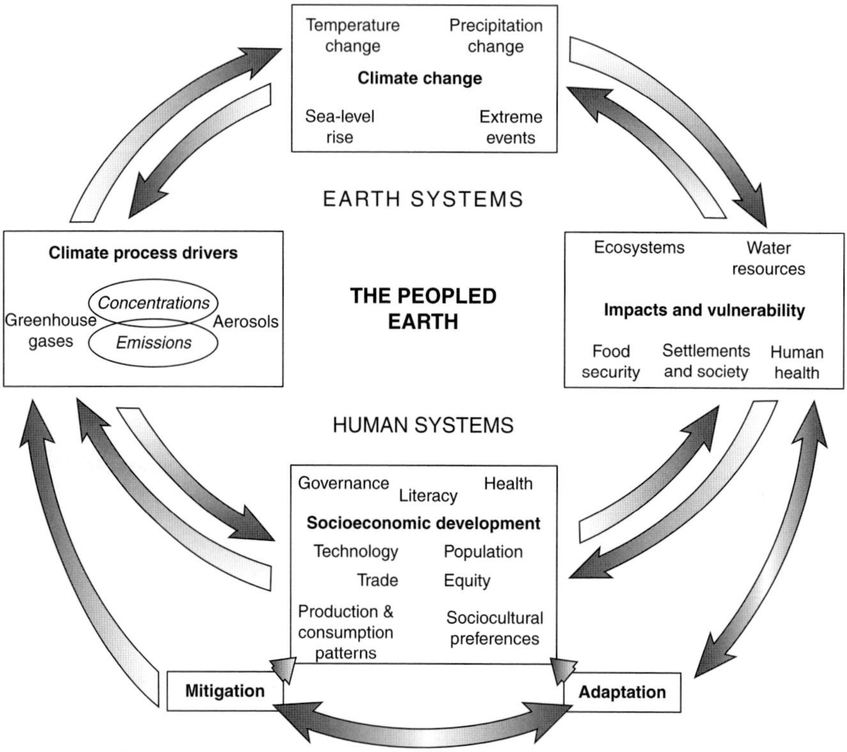

Fig. 1.1. The peopled Earth, showing aspects of the dynamic interaction between Earth systems and human systems pertaining to climate change (adapted from Bernstein et al. 2007: 26).

the context of a community's own history, thereby building upon existing social identity and sense of place; and that this is potentially a powerful device. Drawing lessons from the past should not purport that 'history is replete with replays' (Friedman and Chase-Dunn 2005: 2), but should constitute an informed practice of empowering local communities to deal with changes by learning from the past. Thus, the aim of this book is somewhat more modest than action archaeology (and maybe at the same time more ambitious); it recognizes that only certain elements of the adaptive pathways followed in the past can contribute to building the resilience of communities in the modern world.

This, then, is the essence of what I have called climate change archaeology (Van de Noort 2011a): to see the past as a repository of adaptive pathways, and to gather from this ideas and concepts that can help build the socioecological resilience of communities in the face of rapid climate change.

In this book, I focus on coastal wetlands. The reasons for this are fourfold. Firstly, the impact of global sea-level change will be prominent on coasts around the world; indeed, significant changes are already occurring. Secondly,

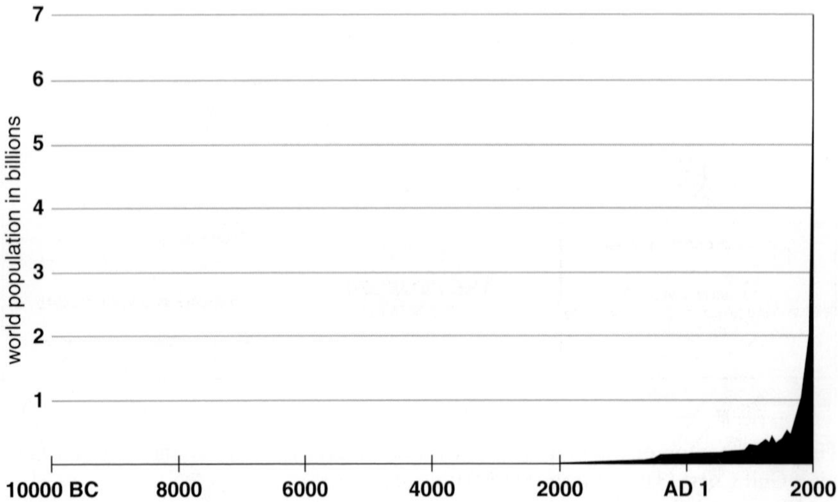

Fig. 1.2. Graph of population growth, 10,000 BC to present. The unprecedented acceleration of the world population since AD 1750 has created a global environment that has no parallels in the past.

coasts and the abundant resources of the sea have always attracted, and will continue to attract, people. Thirdly, many coastal wetlands have been subject to Holocene Eustatic Sea-Level rise since the onset of deglaciation of the West Antarctic Ice, which means that their palaeoenvironmental and archaeological study can reveal long-term perspectives on climate change and adaptation. Finally, wetlands are my specialist area.

This book will focus on four diverse coastal regions which, when combined, will deliver something of a global perspective: the North Sea basin; the Sundarbans in the Bay of Bengal; Florida's wetlands in the Gulf of Mexico; and the Al-Ahwar/Iraqi Marshlands in the Persian Gulf (Fig. 1.3). Past and present research in these coastal wetlands offers complementary perspectives on the adaptive pathways pertaining to each wetland at different times and in different ways. The re-evaluation of evidence for each of these case studies will answer three specific questions:

- How has climatic, environmental, and sea-level change (and neotectonics) during the Holocene shaped the coastal wetlands?
- What pathways were developed by coastal communities to adapt to these environmental changes and what results did these produce?
- What can we learn from these pathways to strengthen the resilience of current coastal communities?

In writing this book, I have been particularly motivated by the statement from the Stern Report (2006) quoted at the beginning of this chapter. Climate

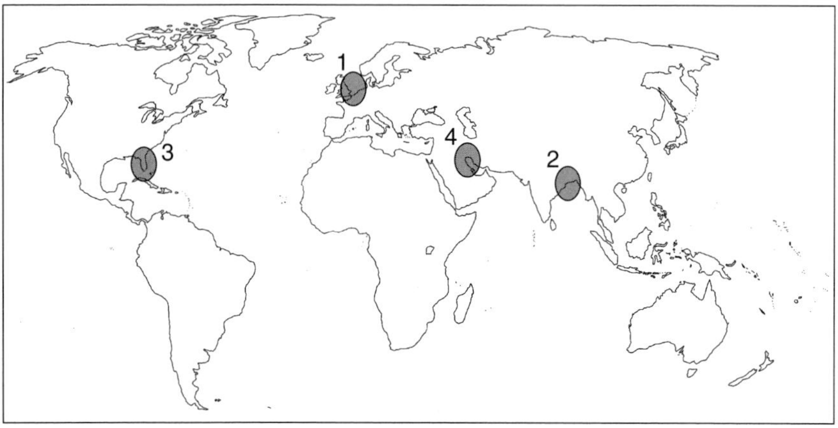

Fig. 1.3. Location of the four case-study areas: North Sea (1); Sundarbans (2); Florida's Gulf Coast (3); and Iraqi Marshlands (4).

change and the consequential transformation of the environment will affect our sense of place, the familiarity with the landscapes that gives communities their roots and their sense of belonging. Landscapes that have been socialized over the last millennia, and which have turned spaces into places, will be rapidly transformed. Global warming may change the familiar wildlife. New tree, plant, and animal species may become dominant, replacing the flora and fauna we have grown up with. Increased precipitation and more frequent extreme weather events may mean that built structures and building techniques that have served people well for centuries may no longer be suitable. To date, most climate change researchers have considered adaptation at 'system' level only, focusing on how humanity or nation states will need to adjust economically, politically, and technologically to the environment in a state of transformation. This book focuses on how this need for adaptation is experienced by local communities, and what archaeology and the broader study of the past can contribute. By looking at the past for inspiration, it may be possible to build up the social, ecological, and technological resilience of local communities in a context that recognizes their sense of place, their heritage, their past. It is in this way that climate change archaeology intends to make a distinct contribution to current climate change research.

CLIMATE CHANGE IN THE TWENTY-FIRST CENTURY

What is the evidence for climate change? This topic, and the research that provides this evidence, will be presented in greater detail in chapter 3, but it seems appropriate to summarize the main findings from the IPCC's *Fourth Assessment Report* here (Bernstein et al. 2007; Metz et al. 2007; Parry et al. 2007; Solomon et al. 2007). The IPCC report provides the most authoritative statement available to us at this point in time, and despite extensive criticism of the work of the IPCC since the publication of the *Fourth Assessment Report*, it is important to note that this was directed principally at the processes of the IPCC, rather than any of the main findings. Nevertheless, climate change research is ongoing, and every year new observations and research findings are produced; this emphasizes the point that climate change research is a relatively young discipline, and that current understandings are bound to alter over the decades to come. These new findings do not always corroborate the findings of the IPCC, but the overwhelming body of new research and observations adds to, rather than detracts from, the realization that the current global climate is warming at a rate not experienced since the end of the last glaciation c. 10,000 cal BC.

The IPCC's role is not to conduct primary research itself, but to bring together published studies on regional climate change in meta-analyses to produce an understanding of global climate change. Archaeological research on climate change in the past has contributed to these meta-analyses. The IPCC defines climate change as 'a change in the state of the climate that can be identified (e.g. using statistical tests) by changes in the mean and/or the variability of its properties, and that persists for an extended period, typically decades or longer. It refers to any change in climate over time, whether due to natural variability or as a result of human activity' (e.g. Bernstein et al. 2007: 30). The notion that human activity is a factor in global climate change is directly related to the burning of fossil fuels, which releases carbon dioxide (CO_2) and methane (CH_4) into the atmosphere, whilst agriculture adds nitrous oxide (N_2O). Deforestation, the decay of biomass, and the oxidation and burning of peatlands add further to the release of CO_2 in the atmosphere. These three long-living gases, together with halocarbons, are the principal GHGs, and basic physics tells us that higher concentrations of these GHGs trap more of the sun's radiation within the Earth's atmosphere (Arrhenius 1896). The IPCC's *Fourth Assessment Report* notes that the atmospheric concentrations of CO_2 and CH_4 in the first few years of the twenty-first century are higher than at any time in the last 650,000 years, whilst concentrations of N_2O are very likely unprecedented for the last 16,000 years (see chapter 3). The trapping of the sun's radiation within the Earth's atmosphere is termed radiative forcing, and increases in the radiative forcing produce higher global temperatures, or global warming. Alongside the concentration of GHGs, there are several other factors involved in changes in radiative forcing, such as the amount of snow coverage on Earth and its ability to reflect sunrays, and these feedback mechanisms will be discussed in chapter 3.

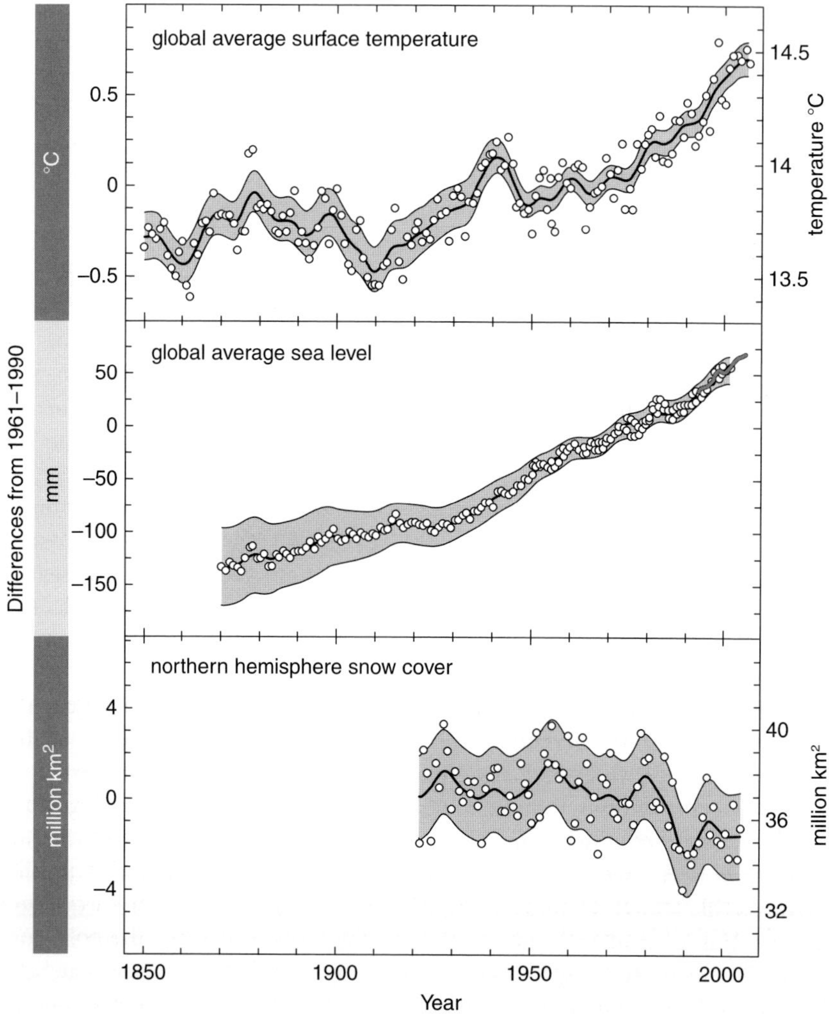

Fig. 1.4. Climate change since AD 1850. Observed changes in global average surface temperature (top); global average sea level from tide-gauge and satellite data (middle); and northern hemisphere snow cover for March–April (bottom). For full caption and source references, see IPCC 2007: 6.

There is no doubt that the climate on Earth is currently warming. This conclusion is based on observations rather than analyses of proxies, and includes records of the temperatures of the atmosphere and the oceans, the rising Eustatic Sea Level, and the retreat of snow and ice as observed from satellite images and aerial photographs (Fig. 1.4). These observations not only show that the global climate is warming, but that the warming trend is accelerating: whereas the global warming trend for the ten decades between 1906 and 2005 was on average 0.074 °C per decade, the figure for the last five decades alone is nearly double, with an average of 0.13 °C per decade. The IPCC concludes that for the northern hemisphere, the temperatures for the second half of the twentieth century were very likely to have been higher than for any other period of 50 years in the last 500; and it is likely that in the last 1,300 years, there were no warmer spells when measured for 50-year periods. The report concludes that there is very high confidence that the warming of the global climate is, at least in part, caused by human activities since c. AD 1750 (Bernstein et al. 2007: 37).

Looking forward, the *Fourth Assessment Report* develops a number of emission scenarios, each with a different storyline of population growth, economic development, and technological innovation and change (see chapter 3). These emission scenarios use the concept of the CO_2-equivalent, or the CO_2-eq, in which the radiative forcing of the other GHGs has been converted into their equivalence for CO_2. The varying emission scenarios indicate a continued global temperature rise in the range of 1.4 to 4 °C by the end of the twenty-first century. Whereas the IPCC report does not attribute likelihood to any of the storylines that underpin the emission scenarios, the lack of progress in cutting global GHG emissions suggests that the likely outcome is towards the upper end of the modelled range of global warming.

So, what environmental impacts can be expected from climate change in the twenty-first century? Obviously, at the global level this is dependent on the actual change in temperature, and we are told to expect significant variations in the global predictions at regional and local levels, so much so that degrees of likelihood and uncertainty cannot be attributed to the impact of climate change on these lower levels. Nevertheless, all emission scenarios predict rising global temperatures, and the Atmosphere-Ocean General Circulation Model (AOGCM) projections for all these scenarios show that the poles and adjacent areas of high latitude will experience greater increase in average temperatures than the tropics and bordering low-latitude regions. This, evidently, will mean that landlocked ice and snow will melt more quickly than is implied by the (averaged) global temperature rise, and this will add water to the world's oceans. Similarly, year-average precipitation is expected to increase at high latitudes, but decrease at middle and semi-arid low latitudes, with as a general rule winters becoming wetter and summers dryer.

It is expected that extreme weather events will occur more frequently. It has been observed that some extreme weather events, such as cyclonic activity, very heavy rainfall, heatwaves, and incidences of very high sea levels have become more common in the second half of the twentieth century and, whilst this is not a sufficiently long time frame to quantify the change in extreme weather events, the prediction is that the twenty-first century will experience more of these.

Together with the rise of the average global temperatures, these extreme weather events will have a number of significant impacts on the environment and the ecosystem services that are important for people. For example, the IPCC report projects that many millions of people will find it increasingly difficult to attain the fresh water they need. Projected changes in precipitation suggest that even less water will be available for people currently living in arid and semi-arid regions and, elsewhere, water use reliant on snowmelt will become less predictable. Many ecosystems around the world will change, and species shift is already a much-noted phenomenon. It is also likely that many more species will go extinct. Wildfires will become more widespread. Many corals will bleach and die. These changes in the ecosystems will have direct consequences for the crops that are grown, and what people will eat. In terms of people's health, challenges include the need to deal with the heatwaves and disease vectors migrating into regions previously unaffected.

An important message from the IPCC report is that societies can adapt to climate change and its environmental impacts, and that adaptation reduces societies' vulnerability, or increases their resilience. Societies can also choose to reduce the emissions of GHGs, and thus try to mitigate climate change. The adaptive capacity of societies is closely linked to social, economic, and political factors, and some nations will be better placed to adapt to climate change than others. In some situations, adaptation is a matter of money, but in others it is limited by the degree of political will to change existing practices, and the willingness or otherwise to cooperate. The report notes that early or proactive adaptation to climate change is likely to be more effective than late and reactive adaptation, and this notion has been echoed by national assessments as well (e.g. Stern 2006). The IPCC report focuses on strategies and policies, as only international cooperation can achieve the reduction of GHG emission, but this means that the impact of climate change and adaptation to it by communities is not considered. However, in very recent years, the non-economic aspects of adapting to climate change have attained greater significance, along with a recognition that individuals and communities can and must play an active role in how societies adapt to climate change and its environmental and economic impacts.

CLIMATE CHANGE IMPACT ON COASTAL WETLANDS

The IPCC report, in particular the 'Paleoclimate' chapter (Jansen et al. 2007), presents an up-to-date knowledge base for understanding the factors determining sea levels over the last 125,000 years, and this provides the information for this section. Climate and sea-level change is already affecting coastal wetlands around the world. What the predicted impact of the future scenarios of climate change will mean for coastal wetlands is detailed in chapter 4, but the topic warrants a general introduction in this opening chapter.

Scientists have developed a range of concepts to quantify the rise of the globe's sea levels over the long term, and for the sake of simplicity these can be referred to in this context as Eustatic Sea Level and Relative Sea Level. The Eustatic Sea Level represents the 'Ice-equivalent Eustatic Sea Level', which is an expression of the contribution of the melting of landlocked ice to sea-level rise and the thermal expansion of the sea, or its reverse effect, at a global level. The Relative Sea Level takes the Ice-equivalent Eustatic Sea Level but sets this against the so-called glacio-isostatic adjustment, which concerns the movements of the Earth's crust in response to the presence or absence of large ice sheets (including the postglacial 'rebound' effect) and the redistribution of water in the ocean basins. As these glacio-isostatic processes vary by region, Relative Sea Level is region dependent. When the Relative Sea Level is measured near the tropics, where there was no measurable glacio-isostatic adjustment, it provides the best equation to the Eustatic Sea Level.

The current geological epoch, the Holocene, is in effect an interglacial stage (or technically, Marine Isotope Stage MIS 1; see chapter 3), and sea levels have been rising throughout this period. Sea level was at its lowest point during the coldest part of the last glacial period, at the Last Glacial Maximum (LGM) around 21,000 years ago, when the Eustatic Sea Level was some 120 m below its current levels. During the first few millennia following the LGM, sea levels rose slowly, but this accelerated significantly after 13,000 cal BC, when the sea levels rose at an average by 10 mm per year, or 1 m during every century, but slowed down towards 7000 cal BC. Subsequently, the rate of sea-level rise decelerated further, with an average sea-level rise of 0.5 mm/year for the last six millennia, and of 0.1–0.2 mm/year for the last 3,000 years (Fig. 1.5). There is no argument that the melting of the polar ice sheets was the main contributor to this rise in the sea level (Church et al. 2001). There remains plenty of ice and snow to increase the sea levels further: the complete melting of the Greenland Ice Sheet alone would add 7 m to the Eustatic Sea Level (Jansen et al. 2007: 457).

The environmental and societal impact of sea-level change has been far-reaching. During the postglacial and early Holocene periods, very extensive and inhabited lowland plains all around the world were submerged, forcing people to find new places to live. Sea-level rise also resulted in the submergence of important land-bridges, separating previously connected landmasses and impeding opportunities for migration to new lands. In the last seven millennia, Eustatic sea-level rise had sufficiently slowed down for coastal wetlands to expand significantly, as the sediments brought down by rivers to their estuaries counterbalanced the decelerating sea-level rise. Mangrove swamps in the tropics and at low latitudes, and salt marshes in regions in mid and high latitudes, expanded and provided new ecosystems for people to explore.

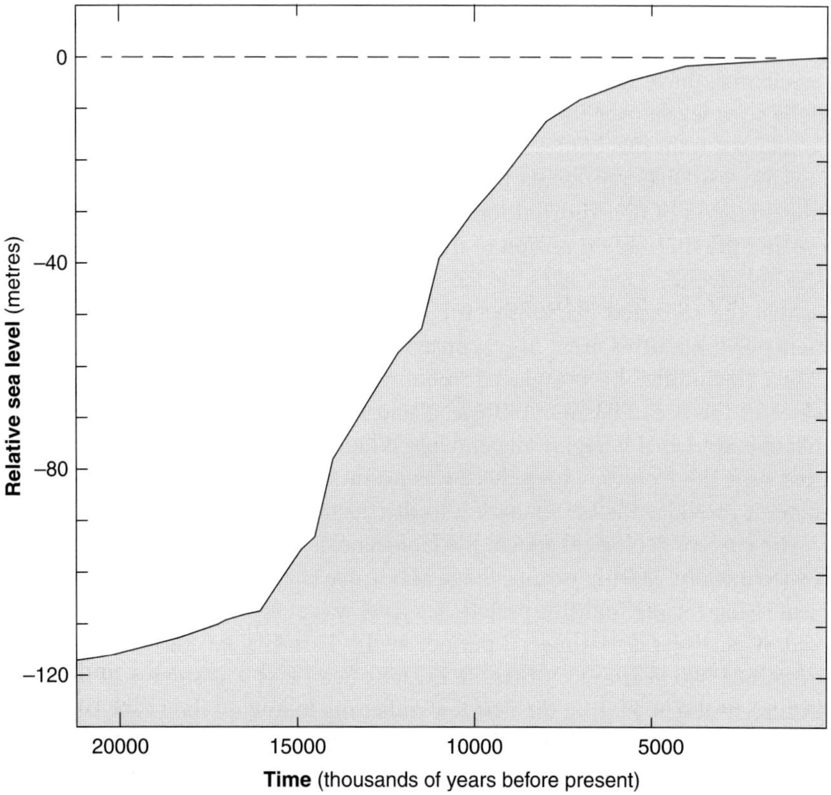

Fig. 1.5. Relative Sea Level reconstruction since the Last Glacial Maximum, based on dated reef corals in tropical regions.

The most recent observations have produced new concerns that the deceleration of the sea levels in the last millennia is reversing. Global mean sea level rose by 1.8 mm/year between 1961 and 2003, a significantly faster rate than the 0.1–0.2 mm/year that have characterized the last three millennia. The most recent data are the most alarming. Satellite altimetry observations between 1993 and 2003 have indicated that the global mean sea level rose around 3.1 mm/year, observations confirmed by the tide-gauge record. However, the IPCC report notes that decadal variations, as represented by the satellite altimetry data, do not always reflect longer-term changes (IPCC 2000: 387).

It has also been observed that the world's oceans are warming. Whilst this has a cooling effect on the globe's atmosphere, warming of the oceans contributes to their thermal expansion (the increased space needed by warmer water), and therefore contributes to sea-level rise. For the period since 1993, it is reckoned that the contribution of the melting of landlocked ice to sea-level change is around 1.2 mm/year, and that of thermal expansion 1.6 mm/year, together giving 2.8 mm/year with an error range of 0.7 mm/year, falling well within the rise of global mean sea level observed through satellite altimetry.

Global warming is likely to have a number of other impacts on the world's oceans, including changes in the salinity and possibly in the main oceanic circulations, such as the Gulf Stream. The time-depth of observations on these marine components is insufficient, and our understanding of their nature not well enough developed, to make any statements about whether and how increases in GHGs and global warming in the twentieth century are causing them to change.

The IPCC has looked back at the previous interglacial period, some 125,000 years ago, to understand what it would mean to sea levels if the global temperatures rose to levels comparable with the last interglacial maximum. Earlier studies of coastal sediments and corals had already shown that the Eustatic Sea Level was up to 4 to 6 m higher during the Marine Isotope Stage MIS 5 than it is today, and climate simulation models suggest that this will be achieved again with arctic summers being between 2 and 5 °C warmer than is the case today—not far off the predictions for global warming by the end of the twenty-first century.

Coastal ecosystems are, by their very nature, shaped by the interrelationship between terrestrial and marine systems (Masselink and Hughes 2003). Sea-level rises, changes in temperature and salinity, as well as the underlying geology and availability of marine and alluvial sediments, are determinants in the flora, fauna, and geomorphology of coastal wetlands. Higher sea levels, and the impact of more frequent extreme weather events on coasts, such as hurricanes, are likely to be among the first undeniable impacts of climate change on our environment (Nicholls, Hoozemans, and Marchand 1999; Adger et al. 2005).

Alongside sea-level change, the warming of the world's oceans, changes in salinity and possibly of the main ocean circulations, the IPCC warns us that in the twenty-first century we will have to learn to cope with increased coastal erosion, more extensive coastal inundation and higher storm-surge flooding, and the landward intrusion of seawater into estuaries and aquifers. In terms of the biological impacts, submergence and increased salinity in freshwater or brackish ecosystems will inhibit primary production processes, affecting both the flora and fauna of coastal wetlands. Coastal wetlands, including estuaries, lagoons, low-lying deltaic coasts, and 'soft rock' and intertidal coasts, are considered by the IPCC to be the ecosystems with greatest risk exposure to the environmental impacts of climate change. It has been estimated that up to 22 per cent of the world's coastal wetlands will be lost by 2080 as a consequence of sea-level change (Nicholls, Hoozemans, and Marchand 1999). This figure includes those areas lost as a result of societies responding to sea-level change through the construction of higher embankments; the existence of such embankments results in 'coastal squeeze', whereby coastal wetlands are

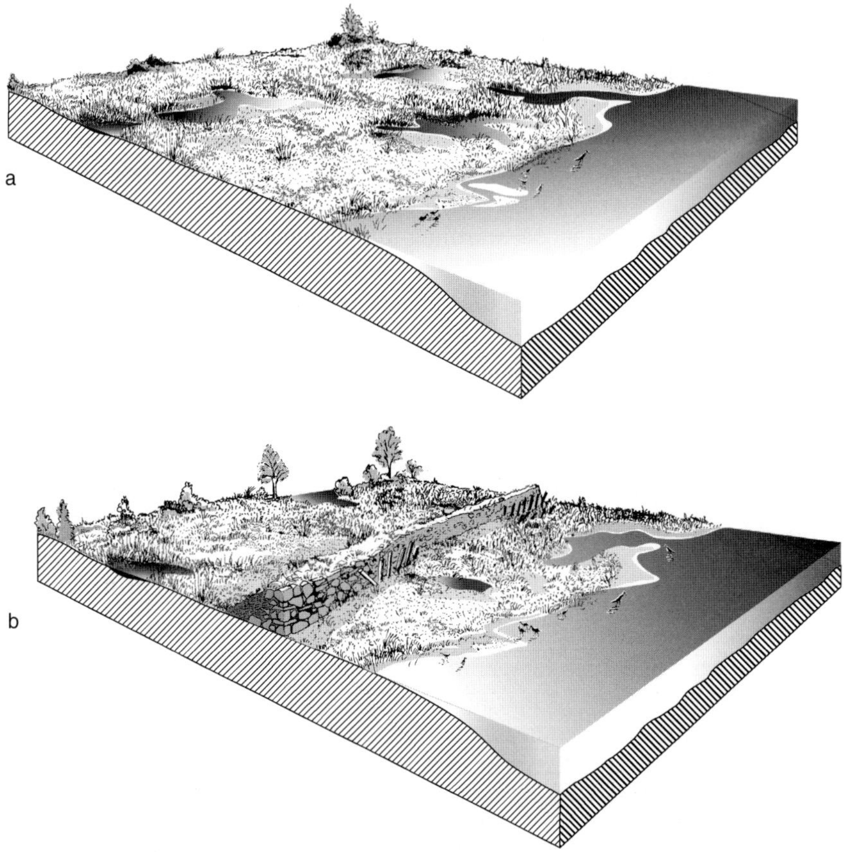

Fig. 1.6. Coastal squeeze: unconfined wetlands respond to sea-level rise by migrating landwards (a), but where seawalls or dikes prevent this from happening, coastal wetlands are squeezed (b).

increasingly squeezed between the man-made coastal defences and the rising sea level (Fig. 1.6; Nicholls and Branson 1998).

Coastal communities will be affected by climate change in several ways. Sea-level rise will threaten their settlements, and the need to construct embankments may require capital and labour that are not always available or affordable. Fish and shellfish populations are likely to change, and saltwater incursion may prove devastating for agricultural activities on the coastal plain. Long-serving ports may silt up, rivers will change course, and established economic systems will need to adjust. In all these ways, the past offers plentiful parallels to the future.

Whilst heritage protection is not a key theme in this book, archaeologists should be aware that climate change will adversely impact on 'living' coastal wetlands, with consequences for the stability of the underlying ancient wetlands that contain the source materials for archaeological and palaeoenvironmental research. Add to this the impact of the anticipated increase in human activity in the form of construction of embankments and coastal barriers, and the likely loss of the essence and integrity of the historical landscape of coastal areas, and it becomes clear that climate change might well have a direct impact on the sources of our livelihoods as well.

ORGANIZATION OF THE BOOK

As argued in the introduction to this chapter, this book seeks to develop ways in which archaeology can become not only relevant to current climate change debates but also add real value to them. Consequently, this book addresses three audiences: the archaeologists who will need to think differently about the contribution their research can make to current climate change debates; the climate change scientists (geographers, environmental scientists, anthropologists) and policymakers who are to benefit from valuing the past and research of the past; and the local communities living on the coasts of the case studies presented here, whose resilience in dealing with climate change is to be strengthened.

The next three chapters aim to provide these three audiences with the contexts that provide the background to this book, and offer the reader an up-to-date understanding of climate change and how it will affect coastal wetlands. In chapter 2, the concept of climate change archaeology is presented in some detail. This includes a short history of the way archaeologists have understood climate change over the last 150 years, and considers how labels such as 'environmental determinism' and 'environmental relativism' have prevented them from developing holistic understandings of the relationships between people and their world that can be appreciated by climate change researchers. The chapter looks at the evolving interdisciplinary study of historical ecology for inspiration, and also explains the concepts of resilience and sustainability.

Chapter 3 concerns climate change in the past, present, and future, presenting a long-term view of climate change, and the principles of palaeoclimate research. This chapter seeks to stress two points that are important for understanding the relationship between the study of past climates by archaeologists and that undertaken by the climate change researchers. These are, first, the relationship between global climate change and climate fluctuations at regional levels and, second, the differentiation between natural and anthropogenic causes of climate change.

Chapter 4 considers how climate change will affect coastal wetlands. It introduces the biogeographical principles and concepts that underpin the relationships between climate change, sea-level change, and coastal evolution. Climate change is the key driver in Eustatic Sea Levels, and sea levels are the most important determinant in the formation of coastal wetlands. Rivers, and especially the sediment outputs from these, also play an important role in coastal evolution, and the impacts of climate change on sediment flows are generally well understood. This chapter will also consider future scenarios for the impact of climate and sea-level change on the coastal communities around the world.

In the next four chapters, comparative case studies will illustrate how climate change archaeology can be made to work. Each of these will consider the influence of climate change on coastal communities and provide examples of adaptation to its environmental impacts. The role of human agency in producing diverse adaptive pathways will be illuminated. In each of the cases considered, Holocene sea-level rise resulted in the creation and loss of coastal landscapes that had been socialized by previous generations, and the manner in which people dealt with this is explored. Adaptation to sea-level change inevitably produces certain feedback mechanisms, and these will be unravelled. The case studies have been selected to present very different coastal environments, with diverse cultural histories, in order to show that adaptive pathways to changes in sea level are culturally specific. These case studies concern the North Sea basin (chapter 5), the Sundarbans in the Bay of Bengal (chapter 6), the wetlands of Florida in the Gulf of Mexico (chapter 7), and the Iraqi Marshlands (chapter 8).

Finally, the conclusions of this study are presented in chapter 9. This chapter will draw out the main themes from the adaptive pathways followed by communities in the past in dealing with climate and environmental change, and will explore how these can contribute to building the socioecological resilience for twenty-first century coastal communities.

In an earlier paper (Van de Noort 2011a), I have argued that the IPCC report presents something of a paradox. Whereas the IPCC provides long-term retrospective contexts for modern and projected climate change, and has been a catalyst in the development of rigorous, global palaeoclimate research, it would seem that when it comes to the subject of adaptation to climate change the past is nearly invisible. As argued in the previous section, I do not suggest that, at a general level, historical antecedents of how societies adapted to (natural) climate change will be of much use in the twenty-first century. What I aim for in this book is to show that the past offers a range of insights into long-term natural processes, human behaviour and values, as well as examples of a more sustainable interaction with the environment. These insights and examples can play a positive role in the building of people's resilience, but so far their potential has been ignored. In this context, we need to acknowledge that many attempts to respond to the effects of climate change are opposed by local communities, with local opposition to the construction of wind farms providing an instructive example (Wolsink 2007). Climate change science has, in such instances, failed to appreciate the importance of the landscape and the sense of place among the local communities affected. In other words, this book is a first step towards addressing the paradox presented in the IPCC report.

2

Climate Change Archaeology: Background, Building Blocks, and Concepts

INTRODUCTION

The aim of this chapter is to provide the background to, and explain the concept of, climate change archaeology. The recent paper, 'Conceptualising climate change archaeology' (Van de Noort 2011a), was very much intended as a 'rallying call' for the archaeological community. In essence, it makes the point that we as archaeologists need to think differently about the interrelationship between societies and their environment in the past if we want our research to play a part in climate change debates. This chapter restates that argument, and in addition seeks to inform both climate change researchers and local communities about the potential benefits of developing long-term perspectives on adaptation to climate change.

As already argued in the introductory chapter, the essence of climate change archaeology is that it sees the past as a repository of adaptive strategies and concepts that can help build the resilience of modern communities in a time of rapid climate change. This position echoes, to some extent, the idea that social memories were and are reservoirs of information from which appropriate adaptive strategies can be selected as and when required, as argued by Stephen Shennan (2002: 80). Climate change archaeology does not suggest that the past offers direct parallels to present and future situations, nor that high-level 'lessons from the past' can be expected to make a significant difference to the way we adapt to climate change. Instead, climate change archaeology is somewhat more restricted in its scope, in that it recognizes that only certain elements of the adaptive pathways followed in the past can help in building up the resilience of modern communities. These elements can include long-term perspectives on natural processes, and occasionally ecological or technical solutions that are no longer part of the social memory of communities. However, the more important elements are social in nature, and are linked directly to people's sense of place and identity in a world that needs to adapt. The past, and long-term perspectives on adaptation to change, can act as a source of inspiration for modern societies. Recognizing that future adaptive pathways can be built on those followed by predecessors who have co-constructed the modern landscape is a potentially powerful source of inspiration.

The first part of this chapter concerns the relationship between archaeology and climate change in the nineteenth, twentieth, and early years of the twenty-first centuries. To a significant extent based on Bruce Trigger's (2006) *A History of Archaeological Thought*, it provides a short history of archaeology and its approach to climate change. It will show how different theoretical perspectives have favoured alternative understandings of how climate and environmental change shaped human history, and what this has meant for the way archaeologists deal with modern climate change. The second part concerns concepts of historical ecology, resilience, and sustainability: these concepts, familiar to many climate change researchers but less so to archaeologists, form the basis of the third part of this chapter, which examines the concept of climate change archaeology itself.

BACKGROUND: ARCHAEOLOGY AND CLIMATE CHANGE

The earliest research into the connections between cultural and environmental change, and the origins of post-antiquarian scientific archaeology, date back to the nineteenth century. The first to link cultural evolution with changes in climate and the environment was the geologist Japetus Steenstrup, whose work in the peatlands in Denmark in the first half of the nineteenth century revealed the succession of postglacial woodlands from aspen, through pine, to oak, and to beech and elm. By the 1840s, Steenstrup had correlated this woodland succession with Christian Thomsen's Three Age System, noting that pine forests were associated with the Stone Age, oak forests with the Bronze Age, and beech and elm woodland with the Iron Age (Morlot 1861: 309–10; cf. Rowley-Conwy 2007). It did not take long for this research to be extended to coastal wetlands. Following the identification of the shell middens, or *køkkenmøddinge*, on Sjaelland and in northern Jutland, an interdisciplinary research commission was established by the Royal Danish Academy of Sciences in 1848. The research linked the consumption of the shellfish and hunted animals whose remains made up the middens with their palaeoenvironments and with sea-level change (Trigger 2006: 132–3; 315).

The findings were presented in *Danmarks Oldtid* (1843; cf. Trigger 2006: 131–2) by Jens Worsaae, a member of the interdisciplinary research commission and also the first professional archaeologist. An English translation was published in 1849. The integration of cultural and environmental research became the hallmark of Scandinavian archaeology from this point onward. In the following decades, significant progress was made in the study of past climate change, in particular in relation to the end of the last Ice Age and its impact on the environment, including sea levels. This included, for example, the pioneering work of the palynologist Lennart von Post in developing pollen diagrams that included non-arboreal species alongside the tree pollen, and which could show phases of forest clearing and be used for the reconstruction of the Earth's climate (1946). The study by Gerard de Geer (1912) of the varve deposits, which formed every spring in front of the retreating glaciers, provided an absolute chronology with calendrical accuracy, against which the rate of cultural and environmental change could be measured.

Despite Worsaae's publication of *Danmarks Oldtid* and its translation into English, and the work of his colleagues, the integration of cultural and environmental research outside Scandinavia happened only after significant delays, due to the predominance of the culture-historical outlook on the past in Europe (Trigger 2006: 314). The environment, however, was not completely ignored by archaeologists outside Scandinavia. Examples of how certain cultural phenomena were linked with environmental conditions included the correlation of the origins of agriculture in the Middle East with the end of the last Ice Age (Pumpelly 1908), the correlation of the settlements of the first LBK (from German *Linearbandkeramik*, or 'Linear Band Ceramic') farmers in Europe with the distribution of the light and easily worked wind-blown loess soils (Gradmann 1906), and the increased use of the heavier clay soils in Britain in later prehistory (Fox 1932). However, none of these studies sought to explain cultural change as a consequence of environmental change, and diffusion and migration retained their importance in this respect.

Trigger (2006: 315–19) notes that the increasing interest in how settlements related to their environmental setting, and the growing realization of the limitations of diffusion and migration for explaining the diversity presented by the archaeological record, sowed the seeds for early functional and processual archaeology. For example, V. Gordon Childe's central tenet in *Man Makes Himself* (1936) is one where 'the scientific knowledge accumulated by human beings gave progressive societies ever greater control over nature and led to the formation of new and more complex socio-political societies', a position very different from his earlier culture-historical preoccupation with diffusion and migration (Trigger 2006: 346). At about the same time, American archaeologists such as Julian Steward called for an ecological context for understanding cultural change (Steward 1937; Steward and Setzler 1938). Thus, when Grahame Clark adopted a more explicit ecological perspective in the 1940s, this had its foundations in the preceding decade.

Nevertheless, the work by Clark heralded significant alterations in the way British archaeologists sought to explain cultural change. In the case of the excavation at Star Carr in North Yorkshire from 1949 to 1951, the connections between the rapidly warming postglacial environment and the economic basis of the Early Mesolithic community were evident (Clark 1954). In 1953, Clark presented a chart that fully integrated cultural (e.g. subsistence, technology, social organization, and trade, but also religion, art, and magic) and environmental (habitat and biome) components into a single system (Fig. 2.1; cf. Trigger 2006: 354–6). Increasingly, functional archaeologists in Britain and northern America sought to emphasize the impacts of climate and environmental change on human behaviour. A number of major works clearly reflected the understanding of archaeology as a discipline that places the study of the human past within the context of environmental change (e.g. Clark 1936, 1961; Willey 1953; Childe 1958; Adams 1965; Butzer 1972; Higgs 1972). This would soon lead to accusations of 'environmental determinism', that is, explaining cultural change exclusively by the environmental pressures on prehistoric communities. However, the impact of the functionalist way of working, which integrates archaeological excavation with an analysis of the palaeoenvironment and the palaeoeconomy, for example through detailed analyses of animal bone assemblages, has had a lasting influence on the discipline in both Europe and northern America. Environmental archaeology was firmly established as a sub-discipline in the 1960s, and one of its most

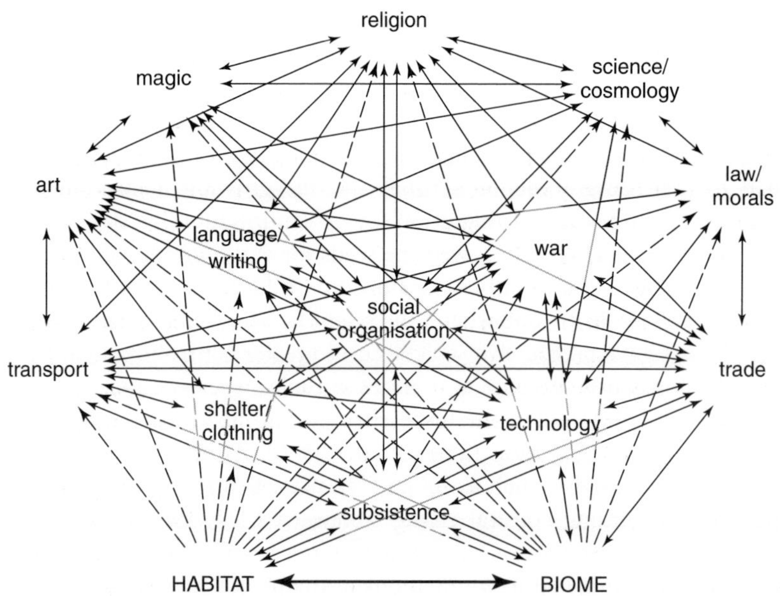

Fig. 2.1. Grahame Clark's integrated human-environment system, first used in the 1953 Reckitt Lecture.

prominent advocates, Geoffrey Dimbleby, was appointed Chair of Human Environment at London University's Institute of Archaeology in 1964.

The New Archaeology or processual archaeology sought to break with the functionalist tradition that singled out environmental determinism as the key cause of cultural change, but the adoption of models of cultural evolution did not exclude climate and environmental change as important aspects of prehistoric life (Trigger 2006: 302–444). Lewis Binford, the most prominent advocate of the New Archaeology, understood the concept of cultural evolution as the cultural adaptation to climate and environmental change (1972: 106); thus a clear role for climate change was retained. Trigger (2006: 395) has argued that Binford assumed 'that prehistoric groups had possessed a nearly perfect knowledge of their environment and therefore were able to calculate the most rational ecological response to any problem', and that Binford believed differences in culture merely reflected the diverse ways in which societies had adapted to these environmental changes. Such concepts were operationalized in Claudio Vita-Finzi and Eric Higgs's (1970) Site Catchment Analysis, where the economic basis of sites was defined by particular patterns of land use within a radius of 5 km, for agriculturalists, or 10 km for hunter-gatherers. Thiessen polygons (which subdivide the landscape between sites of equal status) were also applied, for example to determine the hinterlands of Roman towns (Hodder and Orton 1976).

Whilst Binford maintained this 'ecological' outlook for much of his life, by the end of the 1970s this view of the relationship between societies and their environments had become the exception in processual archaeology (Trigger 2006: 440; 462). Processual archaeologists had by that time adopted alternative perspectives on the causes of cultural change, with an important focus on the cultural evolution of societies, embracing aspects of the work of anthropologists E. R. Service (1962; 1975) and M. D. Sahlins (1968); the latter provided a structure for explaining increasing complexity in his 'band—tribe—chiefdom—state' framework. Systems theory provided an alternative way of describing how societies operated, especially in terms of subsistence strategies, and this theory was favoured by both Kent Flannery (1968) and David Clarke (1972). Alongside ecological factors, such as climate change, the role of sociopolitical and cultural factors in bringing about change—including the impact societies can have on each other—was widely acknowledged from the mid-1970s onwards.

Processual archaeology received progressively more vociferous critiques mainly from English archaeologists who reintroduced the concept of culture to explain the diversity of the beliefs and actions of people in the past (Robb 1998). In 1985 this was labelled 'postprocessual archaeology' (Hodder 1986). The focus of postprocessual archaeology on non-economic aspects of past societies led proponents of this theoretical perspective to turn their backs on the environment (e.g. Hodder 1986, 1990; Thomas 1991; cf. Trigger 2006: 444–83). Whilst not seeking to deny (past) climate change, the interest during the 1980s in both the symbolic and active role of material culture, and the key role attributed in the 1990s and in the first decade of the twenty-first century to human agency in its many different forms as the principal means by which societal change was achieved, rendered the role of climate and environmental change effectively invisible (e.g. Hodder 1990; Bender 1993; Tilley 1994). The influence of the environment was acknowledged mainly in the context of landscape archaeology, where phenomenological approaches had been adopted (Tilley 1994); however, the focus here was on the landscape as a stage for social action, rather than as a factor shaping the economic organization of societies. It is unsurprising, then, that postprocessual archaeologists have been accused of 'environmental relativism'.

Environmental archaeologists continued their palaeoenvironmental and palaeoeconomic work during the last decades of the twentieth century, and a more anthropocentric approach has slowly emerged (e.g. Dincauze 2000). In *Environmental Archaeology and the Social Order*, John G. Evans (2003) offered a comprehensive postprocessual revision of this sub-field. Evans, a renowned environmental archaeologist himself and an expert on the archaeological study of land snails, argued that environments in the past were actively used and manipulated for the purpose of establishing, maintaining, and reproducing the social order. He noted that the 'old' environmental archaeology, where the response to the environment by past societies is seen in terms of producing food or gaining shelter, should give way to a rethought environmental archaeology that focuses on the role of the environment in the development of socialities (p. 19). In tune with this, on the matter of climate and climate change, Evans concluded that 'climate is a human construct which is used as a means of social engagement . . . climate is constituted in its relevance to human action . . . [and] society engages with climate as a means of establishing community and individual identities . . . as in the prediction of climate change' (pp. 118–19).

In his *A History of Archaeological Thought*, Trigger notes that the development of theory in the discipline is not unilinear, nor has this development a directed trajectory or a defined end point (2006: 532–80). This is also true for the archaeological study of climate and environmental change. Against the backdrop of the different theoretical perspectives presented here—which may leave the reader with the impression that, by the first decade of the twenty-first century, environmental relativism had been adopted by all archaeologists, including environmental archaeologists—a significant body of work on the impact of climate and environmental change on past societies continues to be produced. Some of these studies focus on the temporal correlation of significant changes in climate with changes in where and how societies lived, and provide compelling examples of the impact, both negative and positive, of the environment and environmental change on human societies and communities (e.g. Van Geel, Burrman, and Waterbolk 1996; Binford et al. 1997;

Yu et al. 2000; Turney et al. 2006; Fagan 2008; Müller et al. 2011). A particularly significant contribution is delivered by the 1995 'Global change in history and prehistory' conference proceedings, published as *The Way the Wind Blows: Climate, History, and Human Action* (McIntosh, Tainter, and McIntosh, eds., 2000). The two key themes contained within this book are, firstly, the concept that research into the past, including the social memories of current communities as well as the archaeological and historical past, can improve our understanding of climate change; and, secondly, that adaptation to climate change in the past has implications for sustainability in the present.

In recent years, archaeology has experienced a growth in the number of studies focusing on the collapse of societies, and this has direct relevance for the study of climate and environmental change in archaeology in the twenty-first century. Researchers of ancient societies have probably always known that failure and collapse befall even the greatest nations, and Edward Gibbon's *The History of the Decline and Fall of the Roman Empire* (1776–88) offers an early example of this. Whereas archaeologists have noted and studied the finite nature of many ancient societies, the systematic study of failing societies is, surprisingly, of a much more recent date. A phase of 'catastrophe' studies dates to the 1970s and 1980s (Trigger 2006; 412–13). Joseph Tainter offered the first cross-cultural study in *The Collapse of Complex Societies* (1988). In this study he argued that when societies become increasingly complex over time, because of the need to resolve new problems, the demands for ever greater investment into the society's systems will lead eventually to unsustainable requirements of the people within that society, and its inevitable collapse. Several recent studies into the collapse of ancient societies have adopted a more explicit 'neo-environmental deterministic' approach, whereby the unsustainable exploitation of natural resources brings about the end of cultures, with or without climate change playing a role (e.g. Hodell et al.1995; deMenocal 2001).

The best-known collapse study is Jared Diamond's bestseller *Collapse: How Societies Choose to Fail or Succeed* (2005). This is a study in 'ecological suicide—ecocide' where past societies have undermined the ecological basis of their success as a result of 'deforestation and habitat destruction, soil problems (erosion, salinization, and soil fertility), water management problems, overhunting, overfishing, effects of introduced species on native species, human population growth, and increased per capita impact of people' (p. 6). The collapsed ancient societies presented in this book include the Polynesian community on Rapa Nui (Easter Island) who deforested this isolated and ecologically fragile island before its desertion; the Hohokam and the Anasazi in the American South-west, who were successful in adapting to the marginal environments in the short term, but long-term environmental changes and human overexploitation of the marginal environment led to their societies' inevitable collapse; the Maya, whose collapse was brought about by unsustainable population growth, exacerbated by warring, drought, and a disinterested elite; and the Norse medieval settlement of Greenland. This last study provides the most explicit example of how climate change, in this case the early onset of the Little Ice Age after c. AD 1300, played a direct and decisive role in the collapse of this society. The popularity of studies into the collapse of societies resulting from the irresponsible use of their natural sources is, in no small way, linked to current concerns about climate change and the impact this will have on humanity.

Diamond's studies in societal 'ecocide' have not gone unchallenged. For example, the detailed reanalysis of the collapse of the Polynesian community on Rapa Nui (Easter Island) shows that it was not people but rats that were principally responsible for the destruction of the palm-tree forests (Hunt and Lipo 2010). There never was a 'last tree' that was cut down before the island was deserted, but instead the Polynesian population continued to grow till the island was 'discovered' by Europeans; it was the combined effect of colonization, new diseases, and the introduction of sheep that caused the rapid decline of the indigenous population. However, the descendants of the Polynesian inhabitants who carved the massive stone sculptures as embodiments of the spirits of powerful ancestors are alive and well on Rapa Nui. A similar case has been made for the American South-west. Not only was the history of the Native Peoples completely rewritten by American archaeologists and anthropologists, and new ethnicities such as 'Hohokam' for the O'Odham and 'Anasazi' for the Chacoan Pueblos were invented, but no recognition was made of the fact that for centuries the O'Odham and Pueblo had lived rich and 'sustainable' lives, managing water resources carefully and responsibly, and producing such evocative monuments as the Great Kiva at Pueblo Bonito (Wilcox 2010). The collapse of these societies was not the result of unsustainable 'ecocides', but was directly linked to the conquest and colonization of the American Southwest by Europeans.

Collapse studies have had a significant impact on archaeological projects concerned with modern climate change. For example, the IHOPE (Integrated History of People on Earth) project identifies the phenomenon of collapse as its 'central concept' (Costanza, Graumlich, and Steffen 2007: 15). The IHOPE project is an international collaboration involving archaeologists, historians, (palaeo)environmental scientists, ecologists, and modellers, and takes as its point of departure the view that traditional human history has focused on the emergence and collapse of civilizations, largely ignoring the environmental context of these histories. It seeks to redress the balance through the construction of an integrated record of biophysical and human system change. Using this record, the project aims to develop an understanding of the connections and dynamics of the biophysical and human past by testing humans-in-environment systems models, in the anticipation that this will provide a platform for the testing of options and scenarios for the future (Costanza, Graumlich, and Steffen 2007: 4–5). Several research challenges are anticipated, including the quality of data on socioecological systems and the comparability of these with data from environmental systems. The inaugural workshop in 2005 concluded that 'human societies respond to environmental (e.g., climate) signals through multiple pathways including collapse or failure, migration, and creative mitigation strategies' (Costanza, Graumlich, and Steffen 2007: 5), and an environmental-deterministic connection between environmental and societal change is rejected.

The IHOPE project is ongoing, and the publications produced so far have not, as yet, received critical reviews. The innovative nature and laudable intentions of the project, as well as the fact that it incorporates a number of elements that will also be part of the concept of climate change archaeology (see below), cannot be denied. However, the theoretical stance of the project remains closely aligned to a modernist or processual research agenda. This is evident, for example, in the assumption that adaptation of human systems to environmental change can be described in terms of objective quantifiable data. A further example is the emphasis on a singular past where whole societies collapse or survive, rather than an acceptance of the heterogeneous nature and existence of unequal groups in every society, past and present. The central place that is attributed to the collapse phenomenon is, in this context, revealing. The 'collapse' of complex societies is normally to the detriment of ruling elites, but the fate of the poor, land-bound, and dispossessed, however defined, may be very different.

This short history has shown that the archaeological study of climate change can be traced back to the middle of the nineteenth century, and that its place in archaeological research has varied with prevailing paradigms. It offers ideas and inspiration for the development of climate change archaeology. This will be based on a pragmatic synthesis that recognizes, on the one hand, that humans live in a world affected by climate change and its environmental impact, and on the other acknowledges that the responses from societies and communities are culturally defined and diverse, and cannot be simply categorized in terms of success and collapse. The different ways in which societies and communities, exercising their agency, have responded to the warnings and early impacts of fossil-fuel-driven climate change illustrates this convincingly.

BUILDING BLOCKS: RESILIENCE, HISTORICAL ECOLOGY, AND SUSTAINABILITY

Three scientific concepts are presented here as building blocks that can help define climate change archaeology: resilience, historical ecology, and sustainability. Their emergence in the decades after World War II reflected the growing concern that increases in the global population, notably in the developing world, combined with growing consumption levels in the developed world, would lead to a depletion of natural resources. Well-known reports predicting a future of environmental or (neo-)Malthusian catastrophes include Rachel Carson's (1962) *Silent Spring*, Paul Ehrlich's (1968) *The Population Bomb*, and the highly influential *Limits to Growth: A Report for the Club of Rome's Project on the Predicament of Mankind* (Meadows et al. 1972). Apart from their shared sources of origin and inspiration, the concepts of resilience, historical ecology, and sustainability have in common a concern with an understanding of the world in terms of links and connections between its human and natural parts. In addition, attempts have been made to put into practice all three of these concepts.

Resilience

Ideas about the role of resilience in ecological systems can be traced back to the seminal paper by the ecologist Crawford S. Holling (1973), 'Resilience and stability of ecological systems'. In this paper Holling argued that ecologists, up to that point in time, had equated stability to systems behaviour, with stability meaning conditions near equilibrium points. Whilst this had served ecology well, and allowed for the study of behaviour of highly dynamic systems, the persistence and even extinction of particular species were understood and explained as aspects of an ecosystem's stability. Holling argued for a redefinition of these ecological behavioural characteristics, using the terms 'stability' and 'resilience'. Within ecological systems, stability means the 'ability of a system to return to an equilibrium state after temporary disturbance' (p. 17), whilst an ecosystem's resilience is a measure of its ability to absorb change and disturbance, but still persist. In ecological terms, certain communities can have high stability but low resilience, such as fish populations in large freshwater lakes that attain stability over long periods of time (i.e. the size of the community over time remains close to an equilibrium), but prove not very resilient in the face of changes, such as eutrophication of the lake following the introduction or intensification of agriculture on its edge. Conversely, some communities have low stability but are highly resilient, and this is the case for many insect communities that can expand or contract exponentially when conditions are favourable or unfavourable, but rarely disappear completely from an ecosystem. Holling noted the importance of positive and negative feedback mechanisms in determining stability and resilience in ecosystems. These original ideas of resilience have been widely accepted in the field of nature conservation, and to a lesser extent in food production, where the earlier focus on the survival or flourishing of particular species has increasingly been replaced with a concern for the whole ecosystem (e.g. Mitchell et al. 2000).

In Holling's 1973 paper, the role of people in the resilience and stability of ecological systems was already noted, but the extension of the resilience concept to include human societies occurred in the mid-1990s, when the resilience concept was applied to what was termed the socioecological systems (e.g. Gunderson, Holling, and Light 1995; Hanna, Folke, and Mäler 1996; Ludwig, Walker, and Holling 1997; Berkes and Folke 1998; Carpenter et al. 2001). The concept of the socioecological system reflects the understanding that, in the modern world, there are no ecosystems left that have completely evaded the impact of humanity. In many recent publications on resilience and climate change, it has become the norm to talk about socioecological resilience, with the emphasis on how societies, being part of coupled human-environmental systems, can adapt to climate change. For example, in 'Social-ecological resilience to coastal disasters' (Adger et al. 2005: 1036), resilience is said to reflect 'the degree to which a complex adaptive system is capable of self-organization (versus lack of organization or organization forced by external factors) and the degree to which the system can build capacity for learning and adaptation'.

The concept of resilience in ecology has recently been conceptualized within the concept of the 'panarchy'—after the Greek god Pan who wove 'the harmony of the cosmos into playful song'—and it is argued that this concept is applicable to all human and natural systems, regardless of scale (Holling, Gunderson, and Peterson 2002: 63). The panarchy is made up of adaptive cycles, which can be envisioned as the system's equivalent of the stages of life of an organism: birth, growth, death, and renewal. Thus, the adaptive cycle of the panarchy comprises four stages: exploitation, conservation, release, and reorganization. The exploitation stage is characterized by the rapid growth of a community, for example when a new species is introduced in an area without natural competitors or enemies. In the conservation stage, the community reaches the limits of its growth or carrying capacity, and stores energy and material. The stored energy and material are released when the community declines, for example because of changed environmental conditions or the introduction of competitors. Finally, in the fourth stage the community is reorganized through selection, experimentation, and invention, and starts to grow again. The adaptive cycle has been presented as a figure-of-eight loop (Fig. 2.2), and the speed by which certain developments take place has been represented by the length of the arrows. Thus, empirical studies in ecosystems have shown that the storage of energy and material in the conservation stage is a relatively slow process, whereas the release of energy and material in the third stage of the adaptive cycle happens much more rapidly.

Each of the four stages of the adaptive cycle has three properties. The first, potential, defines the (natural) limits to growth and expansion. The second, connectedness, determines the extent to which a population determines it own destiny as opposed to being dependent on external variables. The third,

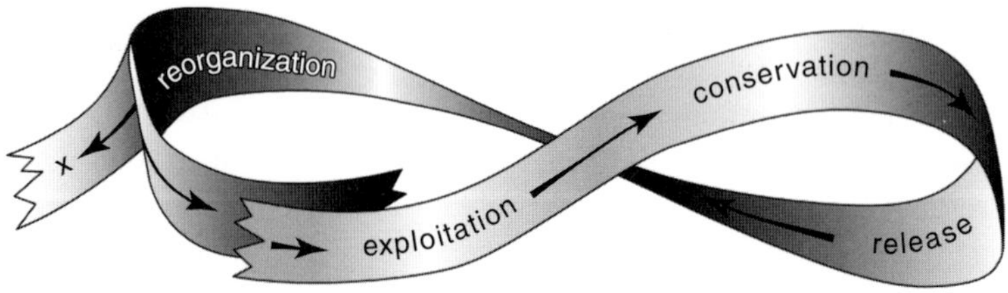

Fig. 2.2. The concept of 'panarchy', a stylized representation of the ecosystem functions of exploitation, conservation, release, and reorganization (after Holling and Gunderson 2002: 34).

resilience, is the capability of communities to absorb changes and unanticipated disturbances. Because the adaptive cycle of any human or natural system is interconnected with other adaptive cycles, the panarchy is sometimes presented as a series of nested adaptive cycles at different levels. Some of these cycles are relatively large and slow, whereas others are smaller and faster. This represents more than a vague resemblance with Fernand Braudel's (1949) concepts of *longue durée*, *conjonctures*, and *événements*.

The most ardent advocate of the use of the resilience concept and theory in archaeological research is Charles Redman (2005). Redman notes that adopting this concept provides archaeologists with a role in multidisciplinary and interdisciplinary research that aims to contribute to 'the management of the globe for a sustainable future' (p. 70). The contribution from archaeology to research with an eye on the future comes, principally, in the form of the long-term (i.e. centennial and millennial) perspectives, which are beyond the fields of view of ecologists' and anthropologists' research. Furthermore, Redman argues that this perspective brings into view the ultimate and proximate causes of the collapse of systems, and that in this way archaeologists can help build sustainable futures. In return, archaeologists are provided with a framework of understanding that integrates social systems with ecological systems in a dynamic setting, which explains phases of (apparent) stability and change. In this paper, Redman illustrates the use of resilience theory in archaeology with examples from the American South-west, notably the 'Hohokam' and their collapse, developing the case along similar lines to those presented by Diamond (2005), and discussed earlier. The same critique is pertinent here: the case of the 'Hohokam's' collapse demonstrates the imposition of external and anachronistic concepts of sustainability and collapse on past societies, rather than an exploration of how a society adapted to external changes.

In short, socioecological resilience, meaning the ability of a society to absorb changes and disturbances whilst retaining essential and defining characteristics, is a valuable concept in understanding how human communities deal successfully with climate change and its environmental impact. The conceptualization of the four stages of the adaptive cycle in the figure-of-eight model (see Fig. 2.1), together with its multilevel extension, the panarchy, provides a model that may be useful for understanding ecosystems, but has not been convincingly applied to human systems or societies.

Historical ecology

At its initial outing, historical ecology was presented as an approach that was complementary to evolutionary ecology, but adding a missing component: history (Brooks 1985: 676). The key methodological difference between evolutionary ecology and historical ecology was the use of 'direct estimates of history', which were obtained from phylogenetic systematics, or the study that reconstructs the pathways behind the present distribution and diversity of flora and fauna. Such pathways are always complex and involve multiple species, so practitioners of historical ecology study many interacting species concurrently, rather than separately. The early excursions into historical ecology were particularly concerned with historical events from the deep past, such as plate tectonics.

The extension of historical ecology into the impact of the more recent past can be largely attributed to Carole Crumley (e.g. 1993; 1994; 2006). Her own fieldwork focused on the landscape of Burgundy, central-eastern France, in the last three millennia, and this required that the impact of people on the region's ecology was fully considered. From this, Crumley defined historical ecology explicitly as a field of research that sought to overcome the divide between, on the one hand, the humanities and social sciences who prefer qualitative narratives, and on the other the biological and physical sciences who base their work largely on quantitative analysis (Crumley 2006: 15–16).

In historical ecology, the unit of analysis is typically the landscape. The term's ambiguity in meaning and scale provides a platform for practitioners from different disciplines to work together. These practitioners are expected to undertake their research using their own methods and techniques, but to place their results in the broader contexts provided by the inter- and trans-disciplinary research. Historical ecology does not presume that environmental change causes cultural change, or vice versa, but seeks to study the interrelationships between biophysical and social/economic/cultural stages and changes.

The way in which the practitioners from different disciplines are brought together into an all-encompassing framework of historical ecology is through the use of systems theory and complex systems theory. In ecological research, the system concept has been established for many decades, as in the term 'ecosystem' which was coined in 1930. Systems theory and thinking have undergone significant changes since then, but the key criticism of systems theory continues to be that it tends to lead to a focus on smaller and simpler systems that are easier to model. Such a reductionist tendency is the opposite of what historical ecology aims to achieve, and it has therefore adopted complex systems theory to bring together the many disciplinary traditions it seeks to fuse. Complex systems theory seems an appropriate vehicle for this, as it understands systems as being made up of components that are interconnected. Unanticipated aspects and properties not present in the individual parts are frequently identified, a process referred to as 'emergence' (e.g. Sawyer 2005). The kind of complex system that specifically integrates the biophysics and humanities/social sciences strands of research is the coupled human-environment system, a version of which is used by the IPCC (Bernstein et al. 2007 and see Fig. 1.1).

Complex systems can be studied at different scales, for example regional landscapes or the entire Earth. These complex systems often have no clear boundaries and can be open to external influences. For example, regional landscapes will have experienced the influx of new species as the climate warmed after the end of the last Ice Age, or as a consequence of the deliberate introduction of new species by people. Complex systems theory notes that because of their dynamic nature, the historical pathways are important. Complex systems are thus thought to contain a 'memory', usually in the form of certain system characteristics that can only be explained by understanding the long term, but are sometimes retained within human populations as social memory and in material culture. The relationships between the components within the complex system are nearly always non-linear, and linear cause-and-effect explanations for real-world phenomena are (nearly) always too simplistic. Complex systems are often nested, meaning that a complex system is made up of other complex systems, which in turn are made up of other complex systems. The way these are organized is not necessarily within hierarchical structures but in a dynamic meshwork where 'the sources of power are counterpoised and linked to values' (Crumley 2006: 25). Historical ecologists work with complex systems theory because nearly all ecosystems and cultural systems display these characteristics, and a common platform and language are created to start exploring the interconnectedness between the two.

A further aspect of historical ecology relevant to the development of climate change archaeology is that it has been successful in applying its findings to the real world of nature conservation and ecosystem management, in what has been termed 'applied historical ecology'. A much-cited example of this concerns the restoration of a highland grassland in northern Mexico, where the invasive trees could be dated using aerial photography, whilst the thinning of woodland and use of fires to manage woodlands were dateable on the basis of reconstructed long-term fire-scar chronologies (Swetnam, Allen and Betancourt 1999). Effective use of the outcomes of historical ecology research has served to strengthen its reputation and the validity of its precepts.

In short, historical ecology provides an important departure from existing research traditions, not just in archaeology but in geography and environmental studies too, where the quantifiable results from biophysical research and the qualitative narratives from humanities and social sciences research continue to exist, to a large extent, in parallel worlds. Historical ecology uses complex systems theory to integrate the findings from biophysical and social systems, produced within discipline-specific protocols, into coupled human-environment systems. To date, historical ecology in both its research and applied form has been most successfully employed to study particular ecosystems, regional landscapes, or specific species. It has yet to produce fully integrated holistic and comprehensive coupled human-environment systems at higher levels of scale or complexity. The diversity of human behaviour at different times and in different places, both within and between societies, defeats such attempts at abstraction.

Sustainability

'Sustainability' can trace its origins to the late eighteenth and nineteenth centuries in the works of Thomas Robert Malthus, notably *An Essay on the Principle of Population* (first published 1798) and others, but the concepts of sustainability and sustainable development first became formalized in the 1980s. The essence of sustainable development was captured in the United Nations *Report of the World Commission on Environment and Development: Our Common Future* (UNWCED 1987), also known as the Brundtland Report after the commission's chair, as follows: 'Sustainable development is development that meets the needs of the present without compromising the ability of future generations to meet their own needs.' Whilst this definition has been considered opaque (e.g. Hueting 1990), and many other definitions have been proposed, the Brundtland formulation of sustainable development has become widely accepted by international organizations, nation states, and non-governmental organizations (NGOs) (Baker 2006: 24). Its explicit long-term intergenerational perspective makes it particularly relevant for the development of climate change archaeology.

The Brundtland Report on sustainable development included a number of assumptions that are now more widely—but certainly not universally—accepted than was the case in the 1980s. For example, the report established a clear link between economic development and environmental degradation. It recognized that the environment has a regenerative carrying capacity with set limits, and when economic development transgresses these limits, the depletion of the available natural resources will be the inevitable consequence. This recognition has had far-reaching consequences for the study of ecology and ecosystems, and concepts such as ecosystem health and ecosystem services are bound up in the same idea. The key developments considered as instrumental in pushing the world towards its ecological limits were the ever-increasing consumption patterns in the developed world, coupled with increasing population, especially in the developing world. The report took an explicitly global approach to these problems. Importantly, *Our Common Future* stressed that the 'needs of the present' concerned in particular the needs of the world's poor, and it supported economic growth as the principal mechanism by which alleviation of poor countries could be achieved. However, that growth needs to be sustainable, and a central tenet of the Brundtland Report is that genuine sustainable development will produce social change. Thus, the three 'pillars of sustainable development' are sustainable environmental, economic, and social development.

To date, considerably more attention has been paid to the sustainability of the environment and economic development than to the needs of the world's poor. Indeed, whilst the interconnectedness of environmental and economic sustainability has become well established, this is not equally true for social sustainability. As a point-in-time illustration of this, a Google search of the World Wide Web in 2011 returned over 4 million results for 'environmental sustainability', 1 million for 'financial sustainability' plus half a million for 'economic sustainability', but fewer than half a million results for 'social sustainability'. Social sustainability is inescapably linked to greater degrees of equality at a variety of scales, and this has political and socioeconomical implications that are not always acceptable to different interest groups advocating environmental sustainability (e.g. Goodland 1995).

Two models have been used extensively to conceptualize sustainability. The first shows the three principal 'pillars' of sustainability—environment, society, and economy—as overlapping ellipses. This symbolizes the nested or interrelated nature of the three 'pillars'. The second conceptualization shows the same three components in a Venn diagram (Keiner 2005; Fig. 2.3), and it is only where the three components of sustainable environmental, economic, and social development overlap that genuine sustainable development is achieved. This model is now common currency, used at a variety of levels ranging from the global down to individual households.

The broader sustainability agenda has been promoted by the UN at international conventions, and the 1972 UN Conference on the Human Environment represents the beginning of this development (Baker 2006: 54). Subsequent conferences include the 1982 conference in Nairobi, where the UN Environment Programme (UNEP) was established. Early concerns about the impact of GHGs on the world's climate led the UNEP, together with the World Meteorological Organization, to establish the IPCC as the scientific body that would investigate these concerns. The *First Assessment Report* of the IPCC, published in 1990, confirmed that the global climate was warming. Despite the fact that the attribution of climate change to human activity was not established with a high degree of confidence till later *Assessment Reports*, the IPCC's early findings were of direct relevance to the sustainable development agenda because they concerned the structural economic conditions that cause climate change. At the 1992 Rio Earth Summit, concerns over climate change led to the establishment of the

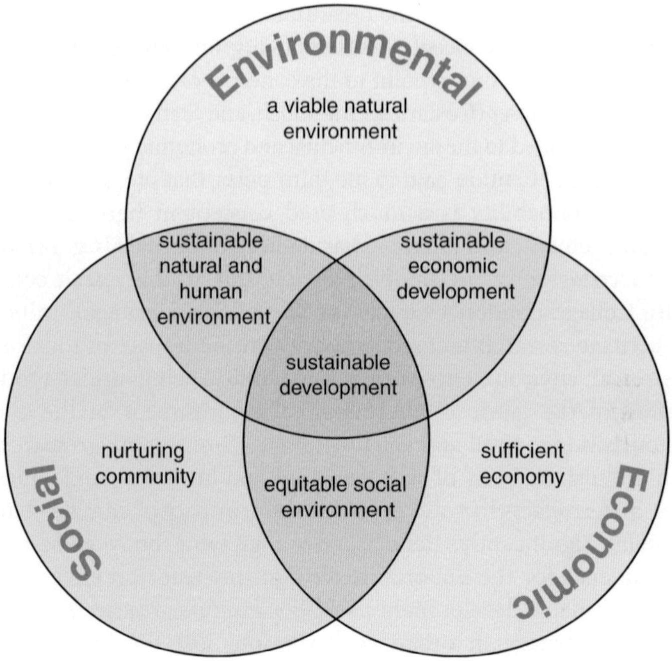

Fig. 2.3. The three 'pillars of sustainable development': environmental, economic, and social sustainability.

UN Framework Convention on Climate Change (UNFCCC), and it has been this organization that has sought to agree national GHG emission targets at its Conferences of Parties, with the third such meeting in Kyoto in 1993 resulting in the Kyoto Protocol (see chapter 3).

The emergence of a science of sustainability and sustainable development is of a somewhat more recent date, and the delay in its development has been attributed to the predominance of the political debate over evidence-based decision making. In 'Sustainability Science' (Kates et al. 2001), a number of key areas of research were presented which have now, to a large extent, been incorporated into the work of the IPCC. For example, the question of how scientists can integrate the study of the Earth system with human development has produced an answer in the form of the coupled human-environment systems model (see Fig. 1.1); and the question of how long-term future trends are presented has an answer in the form of the Special Reports on Emissions Scenarios (SRESs). The IPCC working groups also consider issues such as the vulnerability and resilience of human and environmental systems, and the existence of 'tipping points', especially at the global level.

Sustainability and sustainable development are concepts that are not widely used in archaeological research. Considering the global and forward-looking nature of both the concepts and the organizations involved, this is unsurprising. The deployment of the concept of sustainability in archaeological research is restricted to its use as a counterpoint to the concepts of ecocide or collapse, as in *Sustainability or Collapse?* (Costanza, Graumlich, and Steffen 2007). In most cases these studies are limited to the environmental and economic pillars of sustainability, with little or no attention paid to the third pillar, that of social development.

However, sustainability is a much-used concept in heritage studies and heritage management, and the topic has dedicated journals (e.g. *International Journal of Heritage and Sustainable Development*) and research centres (e.g. University College London's Centre for Sustainable Heritage). Initially, sustainable heritage research focused primarily on the impact of tourism on the built historical environment, with sustainability being understood as the preservation of the fabric of the historical environment in the context of cultural tourism (e.g. Fyall and Garrod 1998). More recently, research associated with the sustainability of archaeological and historical landscapes, buildings, sites, and artefacts has also considered the impact of climate change (e.g. Cassar 2005). Significantly, there is now a growing body of research that provides evidence for the important role that the historical environment has to play in sustainable development itself. For example, research has identified that the built heritage in urban contexts—not just the buildings deemed worthy of conservation but the broader heritage in the sense of the familiar and distinctive built and natural environment—is an essential aspect in determining the quality of life and social development. The contribution of heritage to the quality of life was found to be greatest where the rate of change was fastest (e.g. Tweed and Sutherland 2007).

Evidently, heritage research has progressed further than archaeological research in addressing the concepts of sustainability and sustainable development. The recognition that the absence of heritage leaves people without a sense of belonging, or a sense of place, is important in the development of climate change archaeology. After all, effective adaptation of communities to climate change and its environmental impacts will be achieved where people's sense of place and quality of life are preserved, and alongside the tangible heritage, the intangible history and archaeology of communities can play a role in this process too.

CLIMATE CHANGE ARCHAEOLOGY: DEVELOPING THE CONCEPT

The previous chapter has already outlined the essence of what climate change archaeology aims to offer: a repository of adaptive pathways revealed through archaeological research that can help build the socioecological resilience of communities in a time of rapid climate change. However, the history of archaeological research in relation to the environmental impacts of climate change, presented earlier in this chapter, can have left little doubt that to construct an archaeology of climate change is not a straightforward task. Archaeologists themselves do not agree on a number of fundamental aspects, including: the relationship between nature and culture/society; the use of systems theory and the role of resilience in this; and the application of the findings from archaeological research in the modern world. In the following paragraphs, I will set out the position on these issues for climate change archaeology.

The relationship between people in the past and the environment they inhabited provides one of the longest-running debates in archaeology. The tendency to dismiss or ridicule research as being either 'environmental deterministic' or 'environmental relativistic' has certainly waned in recent years, but this deep-seated opposition can still be discerned beneath the surface of many publications.

Another way of describing this divergence is to consider the relational place of nature in culture and society. On the one hand, for many functional and processual archaeologists who have adopted a biophysical approach, nature is to be found outside culture and society, in the sense that the environment is 'out there', real and defined by its own parameters. This view is, unsurprisingly, shared by many who focus their research on the origins and spread of agriculture, whereby people 'conquered' nature to farm the land. However, ideas of an externalized nature can be found in all areas of archaeological research. A number of postprocessual archaeologists who work in a humanities and social sciences tradition, on the other hand, have explained that nature, as a concept, is wholly internal to culture. In this, they follow the work by geographers such as Dennis Cosgrove (1984) who defined the landscape as, essentially, an ideological concept and characterized by culturally defined attributes. This view has been adopted by many who focus their research on monuments and their place in the landscape, but the idea of an internalized nature can also be found in research on many different periods, regions, and subjects.

In climate change archaeology, the starting point has to be that people and human organizations have an interdependent relationship with their environment, and that this was as true in the past as it is in the present. In this, it follows historical ecology in recognizing the interrelationship without assuming causal connections. The environment provides people with food, water, warmth, and shelter; it contributes to economic security and freedom from threats, whilst at times natural events such as earthquakes, tsunamis, and (rapid) climate change remind people of their dependency on 'nature'. However, what we call 'nature' is a product of cultural perceptions and understanding, how we think about it, what we value, and what names we use to describe it. Even concepts like 'climate' and 'climate change' are human constructs. Recognizing that both perspectives carry validity, and that nature is at the same time external and 'real' but also internal and constructed by culture and society, produces a hybrid concept of the multilayered and multilateral character of the interrelationship between nature and culture/society (cf. Latour 1993). Thus, whilst the terms 'nature', 'environment', 'climate', and 'climate change' are all human constructs, they stand for features and phenomena wholly or partly outside human influence that have direct and indirect connections to the way people live their lives. The way people adapt to changes in these features and phenomena is a matter of exercising agency. For coastal communities this is especially apparent: however we choose to define the sea, and whatever myths and stories we associate with it, the sea itself and the fish it contains are too real to be defined by the cultures of coastal dwellers alone. But where and how people live with the sea is a matter of coastal communities exercising their agency (Van de Noort 2011b).

The use of systems and systems theory has been, to date, the preserve of processual archaeologists. The notion that the way in which societies operate can be explained in terms of systems theory is based on the premise that societies are generally homogeneous and that within societies, individuals concur with the general purpose of the society. Thus, the main operations of 'whole societies' can be presented in the form of a system that serves the individuals within it. Postprocessual archaeologists have taken the point of view that societies are, by definition, heterogeneous and that these are made up of individuals and 'interest groups' with diverse and contradictory views on the purpose of society—such division lines can exist between groups of different race, sex, age, religion, occupation, wealth, and descent. Systems theory is also rejected by postprocessual archaeologists on the grounds that societies are never isolated from external influences, including influences from other societies, and the description of a society as a closed or self-governing system is therefore considered inappropriate. Because the external influences are often from 'Western civilizations'—such as in the examples of Easter Island and the 'Hohokam' presented above—such thinking has been labelled 'colonialism', leading to 'post-colonial critique' within the broader postprocessual paradigm. Systems theory—and complex systems theory which aims to accommodate greater levels of complexity in societies and is advocated in historical ecology and resilience theory—is rejected within postprocessual thinking, and the study of how individuals and groups within society exercise their agency is given prominence instead.

Whilst acknowledging that societies are heterogeneous and include people with diverse and often contradictory views, and that no human society can be described as a closed or self-organizing system, climate change archaeology accepts that the *aggregated* impact of human societies can feature in system models such as the coupled human-environment system used by the climate change scientists (see Fig. 1.1). This is especially the case in the modern, globalized world. For example, the aggregated effect of industrialization, population growth, and car ownership in many different societies in the world has produced the GHGs that are driving modern climate change, and this could be expressed as the impact of the human system on the environmental system. The IPCC, in its *Assessment Reports*, has stressed the global character of climate change. The need to understand the impact of the whole human system on the environmental system is an essential component of modern climate change science. However, it does not follow that individual societies or communities can be studied as systems, nor that the impact of climate change on individual communities can be expressed in terms of a coupled human-environmental system. For example, the IPCC and other commentators have repeatedly made the point that climate change and its impact on the environment will have very dissimilar consequences for different groups within societies across the world and that, in most situations, the poorest communities with the fewest resources will be the ones with the least resilience to adapt successfully to climate change.

The same argument applies to the use of resilience theory and the application of the panarchy to human societies. The adaptive pathway of exploitation, conservation, release, and reorganization may well be applicable to ecosystems and certain species within ecosystems. However, the application of this model to human societies assumes a homogeneity and shared experience of the various stages of the adaptive pathway which are almost certainly never true. Archaeological applications of resilience theory use familiar case studies of collapse/release of the Maya and the 'Hohokam' (e.g. Redman 2005), ignoring the heterogeneous and uneven make-up of these societies, and the influence of external forces such as the arrival of 'Western civilizations'. Similarly, it would be peculiar to describe complex societies such as ancient Egypt, China, the Assyrians, the Roman Empire, the Soviet Union, or any modern nation state as going through these four stages without recognizing the role of specific individuals and groups within these societies who have sought to influence societal change (cf. Reuswigg 2007). More generally, empirical research has yet to prove convincingly that human systems move (repeatedly) through the four stages of the adaptive cycle of the panarchy, without blatant attempts to make societies' development 'fit' the model.

This does not mean that resilience studies have not produced some important insights for climate change archaeology. Resilience should be understood not as a function of systems, but as the ability of communities and societies to adapt to change to create more sustainable futures, both for the environment and for themselves, whilst retaining key aspects of their cultures to avoid loss of social identities. Social memory, learning, and experimentation, and the use of traditional and 'archaeological' knowledge can all strengthen the socioecological resilience of communities (Berkes and Folke 2002). Adaptation can take the form of technical innovation, economic restructuring, and social reformation, and successful adaptation means that the interrelationship between communities and their environment has become more sustainable, ensuring that the success of communities and societies does not damage the prospects of successive generations.

This raises the issue of what archaeological research can contribute to a modern world facing rapid climate change. In the introductory chapter, I referred to Sabloff's (2008) *Archaeology Matters: Action Archaeology in the Modern World,* where it was argued that such a contribution potentially comes either in the form of 'general models of successful and unsuccessful trajectories of sustainable growth over long periods of time' or 'through specific on-the-ground research projects'. The former presupposes homogeneity of past and present societies, which is a simplistic representation of reality. It also ignores the somewhat obvious fact that the modern world—in terms of population density, consumption, technology, rapidity of climate and environmental change, etcetera—is so far removed from the past that it is unlikely that any general models obtained through archaeological research will have any applicability in dealing with the problems faced by the world in the twenty-first century. Therefore, these general models will not feature in climate change archaeology.

However, the use of information from archaeological research in local applications is very much in line with the practice of applied historical ecology, the application of Holling's early work on resilience in conservation management, and the political agenda on sustainable development. For climate change archaeology, it will constitute the informed practice of empowering local communities to deal with changes by learning from the past, through the archaeological study of adaptive pathways to past environmental change. Whilst it may be possible to learn something from the adaptive pathways of other communities, the particular strength from learning from the adaptive pathways of one's own community, or from people who once lived in the same location, comes from the importance of people's sense of place. Let me illustrate this notion with an example.

My county of residence, Devon in the south-west of England, has to date in the main resisted the construction of wind farms, both on land and offshore. A solitary wind farm comprising three wind turbines with a combined capacity of 2.7 MW is currently the total extent of the utilization of renewable wind energy in this county. The opposition from local communities to the construction of wind turbines has influenced county-level politicians and planning officers to discourage or reject nearly all proposals on the basis that such developments devastate the beautiful countryside that is the pride of the county and which is essential for attracting tourists. However, in the adjoining county of Cornwall, equally proud of its attractive countryside and similarly economically dependent on tourism, the development of wind farms has received a much more positive welcome, and the county prides itself on accommodating the first wind farm in the UK and having currently seven wind farms with a combined total of 86 turbines generating 46.05 MW (Renewable UK 2010). Cornwall has also played a pioneering role in the development of wave energy generation.

What explains these different adaptive pathways? Social sciences researchers who have looked into this reject the concept of NIMBY-ism ('not in my back yard') as a constructive explanation, but instead stress the importance of place identity and place attachment—what are now commonly referred to as people's sense of place—as important factors (Devine-Wright 2009). Whereas in Devon that sense of place is essentially an agricultural one, the Cornish landscape reflects a long and rich history of mining, with many of the defunct pumping houses visible in the landscape and now forming a World Heritage Site. Thus, the Cornish sense of place reflects on large machines and the utilization of natural resources with a sense of pride and connectedness, and it is this that has made it possible for renewable energy to become part of Cornwall's adaptive pathway. However, the Devonian sense of place has no such connectedness, and its people have not (yet) embraced renewable energy as part of their adaptive pathway to climate change.

Self-reflected perceptions of the past, or people's sense of place, can play an active role in deciding the nature of the pathways that communities adopt in becoming resilient to climate change and its environmental impacts, which is what climate change archaeology purports to develop. Archaeological research can contribute to the sustainability of such adaptations, because of the long-term nature of the archaeological record, and its ability to consider the social, economic, and environmental 'pillars'. This idea echoes, to an extent, the UNESCO World Heritage Centre Budapest Declaration on World Heritage that 'heritage in all its diversity is an instrument for the sustainable development of all societies' (UNESCO 2002, Article 1), albeit this declaration concerned the physical heritage remains, rather than the intangible heritage. Just as has been argued for climate change archaeology, the Budapest Declaration stresses the importance of involving local communities in the process of deciding how the past should be managed, protected, and used in the present (Article 3f).

The contribution of climate change archaeology to a sustainable future is only a modest one: it is about finding examples of adaptation to the consequences of environmental and climate changes in the past that, in a small way, can help build the socioecological resilience of communities in the present and future. In this quest, I recognize that none of the societies studied by archaeologists had access to an evidence-based understanding of future climate change. Thus, all adaptations to climate change were reactive responses, never proactive endeavours based on predictive scenarios for the future.

However, that should not be a deterrent. After all, poorly informed innovations and even serendipity could produce solutions that provided long-term and sustainable outcomes in the past, and may inspire work in the future.

CONCLUSION

This chapter has sought to develop the concept of climate change archaeology within the context of existing archaeological research traditions. It has taken inspiration from three scientific concepts that emerged in the post-war decades from a concern about the exhaustion of natural resources: resilience studies, historical ecology, and sustainability.

Climate change archaeology, following historical ecology, adopts a hybrid approach to research, seeking to integrate biophysical and human and social sciences in a single framework (Fig. 2.4). In this way, its understanding of the interrelationship between societies and their environments in the past is one

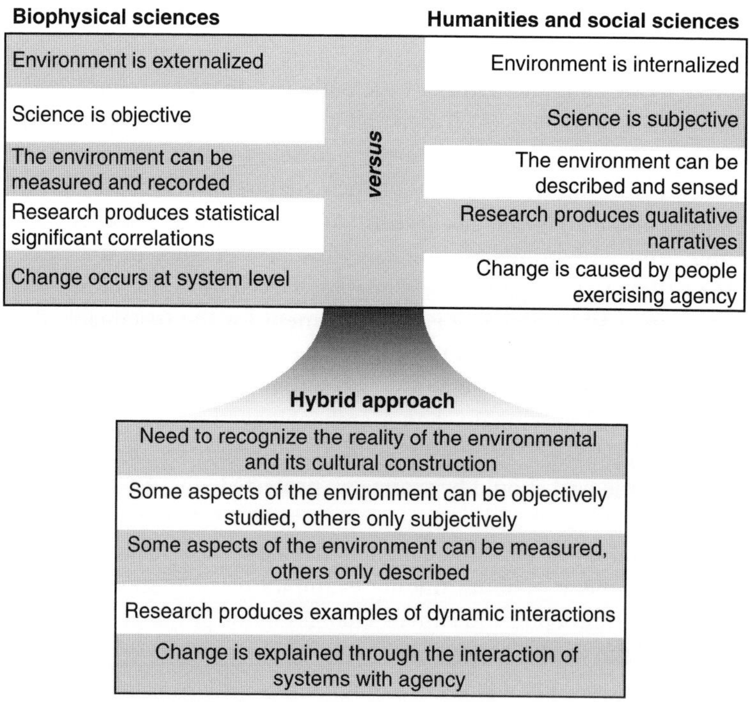

Fig. 2.4. A schematic representation of the hybrid approach to the study of environment-human interaction, bridging the gap(s) between biophysical sciences and humanities and social sciences.

that transgresses the existing, if waning, archaeological traditions of functional/processual and postprocessual archaeology.

Climate change archaeology sees the past as a repository of adaptive strategies and concepts that can help build the socioecological resilience of modern communities in a time of rapid climate change. This repository is best utilized where it strengthens, or builds upon, communities' sense of place. Finally, the long-term developments that are revealed through archaeological research are well suited for revealing long-term, sustainable adaptive strategies that provide alternative scenarios to short-term political opportunism.

3

Past, Present, and Future Climate Change

INTRODUCTION

Climatologists study the Earth's climate in five spheres: the atmosphere (gases surrounding the Earth), geosphere (land surface), hydrosphere (salt- and fresh-water bodies), cryosphere (ice and snow), and the biosphere (living organisms) (e.g. Bradley 1999; Dincauze 2000: 144–58). The Earth's climate is basically controlled by three factors (Le Treut et al. 2007: 96). The first concerns the amount of radiation *received* from the sun, or insolation. The second is the fraction of this solar radiation that is *reflected* back into space, what is called the Earth's albedo. The third concerns the amount of energy derived from the sun *radiated* back from the Earth's surface into space. The combination of these three factors is referred to as absorbed solar energy (Fig. 3.1). None of these factors contributing to the absorbed solar energy has been constant in the past, or will be in the future, and the principal cause of past and future global climate change can be attributed to variations in these three factors.

The shortwave or ultraviolet radiation received from the sun is distributed principally through the atmosphere and the hydrosphere, and this effectively controls the weather. The Earth's albedo is determined by factors in all five spheres, but the cryosphere, geosphere, and biosphere play a significant role in the amount of the sun's radiation that is reflected back into space. For example, volcanic activity can influence the amount of stratospheric aerosols, and reflect significant amounts of radiation back into space (Hansen and Lacis 1990). The radiation that reaches the Earth's surface is converted into heat, which in turn causes the emission of longwave or infrared radiation. This is radiated back into the atmosphere. Some of the longwave radiation passes through the atmosphere unhindered into space, but the remaining radiation is absorbed and re-emitted by gases in the atmosphere. GHGs do not affect the trajectory of the shortwave radiation from the sun to the Earth's surface, but GHGs are able to block the longwave radiation on its way back into space (Jansen et al. 2007). Increased GHGs, therefore, retain more of the longwave radiation emitted from the geosphere and hydrosphere, leading to global warming and climate change.

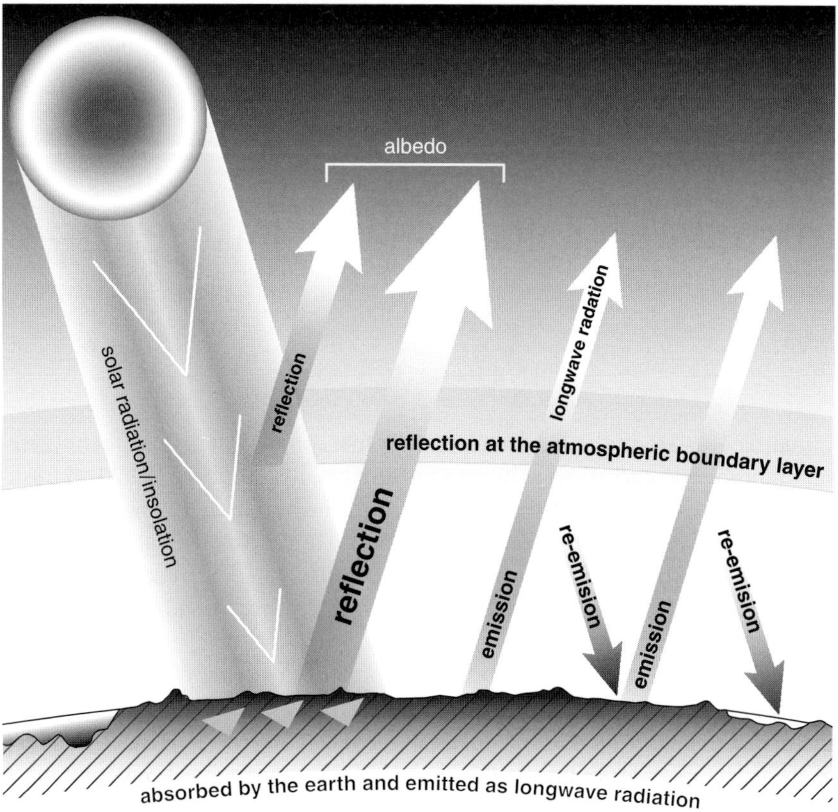

Fig. 3.1. A schematic representation of absorbed solar energy: solar radiation reaches the Earth as shortwave radiation; some is reflected by the Earth's albedo, the remainder is absorbed. Part of this is then emitted as longwave radiation which, in turn, is partly absorbed and re-emitted in the Earth's atmosphere. Higher concentrations of GHGs in the atmosphere lead inevitably to an increased absorption of the longwave radiation, which is the cause of global warming.

Climate on Earth is further determined through autovariation—that is, changes within each of the five spheres that have no external cause—and feedback mechanisms, which can either amplify or reduce the externally induced changes, respectively known as positive and negative feedbacks. A well-known example of a positive feedback is presented by the melting of snow and ice in periods of warming, which reduces the Earth's albedo and the reflection back into space of solar energy, thus accelerating the warming trend as observed during early parts of interglacial stadials (e.g. Shackleton and Opdyke 1973).

This chapter develops the key concepts from climate change science, which will provide a framework in support of the case studies in climate change archaeology. It will consider both natural climate change and climate change driven by increased GHGs, so that the contribution of climate change archaeology to the building of the resilience of modern communities can be placed in the appropriate context. The chapter offers an introduction to palaeoclimate research, noting how new methodologies, concepts, and understandings have been developed in the last few decades. Next, it critically considers the role and claims of archaeological and palaeoenvironmental research. It also presents the most up-to-date palaeoclimate evidence on climate in the deep past, and in some more detail for the period since the Last Glacial Maximum (LGM) c. 21,000 years ago. The evidence for human-induced climate change and the scenarios for climate change in the future complete this exploration of climate change research.

PALAEOCLIMATE RESEARCH

Understanding the amount of radiation that the Earth has received over time provides the foundation for palaeoclimate research. The history of climate change research can be said to have commenced in the late eighteenth century with the realization that the presence of large boulders found at great distances from their natural provenance could only be attributed to large-scale movement of ice, in regions where ice was no longer permanently present. Amongst the first to refer to glacial activity as an agent of geomorphological change was James Hutton who, in his *Theory of the Earth* (1795), discussed evidence for this from his own research in the French Jura. Other pioneers included the mining engineer and geoscientist Jens Esmark, whose work in Rogaland, Norway, showed conclusive evidence for the transport of erratics by ice (1824). A treatise on the origin of Scandinavian boulders in northern Germany was published by Albrecht Reinhart Bernhardi in 1832; the botanist Karl Friedrich Schimper coined the phrase *Eiszeit* in 1837, and in 1840 Louis Agassiz produced the first comprehensive account of glacial activity in his *Études sur les Glaciers*.

The identification of glacial activity as an influence on geomorphology led to the attribution of other phenomena to significant fluctuations in the Earth's climate: evidence for significant sea-level fluctuations, the aeolian origin of the loess covering large areas in the northern hemisphere, and the geological evidence for more than just a single Ice Age, with glacial stages alternating with warmer interglacial stages (Fagan 2009: 32–4). An understanding of the significance of ice as an agent for geological and geomorphological change played a key role in the search for the origins of the Ice Age and, consequently, in the study of the climatic fluctuations that produced the glacial and interglacial stages. The influence of this on the new discipline of palaeoclimatology was such that it remained the primary focus for climate change research till the 1970s (Jansen et al. 2007: 438).

The search for the origins of the Ice Age and the glacial–interglacial fluctuations has included discussions on changes in solar energy and sunspots, on the role of volcanic eruptions, and the subsequent impact of aerosols or airborne dust particles on the world's climate, and on variations in the Earth's orbit around the sun. The latter suggestion originated with Joseph Alphonse Adhémar (1842), and was subsequently promoted by James Geikie (1874) and James Croll (1875), before the Siberian Milutin Milankovitch established the so-called Milankovitch curve. This curve is effectively made up of three variables. The first concerns the shape of the ellipsis of the Earth's orbit around the sun, which follows a 96,000-year cycle of variable distance between the Earth and the sun, resulting in fluctuations in incoming solar radiation. This is known as orbital eccentricity. The second concerns changes in the tilt of the Earth's axis relative to the sun. Over a 41,000-year cycle, this axial tilt fluctuates between 21.5 and 24.5 °C, which means that during this cycle there are prolonged periods during which the Arctic receives less summer sun and the Antarctic more, and periods during which the reverse is the case. This is known as obliquity. The third variable is precession, or the change in the alignment of the Earth's axis as it spins, producing 'wobbles' on a 21,000-year cycle.

Milankovitch calculated the periods over the last 600,000 years where the 65 °N 'equivalent latitude' was located, in terms of the solar radiation received. His graph of the 65 °N 'equivalent latitude' (Fig. 3.2; its first iteration was published in Köppen and Wegener 1924) shows a good correlation of what was, at that point in time, the understanding of the age of the more recent glacial and interglacial stages; and, within these, some recent stadials and interstadials such as the Younger Dryas and the Little Ice Age. Although the Milankovitch curve was initially hailed as the answer to the vexed question, confidence in it was undermined by the application of radiocarbon dating, and later potassium-argon dating. The new dating techniques enabled the dating of glacial and interglacial stages studied on land, but this showed a limited correlation with the Milankovitch curve. However, subsequent research on the existence and age of glacial stages based on sediments from the world's ocean floors revealed a significant number of previously unknown glacial and interglacial stages, alongside stadials and interstadials. This new evidence restored the close link between the Milankovitch curve and the date of the glacial periods in the last 2 million years, and its significance in driving the Ice Ages has now been re-established (Bradley 1999; Jansen et al. 2007). The three spatial variables that produce the Milankovitch curve are now widely accepted as the key long-term drivers determining the total amount and distribution of the solar radiation received over time.

Deep-sea sediment and polar-ice cores have revolutionized our understanding of the palaeoclimate. Deep-sea coring programmes commenced in the 1960s;

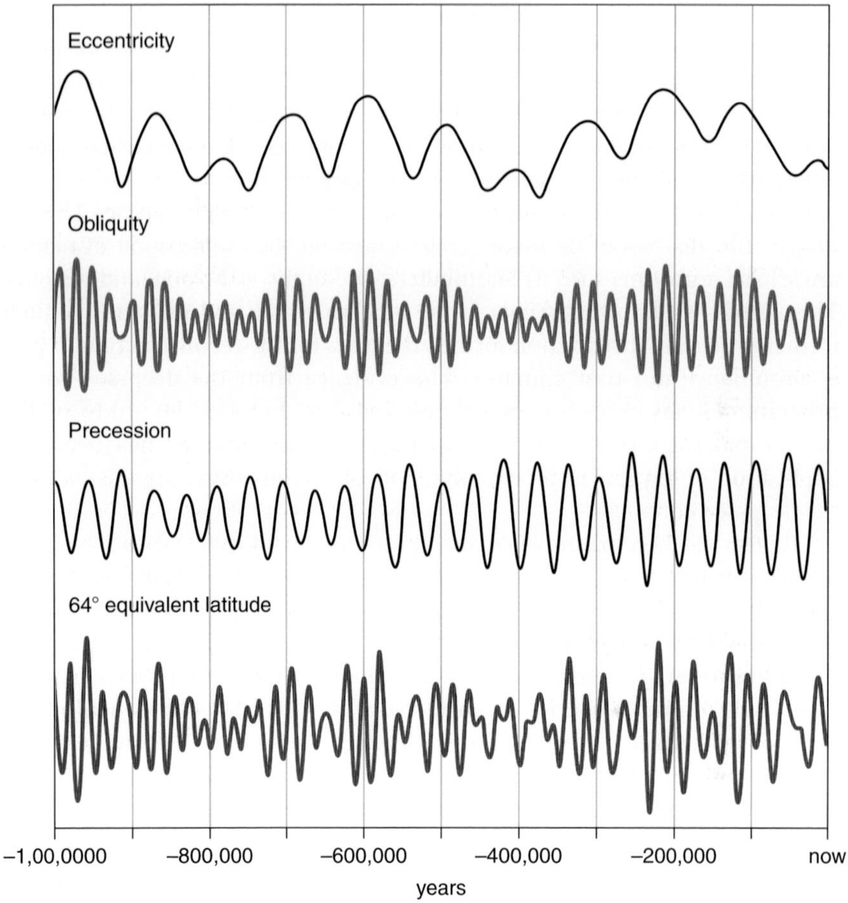

Fig. 3.2. The Milankovitch curve for the last million years: by combining the shape of the ellipsis of the Earth's orbit around the sun (eccentricity), changes in the tilt of the Earth's axis (obliquity), and the alignment of the Earth's axis as it spins (precession), Milankovitch calculated the 65°N 'equivalent latitude', which explains the timing of the Ice Ages.

initially funded by the National Science Foundation, these programmes were developed from the middle 1970s through international research collaborations. Three different projects have been undertaken: the Deep Sea Drilling Project (1968–83), the Ocean Drilling Programme (1983–2003), and the Integrated Ocean Drilling Programme, which had its first operational year in 2004 and is programmed to continue into the 2020s. The three projects have in common the collection of sediment cores from different parts of the hydrosphere for the study of the Earth's history. Their objectives include the study of the deep biosphere, climate and environmental change, and solid Earth cycles and geodynamics including plate tectonics (e.g. http://www.iodp.org). One of the early successes of deep-sea coring was the identification, for the Quaternary, of some 16 previously unknown glacial stages, in addition to the four that had already been recognized on land (Dincauze 2000: 170). The coring programme also elucidated the amount of water held by landlocked ice during the glacial stages, and the corresponding effect this had on the Eustatic Sea Levels (see chapter 4).

Initially, as demonstrated in the first publications of the Deep Sea Drilling Project, the analyses of deep-sea cores focused on the examination of marine unicellular organisms such as foraminifera, coccoliths, radiolaria, and diatoms. These biota had known habitats in the modern world, and the study of their former ratios and distribution formed the basis for the reconstruction of past environments and past climates. The evidence from the deep-sea coring programmes has delivered many new insights, but for the study of the Earth's palaeoclimate the most significant transformation has been the application of oxygen-isotope analysis on calcareous microorganisms from the deep-sea cores.

Oxygen has three naturally occurring isotopes, with fixed ratios. The lightest oxygen-isotope, ^{16}O, makes up 99.725 per cent of the Earth's water and water vapour, ^{17}O represents only 0.0374 per cent, and the heaviest isotope, ^{18}O, contributes 0.2039 per cent. Because of its relative lightness, evaporation of ^{16}O requires fractionally less energy than the heaviest oxygen isotope, ^{18}O. Thus, in glacial periods when much of the Earth's water is held in the form of ice—which is formed from snowfall containing a relatively higher ratio of water with the lightest oxygen-isotope ^{16}O—the hydrosphere will have a relatively higher ratio of water with the heaviest isotope ^{18}O. Conversely, in interglacial periods, much of the ice melts and this releases water with the lightest oxygen-isotope ^{16}O back into the hydrosphere. The analysis of the ratio between ^{16}O and ^{18}O found in the shells and hard tests of marine organisms taken from the deep-sea cores can thus provide a proxy for the history of glacial and interglacial stages (e.g. Epstein and Mayeda 1953). The Earth's water also contains a fraction of HD_2O, where the D stands for an additional neutron in the hydrogen element, or deuterium, and which, through its additional weight, has properties comparable to ^{18}O.

The changes of the ratio of ^{16}O and ^{18}O, or the ratio of ^{16}O and D, are respectively summarized as $\delta^{18}O$ and δD, which represents the difference from a standard known as the 'Vienna Standard Mean Ocean Water' (VSMOW). A graph of either temperature proxy over a long timescale shows the now familiar 'sawtooth' pattern (Fig. 3.3), representing long periods of increasing $\delta^{18}O$ or δD signifying the glacial periods when large quantities of ^{16}O are transferred from the hydrosphere to the cryosphere, interspersed with short periods of decreasing $\delta^{18}O$ or δD which indicate rapid warming during the interglacials, when large quantities of ^{16}O are returned from the cryosphere to the hydrosphere (e.g. Shackleton and Opdyke 1973). These periods are now

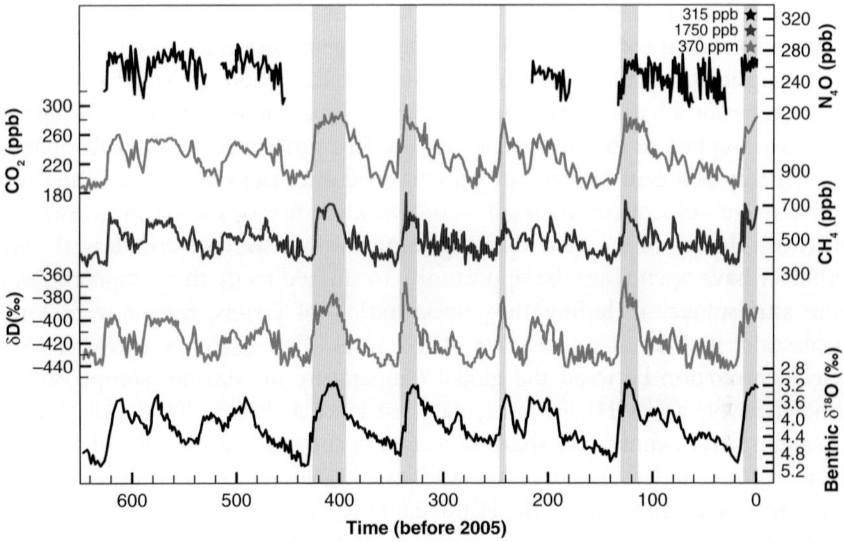

Fig. 3.3. Palaeoclimate: temperature and GHGs over the last 650,000 years. Variations of benthic ^{18}O and deuterium—proxies for global temperature obtained from deep-sea cores—and atmospheric concentrations of the greenhouse gases CO_2, CH_4, and N_2O—derived from air trapped within ice cores and atmospheric measurements—show a high degree of synchronicity. The shading indicates the interglacial periods. Note the current concentrations of CO_2 (370 ppm), CH_4 (1750 ppb), and N_2O (315 ppb) lie outside their natural ranges for the last 600,000 years. For full caption and source references see Jansen et al. 2007: 444.

referred to as Marine Isotope Stages (MISs), with the odd numbers referring to interglacial and the even numbers to glacial stages.

Scientific coring in polar ice has its origins in the 1930s, but the scientific contribution of ice coring to palaeoclimate studies was established in the 1960s and 1970s when cores in excess of 100 m were successfully retrieved and studied. The same isotopic analysis methods as applied to the deep-sea cores could be used for the ice cores, but the temporal extent was limited to the Holocene. However, in the 1980s and 1990s, the ice-coring technique was modified and improved to enable the retrieval of ice from continuous cores several kilometres long. Among the most important of these long cores in the northern hemisphere are those of the Greenland Ice Core Project (GRIP; Peel 1992), the North Greenland Ice Core Project (NorthGRIP; Dahl-Jensen et al. 2002), and the Greenland Ice Sheet Project (GISP; Meese et al. 1997), all of which were more than 3,000 m long and contained palaeoclimatic evidence extending more than 100,000 years back in time. More recent coring in Antarctica, where the lower snowfall rates produce very ancient ice profiles in shorter cores, has been able to extend the temporal range even further back. The Volstok core has a continuous record dating back 440,000 years (Petit et al. 1999), and the European Project for Ice Coring in Antarctica (EPICA) reached a record 650,000 years of ice (e.g. Siegenthaler et al. 2005).

Alongside the achievement of extending the ice core record further back in time, a major advance has been made in the analysis of the tiny air bubbles that are trapped in the compacting snow. This process is not unproblematic, principally as the air bubbles are only trapped once snow compacts under the pressure of subsequent snowfall into crystalline firn, resulting in a time lag between the age of the ice and the age of the air bubbles. Nevertheless, the air bubbles have opened up the opportunity to study directly the composition of the atmosphere, including the concentration of GHGs, over a very long timescale, now stretching back 650,000 years. This evidence shows a very clear correlation between the global temperature or Marine Isotope Stages and the CO_2 and CH_4, in particular (see Fig. 3.3; Brook 2005). All GHGs, including NO_2, show very sharp increases in the most recent part of the ice-core record, and these increases are reaching levels far in excess of their concentrations any time in the last 650,000 years.

Research into the ice cores is more varied than described so far, and a diversity of proxies now complements the original analyses of oxygen isotopes. Analyses have been made, for example, of hydrogen and combined nitrogen and argon isotopes from ice cores (e.g. Severinghaus and Brook 1999). The same is true of the deep-sea cores where alkenone (a phytoplankton biomarker) saturation indices taken from marine organic molecules, for example, have been used to estimate changes in sea surface temperatures (e.g. Ravelo et al. 2004). These and many other proxies have been developed and applied to the ice and sediment sequences obtained through the coring programmes.

Direct and systematic observations of the climate began in the seventeenth century, not long after the invention of the earliest thermometers (e.g. Van den Dool, Krijnen, and Schuurmans 1978). Scientific papers on temperature were published before that century had ended (e.g. Wallis and Beale 1669). The invention of the mercury thermometer by D. G. Fahrenheit in 1714 provided the first reliable instrument for this work (Peterson and Vose 1997: 2,837). Encouraged by the International Meteorological Organization, founded in 1873, scientists from across the world sought to present their observations in a standardized way. W. Köppen (e.g. 1873) is accredited with being the first meteorologist to have achieved this, and his observations from over 100 meteorological stations provide a temperature record that goes back to AD 1840. The number of locations where temperatures are measured has grown rapidly since the middle of the nineteenth century, and it is estimated that ambient temperature, alongside precipitation and atmospheric pressure, is currently measured at over 100,000 stations, including many based in the southern hemisphere and at sea (Wernstedt 1972; Peterson and Vose 1997: 2,840). These observations have, increasingly, been brought and published together, notably in the World Weather Records which have been operating continually since 1923, supporting the study of large-scale and global changes in the climate.

The direct measuring of GHGs in the atmosphere is of a more recent date. The first scientist to collect direct and high-accuracy measurements of atmospheric CO_2 was Charles David Keeling (1961), whose data set from Mauna Loa in Hawaii, which goes back to 1958, is the longest and most valuable. Commencing in 1970, the atmospheric concentrations of other GHGs, notably CH_4 and N_2O, have been measured, and this work is ongoing.

At the end of the twentieth century, and in the first decades of the twenty-first century, the importance of modelling in ever more powerful computer simulation models of the findings from the Milankovitch curve, from deep-sea and ice cores, from direct observations, and the many other proxies and indicators of past climate change, has become the trademark of palaeoclimate research. Most of this modelling utilizes the General Circulation Models (GCMs), or the coupled GCM, which combines the GCMs of the atmosphere and hydrosphere or oceans (AOGCMs). GCMs utilize the numerical basis of modern weather-forecasting models, but are designed to provide long-term outcomes: they are used both for reconstructing past climates and the forecasting of future ones. In the models, specific variables can be altered to see their causal effect on the modelled climate. Climate models allow for the study of the impact of changes on a variety of temporal and geographical scales, for explaining specific climatic phenomena, such as the El Niño-Southern Oscillation (Collins 2000), or for studying the impact on long-term feedback mechanisms, such as carbon cycles (Cox et al. 2000). Because weather-forecasting models form the basis of GCMs, some of the best-known examples are run by national meteorological offices or linked agencies. For example, the UK Hadley Centre HadCM3 model, which underpinned some of the future scenarios presented in the IPCC's *Third Assessment Report*, was designed by the UK Met Office. The GCMs are becoming increasingly complex as new parameters are included. This is made possible by the use of increasingly more powerful supercomputers (e.g. Pope et al. 2000).

This short background of palaeoclimate research has sought to emphasize the need to understand the significance of radiative forcing, and the reflection and radiation of the sun's energy back into space, as the key elements for understanding the climate of the whole Earth. Once these key drivers are in place, the redistribution of the energy of the sun to different parts of the Earth forms the most important element in determining global climate, noting the importance of autovariation and feedback mechanisms. Through these mechanisms, global climate change can have very different impacts on the climates of locales and regions, and this has been the domain of much research by archaeologists and palaeoenvironmental specialists.

ARCHAEOLOGICAL AND PALAEOENVIRONMENTAL PALAEOCLIMATE RESEARCH

Since the middle of the eighteenth century environmental archaeologists and palaeoenvironmental specialists have been undertaking studies aimed at elucidating changes in climate (see chapter 2). What distinguishes their endeavours from those of the palaeoclimatologists is effectively the scale at which they operate: whereas the latter focus on global climate, the former specialists have focused their attention on smaller-scale changes in the climate (Dincauze 2000: 143).

Regional climate change has been studied through proxies such as pollen, whilst significant advances have also been made in the study of microfauna such as foraminifera, fossil insects, and testate amoeba. Information on past climate has also been gained from tree-ring studies, research into tree- and snowlines, cirques, and palaeosols. All these environmental remains can act as proxy indicators of climate change because the success of particular species in establishing presence, growth and expansion, dominance and demise can be linked closely to climatic conditions. Adopting the principle of uniformitarianism—which originated in late eighteenth- and nineteenth-century geology, and is based on the assumption that biophysical processes in the past are the same as in the present—changes in the presence, abundance, and distribution of these proxies can be used to indicate climate change in the past.

Probably the best-known example of a reconstruction of climate change using palaeoenvironmental proxies is provided by the Blytt-Sernander vegetation periods. These were originally defined on the basis of recognizable strata within the accumulated peat in the mires of Scandinavia. The Blytt-Sernander scheme associated the darker bands of peat, which contained pine stumps, with dryer and warmer periods: such a climate would provide suitable conditions for pine to establish itself on the mire surface and also result in a more advanced degree of humification of peat. The lighter peat bands were linked to colder and wetter phases, as these climatic conditions inhibit humification. Axel Blytt named the cold, wet phase 'Atlantic', and the warm and dry period 'Boreal'. Rutger Sernander, on the basis of relative sea-level changes in Sweden, added the sub-Atlantic and sub-Boreal phases of the late Holocene to the scheme. The main phases of the fully developed Blytt-Sernander scheme are shown in Fig. 3.4.

The Blytt-Sernander phases coincide with specific pollen assemblages. These were defined by Lennart von Post (see chapter 2; included in Fig. 3.4), who considered that a classification based solely on macrofossils from peat could not describe complete vegetation histories, and addressed the problem using pollen analysis of cores from peat bogs in southern Sweden. Von Post's initial analysis showed a two-part separation of the woodland cover, one of beech and spruce, the other of mixed oak (1916; 1918). The mixed-oak

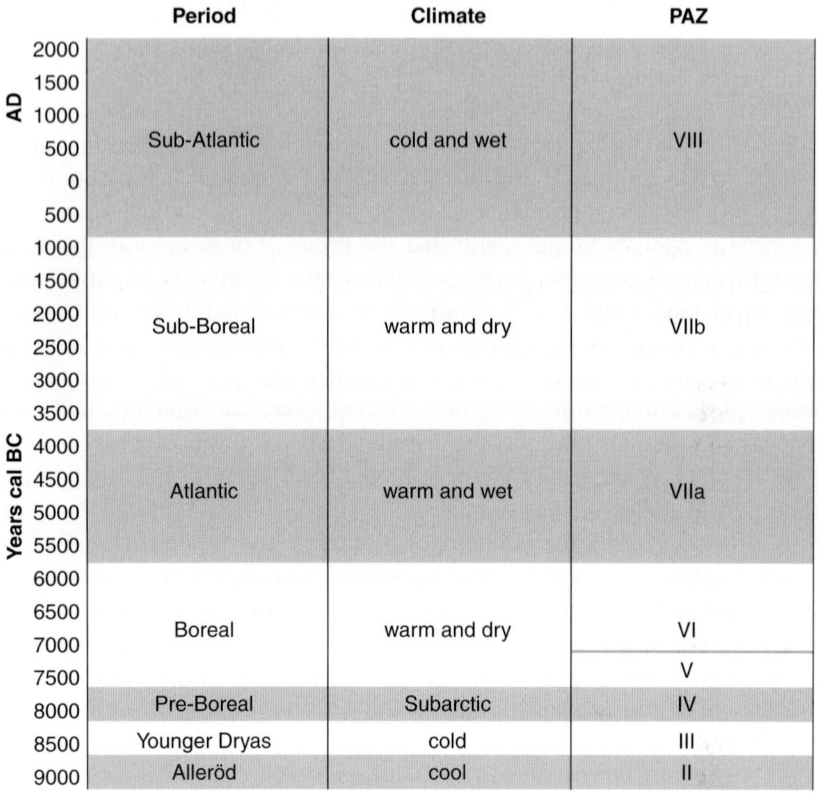

Fig. 3.4. The Blytt-Sernander scheme within a calibrated radiocarbon timescale and with Pollen Assemblage Zones added. More recent research has shown that this scheme does not reflect the full complexity of climate change in the Holocene.

woodland was further divided into a period during which pine and birch pollen predominated, with pollen of alder, elm, lime, and hazel also present, and a period during which oak became the dominant tree species. Von Post noted that the latter division coincided with the *Grenzhorizont* between the Sub-Boreal and the Sub-Atlantic. Whilst the term 'pollen assemblage zone' for the Holocene of northern Europe was not used till later (Jessen 1935), von Post had effectively produced the first example of a pollen diagram, a scheme to date pollen assemblages through their characteristic signatures, and a new paradigm for the study of postglacial climate change through the proxy analysis of pollen (von Post 1946; Fries 1967; Manten 1967; West 1970). This paradigm stood unchallenged for much of the middle part of the twentieth century. Its value greatly increased following the introduction of the radiocarbon technique, which made it possible for the deposits from which the pollen was extracted to be accurately dated (e.g. Godwin 1940; 1956).

Towards the end of the twentieth century the Blytt-Sernander scheme and the use of pollen as a direct proxy of climate change came under pressure, mainly from within palynology and the broader palaeoenvironmental research field. Five broad areas of criticism can be identified. First, the increased use of radiocarbon dating revealed the time-transgressive nature of vegetation changes in the postglacial and Holocene periods. Second, it was found that migration rates of certain plants, and especially trees, were much slower than had been previously thought, and that the presence or absence of particular tree species from an ecosystem was not linked in a unilinear fashion to climate (e.g. Webb and Clark 1977). Third, it was noted that when multi-proxy analysis was undertaken, different proxies often revealed diverging information on the ambient climate. It was observed that animals, including molluscs, beetles, birds, and mammals, respond much quicker than vegetation to climate changes and that, for example, the etymological record was capable of identifying short-term and rapid changes in the local climate that went largely unnoticed in pollen studies (e.g. Huntley and Webb III 1989; Graham and Grimm 1990). Fourth, the influence of people, or 'anthropogenic forcing', on the environmental record became recognized as a factor that could not be ignored, with the period for which this was applicable continually being pushed further back in time (Goudie 1981). Fifth, the principle of uniformitarianism itself was being questioned. For example, the variation of CO_2 in the atmosphere is now known to have an impact on the growth of plants, with individual species affected differently. Observations of biophysical processes in the present are not necessarily the same, therefore, as those inferred for the past (e.g. Street-Perrott et al. 1997).

Environmental archaeologists and palaeoenvironmental specialists have all taken these lessons to heart, and use multi-proxy analyses in their studies of past climate change (e.g. Chen et al. 1997; Hughes et al. 2006). They have also developed an understanding of new proxies, some of which provide different insights into how climates may have changed in the past. For example, the fossil remains of temperature-sensitive chironomids (non-biting midges) from lake sediments can provide an accurate understanding of changes in the July air temperature, thereby contributing to a detailed understanding of local changes in climate (e.g. Larocque, Hall, and Grahn 2001). However, to assume that such observations are manifestations of global climate patterns is simply wrong, as the Earth's climate system incorporates boundless regional variations. Thus, particularly warm July months in one region cannot be used as an indication that the global climate is warming, as at the same time other regions may experience unseasonable cool conditions.

The relatively small geographical scale of nearly all archaeological and palaeoenvironmental research limits its role in global climate change research. The distinguishing feature of modern climate research is the meta-analyses of large numbers of data sets from studies in past and present climate change, which represent the climate of the whole globe as, for example, used by the IPCC in its *Assessment Reports*. For such meta-analyses to be valid in terms of global climate reconstruction it is essential that these data sets encompass both the northern and southern hemispheres, as it is possible that certain short- and medium-term changes in climate may be experienced very differently in the opposite halves of the world.

However, it would be wrong to infer from this that studies into local and regional palaeoclimate change have little relevance to climate change archaeology beyond their incorporation into meta-analyses. Indeed, what makes this research so valuable is the information it potentially provides on the local and regional environmental impact of modern climate change, since it is at this level that societies and local communities will have to adapt. This value has already been recognized. An example is the impact of rapid climate change on the migration of vegetation. Palaeoenvironmental research is now being used to understand the natural migration rates of individual plant species in times of rapid climate change, and it has been argued that certain tree species may be unable to migrate quickly enough into suitable ecosystems (Huntley 1991). Human adaptation to this could include deliberate seeding or planting of these tree species in suitable habitats, or the selection of desirable replacement species. Such adaptive pathways will have been informed by palaeoenvironmental research.

PAST CLIMATE: THE LONG-TERM VIEW

Studying the palaeoclimate for periods for which there are no ice-core data available relies principally on the sediments obtained from deep-sea cores. The proxies from deep-sea core sediments do not include direct indicators of the ancient atmosphere, in the form of trapped air bubbles, as is the case with the ice cores. Instead, a range of proxies provides information on atmospheric levels of CO_2, albeit with a larger error range. In addition, temperatures and the presence of continental ice can be reconstructed using oxygen isotopes (see above) as the key proxy. The information from these proxies has been included in GCMs to reconstruct the palaeoclimate for the pre-Quaternary (prior to 2.5 million years ago). The outcomes from such models consequently lack the high resolution and accuracy now available for the last 600,000 years, for which ice-core data are available. The relevant chapter of the IPCC's *Fourth Assessment Report* (Jansen et al. 2007) offers the most up-to-date meta-analysis of the available scientific evidence on climate in the Earth's past, and unless explicitly stated otherwise, the information on the long-term view of the Earth's climate presented in this section comes from this source.

The relationship between levels of atmospheric CO_2, glaciation, and temperature has been traced back as far as 400 million years ago, when levels of CO_2 were much higher than they are today; no land ice existed and the temperature was significantly above modern levels. Around 330 million years ago, a significant drop in CO_2 and temperature correlates with a major Ice Age period, which lasted to around 270 million years ago. The end of this Ice Age corresponds with rising atmospheric CO_2 levels, their levels peaking around 100 million years ago in the Cretaceous, and a subsequent decline with a new phase of glaciation—in particular in the Antarctic—commencing around 40 million years ago. At this level of resolution, the CO_2 levels remain relatively low till the present, except for a period of about 300,000 years around 3 million years ago, when CO_2 levels were higher and the average global temperature was between 2 °C and 3 °C above the c. AD 1750 level. The greatest differences in temperature when compared to the modern world were found in the polar regions, with fossil pollen and plant material providing clear evidence for coniferous woodland at high latitudes in the northern hemisphere, and for the possible presence of southern beech woodland in the coastal areas of the Antarctic (Thompson and Fleming 1996). During this so-called Middle Pliocene warm period, global sea levels were also considerably higher than they are today, because none or very little of the Earth's water was held as landlocked ice (see chapter 4).

The Earth's climate in the most recent million years has been dominated by glacial and interglacial stages. In the last 430,000 years the interglacial stages closely correspond to the c. 100,000-year orbital variation, which makes up one of the variables used by Milankovitch in his curve. Less pronounced stadials and interstadials can be correlated, to a significant extent, with the two other variables that complete the Milankovitch curve (see above). According to records from the deep-sea cores and the air bubbles from ice cores, it is evident that atmospheric temperature changes, measured through the $\delta^{18}O$ or δD proxies, are closely related to variations in recorded CO_2 and CH_4. The relationship with N_2O is less pronounced but still present (see Fig. 3.3).

Climate researchers have noted that, in the last 430,000 years, the atmospheric CO_2 trails the temperature curve obtained from the Antarctic ice cores by several centuries, and on one occasion by a millennium. In other words, at the beginning of an interglacial stage, the temperature in the Antarctic rises before the CO_2 increases. A clear reason for this delayed rise in CO_2 has not, as yet, been established; but the release of CO_2 stored in the oceans during interglacial periods is an important positive feedback mechanism, accelerating global warming during the early parts of each interglacial stage.

Previous interglacial stages have varied considerably in length, with the shortest (MIS 7) lasting only a few thousand years, and the longest (MIS 11; around 400,000 years ago) lasting 28,000 years. This interglacial stage has been cored in Antarctic ice and in deep-sea cores and was—in terms of duration, temperature, and atmospheric concentration of CO_2—very similar to the current interglacial stage (MIS 1). Dated to between 130,000 and 116,000 years ago, the last interglacial (MIS 5) is thought to have been warmer by about 3 °C to 5 °C than the current one, although evidence is not yet available for the whole globe.

The last glacial stage commenced some 116,000 years ago, and the Last Glacial Maximum (LGM) is dated to around 21,000 years ago. At this point in time, the extent of the ice coverage in the northern hemisphere was at its maximum, and the Eustatic Sea Level was at its lowest (see chapter 4). Both these factors contributed to an increased albedo and reflection of the sun's energy. The atmospheric concentrations of CO_2 and other GHGs were very low at the LGM and the amount of sun radiation radiated back into space was therefore high. The impact of aerosols and vegetation cover are thought to have further contributed to an increase in the Earth's albedo, although these issues continue to be debated. Combined, these factors produced a significant reduction in absorbed solar energy. Inferred temperatures in tropical lowlands were about 2 °C to 3 °C cooler than in the twentieth century, and similar average cooling of the ocean sea surfaces at the equator has been found, albeit with a greater regional variation. The temperatures at the poles were considerably lower than at present. The ice-core data suggest that areas of the Antarctic were 9 °C cooler than today, but temperatures in Greenland were as much as 21 °C colder, with the ice in the northern hemisphere extending over very large parts of the land and sea at middle to high latitudes. Significant changes in the Atlantic circulation during this glacial stage are the probable cause of this difference in cooling between the Earth's poles.

The IPCC *Fourth Assessment Report* places the start of the current interglacial, MIS 1 or the Holocene, at 11,600 years ago, on the basis of changes in the Earth's orbit around the sun (Jansen et al. 2007: 459). The retreat of land and sea ice, and the consequent reduction of the Earth's albedo, had already commenced by this time. Ice-core analysis provides important insights into the concentration of atmospheric GHGs in the early millennia of the Holocene: these show an initial reduction between 11,600 and 8,000 years ago for CO_2 and N_2O, and a reduction between 11,600 and 6,000 years ago for CH_4, followed by moderate increases of all GHGs. This initial reduction of GHGs in the early part of an interglacial stage has been attributed to their storage, in particular the storage of CO_2, resulting from the expansion of forests into higher-latitude regions. The subsequent increase of GHGs can be explained by their release from the oceans, which warmed up more slowly than the land.

During the Holocene, global temperatures increased, as has been evidenced from deep-sea cores, and this progressive warming was experienced in tropical and near-tropical regions. Regional variations at higher latitudes were significant. In the North Atlantic and Antarctic, for example, the warmest period is considered to fall between 10,000 and 8,000 years ago. But for northern Europe and north-western America the warmest period, also referred to as the Mid-Holocene Thermal Optimum, occurred between 7,000 and 5,000 years ago. These regional variations were caused by the distribution of the sun's radiation through the atmosphere and hydrosphere, and also by the effect of local ice on the reflection of the sun's energy. However, regional variations such as the Mid-Holocene Thermal Optimum are not necessarily indicators of global climate conditions.

The dynamics of climate change include a number of abrupt changes, such as the '8.2 ka' event, which is characterized by a rapid cooling in the northern hemisphere at higher latitudes by between 2 °C and 6 °C. This event is, most likely, attributable to an outburst of the meltwater containing Lake Agassiz in northern America, which disrupted the North Atlantic Meridional Overturning Circulation, and so produced a rapid regional cooling. This cooling had a significant impact on northern America and Europe (e.g. Marshall et al. 2007), and has been identified in palaeoenvironmental research as far away as China (e.g. Wang et al. 2002) and southern Africa (e.g. Powers et al. 2005).

In the last 2,000 years, we know of a number of climate fluctuations from the northern hemisphere, especially for Europe, where there is an abundance of historical accounts. Periods such as the Medieval Warm Period and the Little Ice Age were noted at the time of these changes happening, and such variations in regional climates had significant impacts on the societies and communities that experienced them first-hand (e.g. Fagan 2008). Climate reconstructions based on this historical data, and enhanced with data from tree rings and other proxies, provide ample evidence for the existence of decadal variations in temperatures, and also of longer regional variations. The available evidence from the northern hemisphere suggests that the warmest period before the twentieth century was between AD 950 and 1120. This implies that the Medieval Warm Period, with global temperatures raised by 1 °C to 2 °C compared with earlier and later centuries, was a reality. Similarly, tree-ring data from various locations in the northern hemisphere have shown that the fifteenth century was amongst the coolest in the last two millennia. Multi-proxy analysis suggests that a generally cooler climate existed into the sixteenth and especially the early seventeenth century, and what is known as the Little Ice Age is reflected in the proxy temperature record. There is, however, very little evidence for corresponding colder and warmer periods for the southern hemisphere, where tree-ring data provide the longest proxy record for climate change. One such sequence, from Tasmania, extends back 1,100 years, and suggests a relative warm period between AD 1300 and 1500. Tree-ring sequences from southern Africa and South America are too short to provide information on climate changes in the last two millennia.

This picture changes for the last 200 years or so. Combining the observations of the Earth's climate, it is possible to start to reconstruct changes in the global climate. It should be noted, however, that this picture is still dominated by observations from stations in the northern hemisphere, and that the early part of the record of observations is, in turn, dominated by stations in north-western Europe. Nevertheless, these observations show that global temperatures have been continually rising since AD 1800, initially from a low basis, and that by 1950 the global temperature started to exceed the range of temperatures measured in the preceding millennia. Temperature proxies in both the northern and southern hemispheres, including pollen records, tree-line and tree-ring studies, and glacier retreat, show good correspondence with the direct temperature observations (see Fig. 3.4).

The discussion on palaeoclimate in the IPCC *Fourth Assessment Report* places human impact on the global climate (see below) in a long-term context (Jansen et al. 2007: 435). Amongst the most notable contributions are the observations concerning the atmospheric concentration of GHGs in AD 2000, which is highly likely to exceed its natural range. Evidence for this observation comes from air bubbles trapped in ice for a period in excess of 650,000 years. The rate of increase of radiative forcing during the twentieth century has very likely no precedent in the past 16,000 years. The reason for the high levels of GHGs lies, principally, in the extraordinary increase of fossil-fuel burning since the eighteenth century.

HUMAN IMPACT ON CLIMATE: ATMOSPHERIC CONCENTRATIONS OF GHGs

It has been argued by some climate change scientists that the first noticeable impact of anthropogenic activity on the Earth's climate dates to the onset of farming. William Ruddiman argued that early agriculture produced increased levels of CH_4, to the extent that the onset of the next Ice Age was prevented (2003; Ruddiman and Ellis 2009). This view was not supported by the IPCC, however, in its *Fourth Assessment Report* (Jansen et al. 2007: 460). The report noted that the recorded impact of early agriculture as observed in air bubbles trapped in the ice and recovered from ice cores was simply insufficiently large to have had a noticeable impact on the global climate. The differences, in terms of GHG emissions or changes to the Earth's albedo, between pre-agricultural woodland or grassland vegetation and early agricultural crops are very small, and certainly not so big as to have had an impact on the global climate. Furthermore, the IPCC report considers the current interglacial, MIS 1, mostly analogous with MIS 11, which lasted in excess of 28,000 years and had high CO_2 levels throughout. Based on the Earth's orbit around the sun, and under a natural GHG regime, the next glacial period would not be expected to start in the next 30,000 years, and therefore the claimed anthropogenic forcing of the Neolithic is not accepted.

The basic principle of GHGs and their impact on the Earth's temperature has been known for a very long time. As noted earlier, solar radiation reaches Earth as ultraviolet or very short wavelengths. The Earth's albedo ensures that about one-third of this solar radiation is reflected back into space, although the albedo is subject to significant variations over time. The remaining two-thirds of solar radiation is absorbed by the Earth's hydrosphere and geosphere, and in a small part by the atmosphere, from which a component of energy is radiated back in the form of long wavelengths which lie predominantly in the infrared part of the spectrum. On its way back into space, most of this energy has to pass through the atmosphere, where these longer wavelengths are in part absorbed and subsequently re-emitted by GHGs (see Fig. 3.1).

This basic understanding of GHGs and the Earth's greenhouse effect can be traced to the late seventeenth century. In 1681, Edme Mariotte observed that sunlight and heat passed through glass more easily than heat from other sources; Horace Benedict de Saussure conducted a simple greenhouse experiment in 1760; Fourier and Pouillot in the 1830s came to an understanding that the atmosphere absorbed more radiation from the Earth than from the sun; and in 1859 the absorption of thermal radiation properties of complex molecules such as CO_2 was established by John Tyndall (Le Treut et al. 2007: 103). Since these early pioneers, the field of climate research has made many advances in understanding how different complex molecules, including CH_4 and N_2O but also others such as H_2O or chlorofluorocarbons (CFCs), impact on absorbed solar energy, and research continues to progress rapidly. The key point here is that the concept of increasing atmospheric concentrations of GHGs leading to increased temperatures on Earth is a matter not of conjecture but of basic science.

It has already been noted that during previous glacial–interglacial transitions, global warming preceded the increase of GHGs. Based on this, and on other observations, it is not suggested by climate science that the increase of the atmospheric concentration of GHGs is the key cause for the onset of interglacial stages: it has been noted previously that this effect is a function of the Earth's orbit around the sun. However, the release of GHGs from the hydrosphere, which warms up more slowly than the geosphere, acts as a positive feedback mechanism through the increased solar forcing it brings about. The positive feedback of increased levels of GHGs explains, to a large extent, the rapid increase of global temperature in the early stages of an interglacial.

Both recent palaeoclimate research and observations made over the last 150 years indicate that the temperature in the polar regions fluctuates more than the temperature of the tropical regions. This can be explained by changes in the GHGs. Water vapour, H_2O, is the most common and influential (albeit short-lived) GHG, and high concentrations of water vapour in the form of clouds over tropical regions ensure that much of the radiation of short-wavelength energy from the Earth is reflected back here. Changes in other GHGs influence the greenhouse effect over the tropics in relatively small degrees only, as the relatively high concentrations of water vapour in both warm and cold periods attenuated the impact of other GHGs. In the polar regions, the opposite is the case. The atmosphere in these parts of the world contains very low concentrations of water vapour, and changes in other GHGs, notably CO_2, have a proportionally much greater greenhouse effect. The resulting rapid temperature shifts in the polar regions produce positive climate feedbacks through a reduction of the albedo (Le Treut et al. 2007: 115). This rapid warming of the polar regions in the early parts of each interglacial stage also explains the relatively rapid rise of the global sea level during these periods (see chapter 4).

Unlike the GHGs CO_2, CH_4, and N_2O, atmospheric concentrations of water vapour in the atmosphere vary as a direct result of changes in the climatic conditions on the surface of the geosphere, biosphere, and hydrosphere, and this GHG is thus considered a feedback effect (Solomon et al. 2007: 23). Variations in the atmospheric concentrations of CO_2, CH_4, and N_2O can have a natural origin, as is clearly illustrated in the graphs based on the air bubbles taken from ice cores, but the very rapid increase of these GHGs in the last 250 years is largely attributed to human activity, leading to anthropogenic forcing of the climate. This argument is, to a large extent, based on a comparison of what happened over the last 650,000 years. The air bubbles trapped in ice show that, between 650,000 and 250 years ago, the atmospheric concentrations of CO_2 varied between 180 and 300 ppm; that of CH_4 between 300 and 700 ppb; and the concentration of N_2O between 200 and 280 ppb. The AD 2005 measurements of the GHGs in the atmosphere produce results that lie significantly outside this 650,000-year-long variation: the measured atmospheric concentration of CO_2 was 370 ppm, that of CH_4 was 1,750 ppb, and N_2O was 315 ppb (see Fig. 3.3). In short, current atmospheric concentrations of CO_2 and CH_4, and N_2O to a lesser extent, are far in excess of their concentrations during any of the interglacial stages in the preceding 650,000 years, and are therefore considered to be almost certainly unnatural and the effect of human industrial activity.

Further detailed analysis of atmospheric levels of GHGs illustrates the accelerating pace of industrialization over the last 250 years. Take the case of CO_2. During the Holocene, atmospheric concentrations of CO_2 increased. In the 8,000 years before AD 1750 this increase was about 20 ppm, but in the 255 years following AD 1750 this increase was just short of 100 ppm, or just under 0.4 ppm CO_2 per year on average. This annual average has increased to 1.4 ppm for the 55 years up to AD 2005, and to 1.9 ppm for the ten years leading up to AD 2005 (Le Treut et al. 2007: 25).

This human industrial activity included the burning of fossil fuels from the late eighteenth century onwards, which released very large amounts of CO_2—stored over millions of years in the form of coal, oil, and natural gas—as well as the clearing of forests and expansion of agriculture, which have added large amounts of CO_2 and CH_4 to the atmosphere. Increases in atmospheric CH_4 are thought to be linked to emissions from peatlands, which naturally capture CO_2 but release CH_4, and from intensive agricultural practices, including stock breeding, rice planting, and the burning of biomass. The recent increase of N_2O is linked to agricultural expansion and related land-use changes (Le Treut et al. 2007: 27). GHGs such as CO_2, CH_4, and N_2O are long-lived gases which are not readily removed from the atmosphere through natural processes such as oxidation. Many other aspects of the modern industrialized world have an effect on radiative forcing, such as short-lived GHGs which have a temporary greenhouse effect; aerosols, including those that have been produced by the deliberate burning of biomass; the impact of urbanization and change of land use on the Earth's albedo; and the greenhouse effect produced by contrails from jet planes. The radiative forcing of many of these factors is poorly understood and not well qualified. However, taking a balanced view that considers both the warming and cooling effects of human actions, it is estimated that, at 90 per cent confidence, increased radiative forcing attributable to human activity adds between 0.6 and 2.4 watts of energy to each square metre of the Earth's surface, with a midpoint around 1.6 watts per square metre (Le Treut et al. 2007: 32). The Earth's dynamic relationship with its atmosphere is recognized, and about half of the atmospheric CO_2 attributed to human activity has been removed through a range of natural processes, including the increased growth rate of plants and trees and the uptake of CO_2 by the world's oceans (Le Treut et al. 2007: 26). Volcanic eruptions also play a role, as explosive eruptions add sulphate aerosols to the stratosphere, with an overall cooling effect.

Nevertheless, the impact of these long-lived GHGs on the Earth's climate will last for decades and even centuries. Its key result will be global warming. As noted in chapter 1, the observed changes in average global temperatures show that the Earth is warming and that the rate of warming is accelerating. Thus, linear regression trends through the observed average global temperature changes are 0.045 °C per decade over the 150 years before 2005, 0.074 °C per decade over the last 100 years, 0.128 °C per decade over the last 50 years, and 0.177 °C per decade for the last 25 years before 2005 (Le Treut et al. 2007: 37). The *Fourth Assessment Report* noted that 11 of the last 12 years before 2007 ranked amongst the warmest in the record of observed temperatures. Since then 2007, 2009, and 2010 can be added to this list, so that now 14 out of the 16 years before 2011 rank amongst the warmest on record of average global temperatures that goes back to 1850 (e.g. Kennedy, Morice, and Parker 2011).

The concept of average global temperatures hides spatial differences. These include the difference between the geosphere and the hydrosphere, with the former becoming warmer more quickly than the latter. The polar and higher northern-latitude regions also become warmer at a faster rate than tropical and lower-latitude regions, and the relative impact of water vapour and CO_2, and the distribution of the land mass across the Earth explain these differences. Changes have also been observed in regional climates, most notably the increased frequency of El Niño-Southern Oscillation events, or the increased occurrence of heatwaves experienced in Europe (Barriopedro et al. 2011). At smaller geographical and shorter timescales, climate change is influential in different ways. For example, patterns of rainfall are changing, leaving some areas that are already short of water, such as the Sahel in northern Africa, the Mediterranean, and parts of southern Asia, with even less water in the future than in the twentieth century. Global warming is likely to cause many more heavy precipitation events and increased tropical cyclone activity (Solomon et al. 2007: 41).

FUTURE SCENARIOS OF CLIMATE CHANGE

The IPCC makes projections of future changes in the climate, and in the *Fourth Assessment Report* these projections are based on the latest coupled Atmosphere-Ocean General Circulation Models (AOGCMs). These projections have two components. The first concerns the impact of the increased GHGs already in the atmosphere. The impact of these long-living GHGs is referred to as 'committed climate change'. The second concerns the impact of future global economic development and its impact on GHGs. A number of future scenarios were encapsulated in the *Special Report on Emission Scenarios* (SRES; IPCC 2000), which have been used by the IPCC in its *Third* and *Fourth Assessment Reports* in their forward scenario planning (Fig. 3.5). SRES developed different storylines within scenario families. Scenario families A1 and A2 are projections of a world with more economic focus, whereas B1 and B2 project a world with a greater concern for environmental issues. Paraphrasing this SRES report (IPCC 2000: 4–5), the principal different scenario families are:

- A1: a future with very rapid economic and population growth which reaches its peak around AD 2050 and falls thereafter, but also one wherein new and efficient technologies are introduced. Regional differences in per capita income are substantially reduced through collaborations. Within this scenario family, three different storylines are presented: A1FI represents the fossil-intensive storyline; A1T the non-fossil energy storyline; and A1B the balanced use of a range of renewable and non-renewable energy storyline.

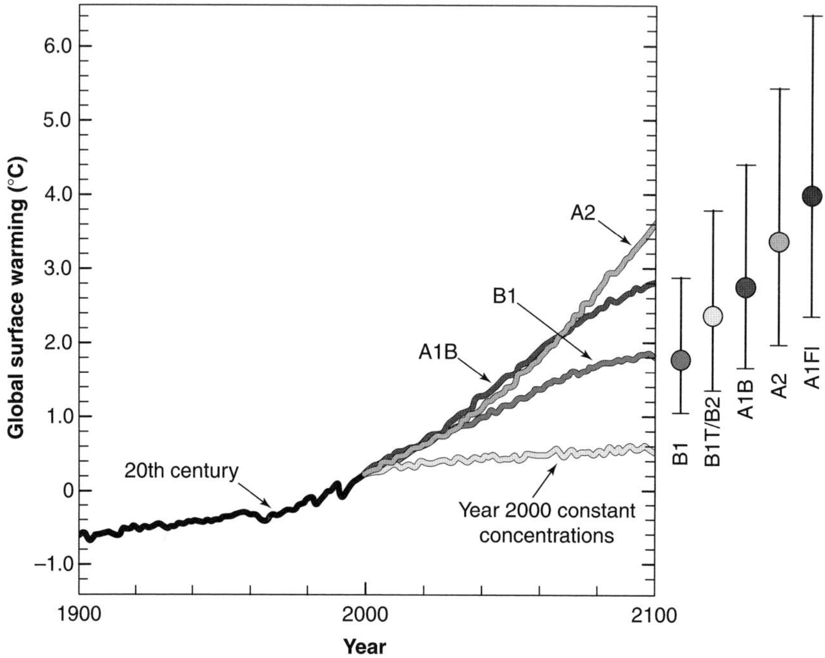

Fig. 3.5. IPCC's SRES scenarios. Each scenario concerns a different storyline for global development in the twenty-first century (see main text for definitions). The 'Year 2000 constant concentrations' shows the committed climate change: an increase of global temperature of 0.6 °C. Best estimates for scenario B1 is an increase of 1.8 °C (likely range: 1.1–2.9 °C) and for scenario A1FI of 4.0 °C (likely range: 2.4–6.4 °C). For full caption and source references see IPCC 2007: 14.

- A2: a future with economic and population growth, but one wherein regional differences and self-reliance are maintained, and where the global population continues to grow after AD 2050 and where new and efficient technologies are introduced at a much slower pace than in A1.
- B1: a future with an increasingly integrated and cooperative world seeking global sustainability, where population peaks around AD 2050 but falls thereafter, and where rapidly changing economic structures enable the emergence of clean and resource-efficient technologies.
- B2: a future with an emphasis on regional and local sustainability, with a growing population and where new and resource-efficient technologies are developed, but introduced at a slower pace on a global scale because of the fragmented nature of this scenario family.

The best estimate of the impact of committed climate change is that global average temperature for the last decade of the twenty-first century will be 0.6 °C (±0.3 °C) warmer than the last decade of the twentieth century. The best estimates for the impacts on the global average temperature for the different scenarios, which should be added to the impact of committed climate change, are as follows:

- A1FI: 4.0 °C (likely range: 2.4–6.4 °C)
- A1T: 2.4 °C (likely range: 1.4–3.8 °C)
- A1B: 2.8 °C (likely range: 1.7–4.4 °C)
- A2: 3.4 °C (likely range: 2.0–5.4 °C)
- B1: 1.8 °C (likely range: 1.1–2.9 °C)
- B2: 2.4 °C (likely range: 1.4–3.8 °C).

Taking into account recent developments in the field of climate change policy, the majority of scientists working on how the scenarios will affect climate change have concentrated on scenario families A2 and B2. The slow pace at which the 1997 Kyoto Protocol was ratified, accepted, approved, or accessioned by the parties to the UNFCCC has shown that in the first decade of the twenty-first century national and economic concerns retain higher political priorities than global environmental ones. It took more than seven years before enough parties had signed up to the Protocol for it to be entered into force, and the combined CO_2 emissions of these parties represented only 63.7 per cent of the total global emissions.

On this basis, the increase of global temperatures by the end of the twenty-first century is predicted to be between 3 °C and 4 °C higher than that at the end of the twentieth century, adding the committed climate change to the A2 and B2 scenario families. Temperature rise is expected to be most pronounced over the polar regions and the continents of the northern hemisphere, especially in the winter, whilst the slower warming of the oceans will attenuate the warming effect in coastal areas to some extent. Global warming will lead to further sea-level rise, both through thermal expansion of the oceans and the melting of land ice (see chapter 4).

CONCLUSION

The purpose of this chapter has been to provide archaeologists with an understanding of the key concepts from climate change science, focusing in particular on the findings contained in the IPCC's *Fourth Assessment Report* (2007). Modern palaeoclimate research has been described, and particular emphasis given to the global scale and how research in the deep sea and polar ice has provided new sources for reconstructing the Earth's climate over a period of millions of years, with the air bubbles trapped in ice providing direct information on the atmospheric composition for the last 650,000 years.

Next, this chapter has sought to contrast the long-standing archaeological and palaeoenvironmental research on regional and local palaeoclimates with the global approach which is the key concern of climate change science. Palaeoclimate evidence from the deep past has shown how significant changes of the Earth's temperature, resulting in the glacial and interglacial stages of the Quaternary, can be correlated to the Milankovitch curve, which is made up of three variables each describing the spatial relationship between the sun and the Earth. These temperature changes can be shown to be correlated in time with the concentration of GHGs, and before AD 1750 this represents a wholly natural relationship. Since AD 1750, however, the atmospheric concentration has increased, and this increase continues to accelerate as a direct consequence of the emission of GHGs through human activity, in particular the burning of fossil fuels, deforestation, and the expansion of intensive agriculture. Higher GHGs produce the greenhouse effect, and the observed global warming at the end of the twentieth and beginning of the twenty-first century is almost certainly happening because of anthropogenic forcing. This chapter has, finally, considered how the IPCC makes its projections for future climate change.

4

How Climate Change Will Affect Coastal Wetlands and Coastal Communities

INTRODUCTION

In the previous chapter the scientific evidence for global climate change was presented. But how will this affect coastal communities? This chapter introduces the principles and concepts that underpin the chain of relationships that links global climate change with the lives of coastal communities around the world.

In essence, climate change has a direct relationship with sea levels, since sea levels rise during periods of global warming. Furthermore, the predicted increased frequency of extreme weather events is relevant here. Higher sea levels and more storms will affect coastal wetlands in myriad ways. Most types of coastal wetlands will respond autogenically—without human interference—by accreting sediments and raising the surface of the wetlands, and by migrating landwards. However, coastal wetlands currently in existence have developed during the middle and late Holocene, a period during which the rate of sea-level rise decelerated, whilst the rate of sea-level rise is set to accelerate in the twenty-first century. Without clear precedents for such an environmental change, we cannot draw on palaeoenvironmental studies nor predict with accuracy what this will mean for coastal wetlands. Some wetlands are expected to survive practically unchanged, but others may drown beneath the rising sea. The growing presence of human activity in the coastal zones is expected to aggravate the impacts of sea-level rise, for example where artificial sea defences prevent the landward migration of the coastal wetlands. Coastal wetlands provide coastal communities with a range of socioeconomic benefits or 'ecosystem services': fresh water, food from agriculture, fishing, and aquaculture, protection from floods, living space, transport routes, and income from tourism, to name the most important ones. These services will be affected when coastal wetlands diminish in extent, or alter their nature, as sea levels rise and more frequent storms occur.

This chapter will explain this chain of relationships in some detail. It commences with a section on how future scenarios of climate change will affect sea levels. The second part will explain how higher sea levels and more frequent extreme weather events will impact on the different types of coastal wetlands. It will also consider the additional stress placed on coastal wetlands by the way societies will adapt to the environmental impacts of climate change. The consequences of this on communities living on the coasts will be presented next, focusing on the various socioeconomic services currently delivered by the coastal wetlands. The final section introduces the concept of the Integrated Coastal Zone Management (ICZM), as a tool that can guide the way in which groups and individuals with an interest in the coast can work together.

CLIMATE CHANGE, SEA-LEVEL RISE, AND COASTAL TRANSGRESSION

Climate change affects ESL in two key ways. First, the amount of water available to fill the oceans is determined by the extent of the land-based ice sheets: the more water locked up in ice, the lower the ESL will be. When land ice melts, it adds water to the oceans. Changes in the volume of sea ice, for example in the form of frozen seas in the polar regions or as icebergs, have no impact on the sea level. Second, when the oceans become warmer the water contained within is subject to thermal expansion. Thermal expansion refers to the tendency of nearly all materials to expand when heated. The thermal expansion of ocean water may appear relatively small: it needs to be heated by 40 °C to expand by 1 per cent. However, the vast amounts of water contained within the Earth's oceans—estimated at around 1.37 billion km^3 (e.g. Duxbury 2000: 39)—produce a noticeable increase in the volume of water when the oceans warm up. For example, an increase of 0.6 m in the level of the sea is achieved when an ocean over a depth of 4,000 m is increased by just 1 °C (Masselink and Hughes 2003: 25). In reality, the thermal expansion of the Earth's oceans, or steric sea-level change, is not a linear process because the water in the oceans is not well mixed. Thus, thermal expansion of the oceans is the result of a complex interplay between warm and cold water, and between water at higher and lower pressures, producing sometimes diverging trends in the temperature changes of the oceans. Because these two principal ways in which climate change affects global sea levels complement each other, ESL rises during warm interglacial stages, and falls during cold glacial stages (Bindoff et al. 2007: 408).

The record of sea-level change for the Holocene has already been introduced in chapter 1. Reconstructing past sea levels involves today a very broad range of palaeoenvironmental proxies obtained from cores of sediments from seabeds or from coral reefs, combined with isotopic analysis and radiometric dating (see chapter 3). The Relative Sea Level (RSL) reconstructed from research into the coral record off Barbados is considered to offer a good approximation to the Eustatic Sea Level (ESL), as this island near the equator was never subjected to any tectonic alteration that can be attributed to the presence or absence of ice (see below). The evidence from Barbados shows that at the Last Glacial Maximum (LGM), 21,000 years ago, very large quantities of water were held in land ice, not just in the polar regions but also on the continents, and the sea level was some 120 to 130 m lower than it is today. With the onset of the Holocene/Marine Isotope Stage 1, ESL started to rise. Between 23,000 and 16,000 cal BC, the rate of this rise was relatively slow, especially compared with the subsequent 7,000 years, when the ESL rose from about −110 to −10 m. Over the last 8,000 years, the rate of sea-level rise has decelerated, and has made up the last 10 m to reach the current levels (see Fig. 1.5).

In the twentieth century, direct observations were made possible by two major advances in the estimation of sea levels. The first was the network of linked tide-gauge observation stations, part of the Global Sea Level Observing System (GLOSS; www.gloss-sealevel.org). There are now over 1,400 stations across the world, with both the northern and southern hemisphere well represented. The second advance concerned the development and application of satellite altimetry. This technique now provides a reliable means of directly observing sea levels. These direct observations indicate that the contribution of land ice melt to sea levels has been 0.5 mm/year between 1961 and 2003, and 0.77 mm/year for the decade between 1993 and 2003 (Bindoff et al. 2007: 418). The melting of the ice from the polar regions 'freshens' the oceans at high latitudes because meltwater is added here in large quantities. Estimates for the sea-level rise due to thermal expansion reflect its dynamic nature. The measured warming of the oceans up to a depth of 700 m is calculated at contributing in excess of 20 mm to the ESL through thermal expansion in the two decades before 1980. This was followed by a five-year period of sea-level fall, and a subsequent period of 25 years of rising sea levels (see Fig. 1.5). Overall, the measured warming of the oceans accounts for a net rise of some 17 mm between 1961 and 2003, or an average rate of 0.32 mm/year. If the oceans below 700 m depth, which have not been measured, warmed in a congruous manner, then the rate of sea-level rise for this period increases to 0.42 mm/year. However, sea-level rise due to thermal expansion appears to be considerably higher for the last decade of the twentieth century, with an estimated annual rate of 1.5 mm/year, but the time period over which this observation holds true is rather short (Bindoff et al. 2007: 415).

In the IPCC's *Fourth Assessment Report* no conclusions are drawn as to which of these two factors has had the greatest impact on the ESL for the period from 1961 to 2003, but it appears that for the period 1993 to 2003 the impact of thermal expansion is the greatest contributor to sea-level rise. Combining the effects of the melting of land ice and thermal expansion, ESL rose by an average of 1.7–1.8 mm/year in the twentieth century, accelerating to c. 3 mm/year in the final decade. Regionally and locally, wave heights have become higher and the frequencies of storm surges and extreme high-water levels have increased in the last decades of the twentieth and in the early twenty-first centuries (Rosenzweig et al. 2007: 92–3). It is, however, not always clear to what extent these changes can be attributed to the impact of climate change or human adaptation to these impacts, or are happening independent from global climate change.

The IPCC has calculated the impact of twenty-first-century climate change according to the SRES scenarios (see chapter 3) on the contribution of thermal expansion and the melting of land ice to ESL. The annual sea-level rise by the end of the twenty-first century is calculated to be in the range of 1.5–3.9 mm/year for the global sustainability scenario, SRES B1, and in the range of 3.0–9.7 mm/year for the fossil intensive storyline, SRES A1FI. The impact of climate change on the ESL, comparing the last decade of the twentieth with the last decade of the twenty-first century, is calculated as an increase of 180–380 mm for SRES B1 and 300–590 mm for SRES A1FI (Meehl et al. 2007: 820). Concern that these estimates are too low, and should be doubled at the very least, is widespread amongst the scientific community (e.g. Rahmstorf 2010).

If global warming were to continue in future centuries, how much higher could the ESL rise? The past provides potential answers. Research on coral remains near the equator, and on coastal deposits that are beyond the influence of glacial ice and glacio-tectonic land movements, have shown that at the end of the Marine Isotope Stage 5 interglacial (around 125,000 years ago), ESL was 4 to 6 m higher than it is today. Further back in time in the Pliocene (5.3–2.6 million years ago), sea levels were as much as 25 m higher than today, but the configuration of the tectonic plates was somewhat different from the present one and direct comparisons are not necessarily appropriate. Nevertheless, with Pliocene atmospheric concentrations of CO_2 between 360 and 400 ppm, and temperatures between 2 °C and 3 °C higher, there appears to have been a clear correspondence between climate, the extent of land ice, and ESL. Whilst this does not constitute proof that the ESL will rise to such heights in future centuries if GHGs are not reduced, it illustrates the point that the potential for sea-level rise exceeds the IPCC calculations for ESL for the twenty-first century.

So far in this section, the impact of climate change on sea-level rise has focused on the ESL. However, of more significance to local communities is the change of the Relative Sea Level (RSL), which takes into account the fact that a land mass is not static. The best-studied aspect of this is known as glacio-isostatic adjustment, which is driven by the transfer of water between the hydrosphere and the cryosphere. This transfer produces changes in the loading of the Earth's crust and mantle. Thus, the weight of the ice sheets deforms the crust and presses it down into the underlying asthenosphere, and the land will fall. The extent of the fall is approximately one-quarter of the depth of the ice sheet on the land surface. However, the asthenospheric material is redistributed, lifting areas that are not loaded by ice sheets and producing the so-called forebulge. When ice melts, as with the transition from glacial to interglacial stages, the isostatic balance between the Earth's crust and the asthenosphere is restored, and this is referred to as the glacio-isostatic rebound (Fig. 4.1). This rebound occurs over very long time spans. For example, the glacio-isostatic rebound of the middle of the uplifted area of eastern North America measures nearly 140 m to date, but the ongoing rate of rising is 20 mm/year (Walcott 1972).

The glacio-isostatic rebound has profound effects on how ESL change is experienced on land. Because the ice sheets were most substantial in weight and size in northern parts of the North American and Eurasian continents, the difference between ESL and RSL is most noticeable in these regions (Steffen and Wu 2011). For example, in the southern Baltic region, where the glacio-isostatic uplift was negligible, RSL rise was a consistent feature throughout the Holocene, but in the northern part of the Baltic, where the maximum glacio-isostatic uplift to date is in excess of 300 m, RSL has fallen for the last 8,000 years (Harff et al. 2005). Movements of the Earth's crust are not limited to the glacio-isostatic rebound, and other 'neotectonic' activities can raise or lower land masses. For example, shelf-edges are prone to sinking where excessive sediment loading takes place, as is the case in the Ganges–Brahmaputra delta in the Bay of Bengal (Stanley and Hait 2000; see chapter 6), and earthquakes can lift or lower land masses, as has been claimed for the Pacific coast of Washington State (Atwater 1987).

What is the impact of a rising RSL on coasts? Generally speaking, the impact will be a transgression of the marine environment onto the receding coast, although locally this impact may be attenuated, prevented, or even temporarily reversed due to the riverine transport of sediments to the coast (known as progradation; Fig. 4.2). An example of the latter is where clastic sediments eroded from a particular stretch of the coastline are deposited along the path of the prevailing wind and wash to extend beaches and dunes, a phenomenon known as longshore drift. There is much evidence from around the world for recent marine transgression and coastal regression linked to higher sea levels (Leatherman, Zhang, and Douglas 2000). For example, 67 per cent of the UK's eastern coastline has retreated landward since the middle of the nineteenth century (Taylor, Murdock, and Pontee 2004); the rate of erosion of the Arctic coast of Alaska doubled in the second half of the twentieth century (Mars and Houseknecht 2007), and coastal erosion and an increase in the frequency of flooding of the Tangier Bay hinterland in Morocco is predicted for the first half of the twenty-first century (Snoussi et al. 2009).

At a high level of abstraction, there are several theories and formulas that seek to calculate the impact of future sea-level rise on coastlines.

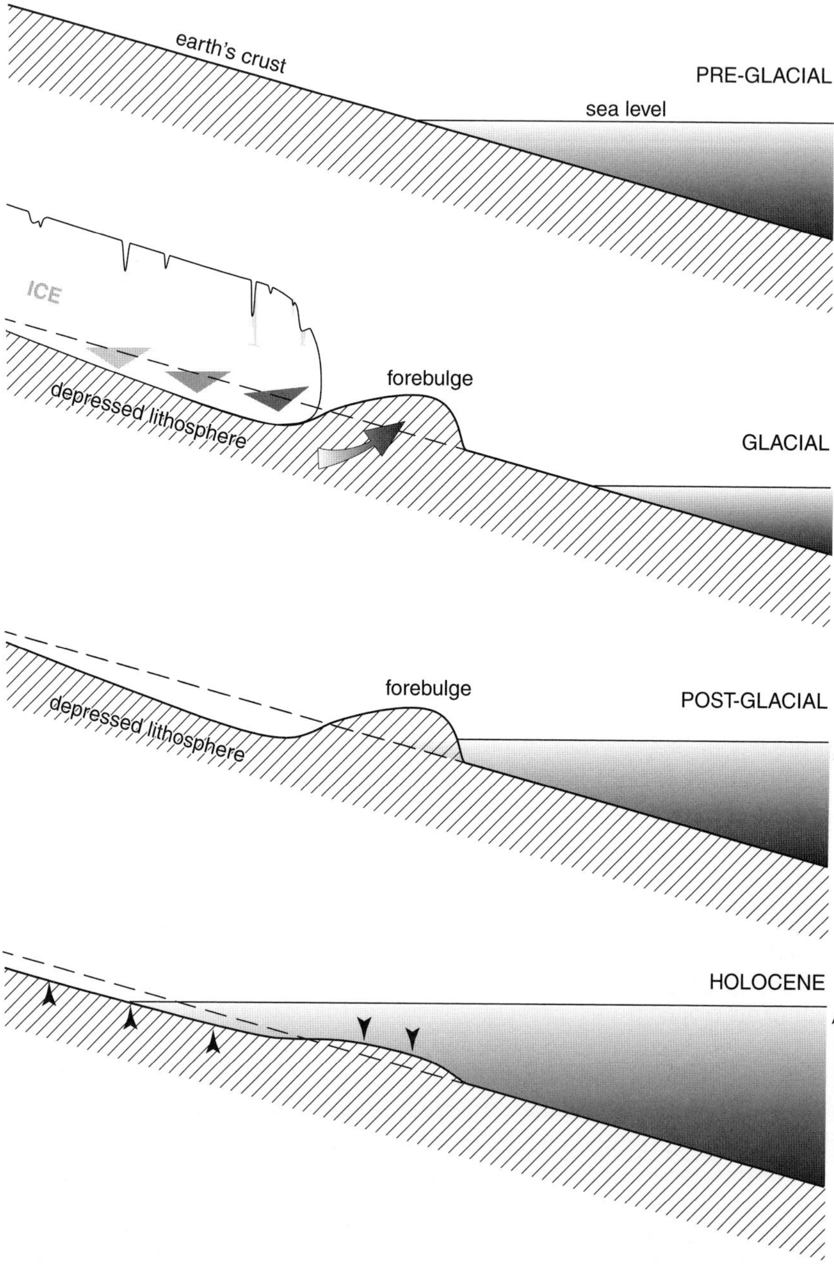

Fig. 4.1. A schematic representation of glacio-isostatic adjustment and sea-level rise. During the glacial period, ice depresses the lithosphere as it puts pressure on the underlying asthenosphere; displacement of the asthenosphere creates a forebulge. In the postglacial period and the Holocene, the ice disappears and the redistribution of the asthenosphere returns the lithosphere to its preglacial position; however, this is a very slow process, but sea levels respond instantly to the amount of water held in the world's ice sheets.

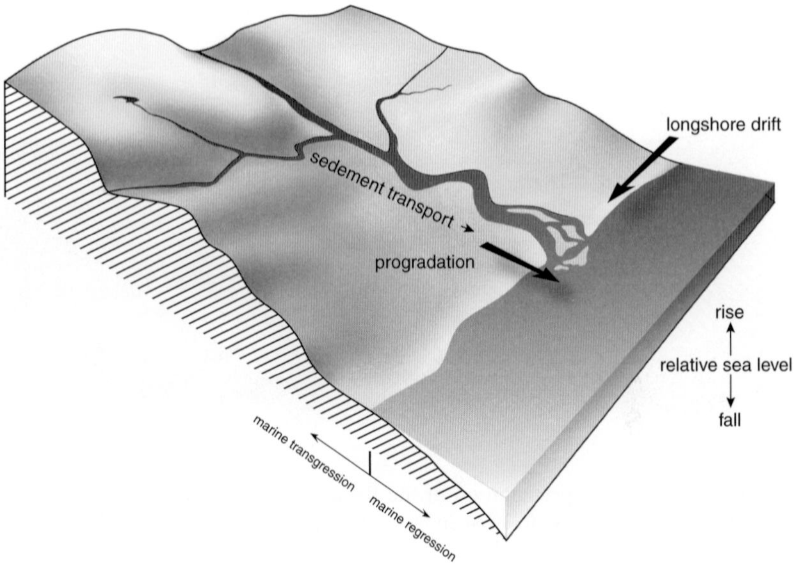

Fig. 4.2. A schematic representation of coastal responses to sea-level change. A rise in the RSL brings about marine transgression, and a fall results in marine regression. Rivers transport sediments to the coast, and the ensuing seaward migration of the coast is called progradation. Sediment transport parallel to the coastline is referred to as longshore drift.

The well-established Bruun theory (Bruun 1962, Schwartz 1967) states that following a rise in the RSL, the beach profile recedes landward as the upper part of the beach is eroded; the volume of the eroded material is equal to the material deposited in the near-shore, and the resultant rise of the near-shore bottom matches the sea-level change (Fig. 4.3). Tested in a range of natural settings, this theory has been found to work well in coastal areas with predominantly medium-fine sediments, such as sandy beaches.

Others have sought to understand the impact of future sea-level change by looking at the record of past coastal change. One such formula was designed to apply to soft-rock shores with either a shallow beach or no beach (Walkden and Dickson 2008: 83). The formula predicts the future impact as follows:

$$\text{future recession rate} = \text{historical recession rate} \sqrt{\frac{\text{rate of future sea-level rise}}{\text{rate of historical sea-level rise}}}$$

Thus, if the sea-level rise in the twenty-first century is 2.4 times the rise measured in the twentieth century—IPCC's best estimate (Nicholls et al.

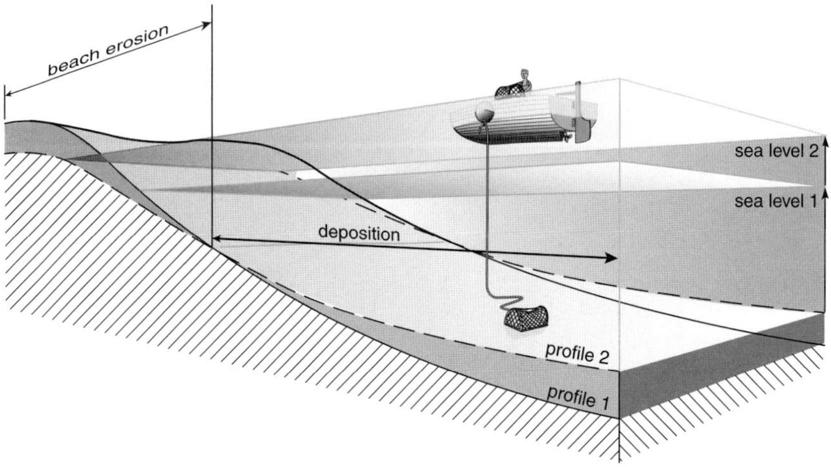

Fig. 4.3. A schematic representation of the Bruun theory on the impact of sea-level rise on beaches. In response to a rise in the RSL, beach profiles recede landward as the upper part of the beach is eroded. The volume of the eroded beach material is equal to the material deposited in the near-shore. The resultant rise of the near-shore bottom from profile 1 to profile 2 matches the sea-level change from level 1 to level 2.

2007: 324)—then the rate by which the coastline recedes in the twenty-first century is 1.55 times the rate observed in the twentieth century.

Bruun's theory and Walkden and Dickson's (2008) formula focus on the establishment of new geomorphological equilibria following a period of sea-level rise or adjustment to 'dynamic equilibria' respectively. Other researchers have stressed the complex and dynamic nature of coastlines and coastal systems. Their research has addressed, for example, the importance of the transport of sediments through rivers and estuaries, the formation of dune systems, and the existence of natural thresholds that produce non-linear responses to change. At their most extreme, coastal systems can drown where the rate of sea-level rise exceeds the geomorphological upward growth rate. Furthermore, extreme events such as storms and hurricanes are now recognized as formative forces in the development of shorelines and coastal systems (cf. Edwards 2008). An important element in determining coastal erosion is the human element, not only in the sense of the defence of the coast against marine transgression, but also through unforeseen consequences of human actions, ranging from the damming of rivers which alter natural sediment flows, through to the cutting of mangrove forests which act as natural protectors against coastal erosion (see below).

Climate change is responsible not only for altering the temperature and level of the oceans, but also for a range of observed chemical and biological changes. One example is the acidification of the oceans, with an average decrease in pH of

0.1 units since AD 1750 being directly attributed to the oceans acting as CO_2 sinks (Nicholls et al. 2007: 320). Other changes include changes in oxygen levels and circulation patterns, although there is as yet no evidence that the major circulation patterns, such as the North Atlantic Oscillation, are altering as a direct consequence of climate change. These biophysical changes impact directly on the biological nature of the oceans. For example, primary producers such as plankton expanded into seas at higher latitudes by about 10° in the North Atlantic in the second half of the twentieth century, and this poleward movement will affect the behaviour of higher-order producers (Rosenzweig et al. 2007). The impact of these changes on human communities will be felt most strongly by fishermen and their families. Amongst researchers there is broad agreement that climate change will have significant impacts on global fish production, but the impact of regional climate changes such as the El Niño-Southern Oscillation dominate the current research (Brander 2007; see below). These changes are likely to have an impact on coastal wetland, in particular where adaptive strategies in fragile landscapes, such as mangrove forests, include the extension and intensification of aquaculture and agriculture to offset reduced fish catches. The impact of human adaptation to climate change and sea-level rise on coastal wetlands will be discussed later in this chapter.

This section has sought to explain the way in which global climate change drives sea-level change, through the transfer of water from the cryosphere to the hydrosphere and through the mechanism of thermal expansion. The world's oceans have never been static, and significant changes in the ESL are a key characteristic of the Quaternary. The regional glacio-isostatic adjustments mean that few coastal regions across the world will experience the same sea-level rise, and for this reason the concept of RSL is important when considering the impact of climate change-driven sea-level change on local communities. Nevertheless, the rate by which the ESL is set to rise in the twenty-first century will have a significant impact on coastal wetlands and the communities that are, for economic, social, or cultural reasons, dependent on them.

COASTAL WETLANDS AT RISK FROM CLIMATE CHANGE

Coasts are dynamic systems, and their point-in-time manifestation is the result of terrestrial and marine geomorphology and processes. Sea-level change is a marine process that has been a factor throughout the Quaternary and the Holocene. However, the impact of global climate change on the marine environment will come in the form of an acceleration of the rise of the ESL after a 7,000-year period of deceleration, and there are no palaeoenvironmental precedents for the impact of this on the coastal wetlands around the world. Climate change is also set to produce increased frequencies of storminess and extreme weather events, and these will impact directly on coastal wetlands, and indirectly through changes in the flow of, and sediment transport through, rivers and estuaries. Whether coastal wetlands receive or lose sediments—whether they have a positive or negative sediment budget—is a key factor in the sustainability of many coastal wetland types. Finally, climate change will have an impact on the temperature and chemistry of the oceans, which may have an impact on coastal flora and fauna (Nicholls et al. 2007). This section will consider the environmental impact of climate change on coastal wetlands generally, and also on the different coastal wetland types, as defined by the Ramsar classification system (www.ramsar.org). It concludes by considering the impact of societal adaptation to climate change and the feedback from this adaptation on coastal wetlands.

Climate change impacts on coastal wetlands

The aggregated impact of global climate change on coastal wetlands is considered very significant in the IPCC'S *Fourth Assessment Report*. The report refers to a paper by Lorraine McFadden, Tom Spencer, and Robert Nicholls (2007), which is based on their Wetland Change Model. This model was designed to quantify the impact of sea-level rise on coastal wetlands and characterizes coastal wetlands in 85-km segments by ecological type and by a set of controlling factors. This model uses three controlling factors, which enable coastal wetlands in the model, as in the real world, to respond to sea-level change through vertical adjustment of the surface and/or species composition, and horizontal/landward recession (e.g. Allen 1990; see also below). The first controlling factor is the ratio of sea-level rise to tidal range. In coastal situations, the sea level is subject to tides, and a rise in sea level is experienced as a change in the tidal range—the height difference between high and low tide—and the congruent time that certain parts of the intertidal landscape are under water. If intertidal zones rise with the sea level, they are more likely to be resilient to change, and coasts with the largest tidal ranges appear to have the greatest resilience to sea-level rise. The second factor concerns the availability of sediments. Sediment availability—especially that of fine-grained sediments such as sands which are highly mobile in intertidal zones—is essential for coastal wetlands to raise their surface in periods of sea-level rise, and to maintain their ecosystems. The third controlling factor concerns the presence or absence of so-called accommodation space, or the ability of coastal wetlands to migrate landwards in response to rising sea levels. These three factors are weighted in the Wetland Change Model as follows:

0.5 for the ratio of sea-level rise to tidal range; 0.3 for the sediment availability; and 0.2 for the presence of accommodation space.

Calculating the global effect of sea-level rise on coastal wetlands using the Wetland Change Model shows that with a 0.5 m increase of the ESL by 2080, some 32 per cent of coastal wetlands will have been lost. If the ESL was to rise by 1.0 m, the loss of coastal wetlands would increase to 44 per cent. These figures are widely quoted, including in the IPCC's *Fourth Assessment Report* (Nicholls et al. 2007). These figures estimate the loss of coastal wetlands to open water, or the drowning of the coastal wetlands, but this will affect the low-lying mudflats and salt marshes proportionally more than higher mangrove swamps and coastal forests.

More recent research has stressed the role of plant communities in the adaptation of ecosystems to sea-level change. This research notes, for example, that coastal wetland vegetation including glasswort and mangroves are capable of trapping increased amounts of sediments during periods of sea-level rise, and the accretion of organic matter in the form of organic mud or peat outpaced the rise of sea level in the twentieth century. Thus, mathematical models need to recognize the vertical growth 'threshold' rates for the different types of coastal wetlands, taking into account the tidal range and sediment availability (as above). Coastal wetlands will be drowned only if sea-level rise exceeds these thresholds. The research concludes that the threshold of many coastal wetlands, including the Plum Island Estuary in New England and the Albemarle-Pamlico Sound marshes in North Carolina, is around 5 mm/year, above the ESL rate for the most optimistic SRES scenario B1 but towards the lower end of the range of sea-level change for the most pessimistic SRES scenario A1FI (Kirwan et al. 2010). Thus, the majority of coastal wetlands will survive the impact of global climate change if humanity can reduce the increase of GHGs, but coastal wetlands will be lost to a very large extent if the output of GHGs from fossil fuels continues to grow.

Climate change impacts on Ramsar-type coastal wetlands

Turning to the different types of coastal wetlands as defined by the Ramsar convention, the following predictions can be made for their future.

Marine subtidal aquatic beds; includes kelp beds, sea-grass beds, tropical marine meadows

The IPCC's *Fourth Assessment Report* notes the decline of this type of coastal wetland as a direct response to sea-level rise and other elements that have an impact on the sea grasses, including warming of the sea and increases in salinity, atmospheric CO_2, and ultraviolet irradiance, as well as increased storm activity which is capable of damaging the vegetation (Nicholls et al. 2007: 329). It is anticipated that kelp beds may become less resilient in warmer waters that facilitate their growth but reduce competitive interaction between juvenile kelp, leading to diminished resilience of mature kelp beds to storm activity (Wernberg et al. 2010). The composition of seagrasses has been shown to have changed significantly in the last decades of the twentieth century as a direct consequence of higher sea levels and warming of the oceans. This is likely to lead to the loss of important habitat functions (Micheli et al. 2008). Climate change has had some surprising and unexpected impacts on certain coastal wetlands included in this group. For example, a case study of the Tasmanian kelp beds has shown that warming of the sea at four times the global average enabled the long-spined sea urchin to extend its habitat range into the kelp beds, with as a consequence the catastrophic overgrazing of the flora. This has coincided with over-fishing of the spiny lobster for human consumption, removing a key predator from the ecosystem and a further acceleration of the overgrazing (Ling et al. 2009).

Coral reefs

It seems unlikely that coral reefs, which survived the rapid sea-level rise of the postglacial and early Holocene periods, are at danger of drowning as long as other stresses can be avoided. Stress on corals around the world comes in the form of coral bleaching, a widespread phenomenon by which corals appear to lose their colour due to the loss of essential symbiotic algae and/or their pigment. Coral bleaching has been observed since the 1980s, and is taken as an indicator of the potential for corals to grow. There is a direct correlation between the temperature of the sea surface and coral bleaching, which happens when the sea surface temperature is raised by 1°C. Prolonged coral bleaching, or when the sea surface temperature is raised by more than 2°C, results in corals dying. Some coral types, such as fanning corals, are more susceptible to bleaching than others, like massive corals (Nicholls et al. 2007: 321–2). More recent research has correlated, through palaeoenvironmental studies, past periods of ocean warming with periodic decline of coral reefs. It emphasizes the importance of ocean acidification as an important contributing factor to the ability of coral reefs to grow and to maintain strong coral-reef structures, alongside the impact of warming of the oceans. It concludes that coral reefs are indeed at grave risk from the impacts of climate change, but that earlier statements on the imminent decline of the coral reefs did not reflect the regional diversity in the response of coral reefs to ocean warming and acidification (Pandolfi et al. 2011). The forecasted increase in extreme weather events such as hurricanes and storminess is set to result in the further weakening of coral-reef structures (Nicholls et al. 2007: 330; Lough 2008).

Rocky marine shores; includes rocky offshore islands, sea cliffs

Rocky coasts, and especially hard-rock coasts are, of all the coastal wetlands types, the most resilient to the impact of sea-level rise. The impacts will come in the form of increased toe erosion, and erosion resulting from the increased frequency of extreme weather events. Soft-rock cliffs may, in addition, be affected by higher groundwater levels following increased precipitation and the impact of frost, which fractures the geology. The earlier mentioned formula by Walkden and Dickson (2008) was developed, in part, on the basis of its application to the soft-rock (chalk) cliffs of Norfolk, UK, which suggested a linear relationship between sea-level rise and coastal regression (see above).

Sand, shingle, or pebble shores; includes sandbars, spits, and sandy islets; includes dune systems and humid dune slacks

A general trend of retreating sandy shorelines has been observed for the second half of the twentieth century, with sea-level rise being one underlying cause (Nicholls et al. 2007: 324). The Bruun model may be applied here, implying a direct link between the increased level of the sea and the amount of material that will be removed from existing beaches (see above). This effect may be exacerbated if coastal embayments such as estuaries and lagoons seek to maintain their equilibrium by building up their beds, thus starving the beaches of sediments. The expectations of more frequent extreme weather events may disproportionally affect beaches and dune systems. The low-gradient dissipative beaches are the most at risk from erosion, and this type of shore contains the greatest biodiversity. Global warming may have unforeseen impacts on the beaches' ecosystems, either through higher ambient and sea temperatures or through the interaction with the sea, and acidification of the sea is expected to affect certain sandy-beach crustaceans and molluscs in particular (Defeo et al. 2008). Pebble and gravel beaches are the most resilient of this shore type, but the increased frequency of extreme weather events may have as yet unexpected impacts on these beaches. The impact of climate change on dune systems is thought to come principally in the form of increased wind strength and prolonged periods of drought which, when combined, can reactivate fixed dunes into active dunes, a process that is considered irreversible (Yizhaq, Ashkenazy, and Tsoar 2009).

Estuarine waters; permanent water of estuaries and estuarine systems of deltas

Because of their intertidal nature, estuaries are vulnerable to drowning during periods of rapid sea-level rise, and this is exacerbated when the sediment budget is negative. The predicted increase in the frequency of extreme weather events may damage existing estuaries beyond (natural) repair, as has been shown by the impact of hurricanes Katrina and Rita on the Mississippi delta (e.g. Day et al. 2007). Meta-analysis of some of the world's largest deltas has shown that this type of coastal wetland diminished in extent and biodiversity during the later parts of the twentieth century, with human development accounting for just under half this loss (Coleman et al. 2008). The interaction between climate change and human modification or alteration of the natural sedimentation processes in estuaries and deltas is particularly important. For example, the construction of dams upstream from the estuaries was found to be a major factor in reducing the sediment input for estuaries, with the depletion resulting in the increase in drowned estuaries and deltas (Day et al. 2008; see also below). The ecogeomorphology of this type of coastal wetland is also affected by climate change through changes in air and water temperature and salinity changes. Many estuaries around the world are important spawning grounds for fish, and there is a growing body of evidence which indicates that the warmer waters with reduced levels of dissolved oxygen, or 'hypoxia', are not well suited as spawning grounds when compared to waters with high levels of dissolved oxygen (Nicholson et al. 2008). As hypoxia is expected to become more common with higher ambient and water temperatures, the sustainability of some fish species may be at risk as a direct effect of climate change.

Intertidal mudflats, sand or salt flats, and intertidal marshes; includes salt marshes, salt meadows, saltings, raised salt marshes; includes tidal brackish marshes

These types of coastal wetland are at high risk from drowning following sea-level rise because of their low-lying intertidal nature. The ecological systems of mudflats, sand and salt flats are thought to be at risk from the effects of climate change in particular because of the rise of ambient and water temperature, which may affect the primary producers within the mud and the higher-order organisms such as birds and fish which feed on these. The geomorphological vulnerability of intertidal marshes lies especially with the plant species such as the halophytic grasses that characterize these saline coastal wetlands. If changes in the sea level, temperature, or chemistry of the water occur faster than the resident plant communities can adjust to, then the nature of the intertidal marsh will change. However, the ability of salt marshes to grow upwards through the vertical accretion of sediments, especially in areas with large tidal ranges, and to retreat horizontally where there is sufficient accommodation space, is well known. This ability makes coastal wetlands in this group relatively resilient to sea-level rise (Allen 1990). Other potential threats include the increase of river flow following the increased occurrence of extreme weather events, and the 'freshening' of estuaries following increased precipitation, which could affect phytoplankton and fish species for which the estuary acts as a nursery (Nicholls et al. 2007: 328). Research across the world has shown that many salt marshes have existed from the middle Holocene, and their early formation is linked to the deceleration of the rise of the ESL. Over the last millennia, salt marshes have accreted vertically and regressed horizontally in response to sea-level rise. However, the idea that salt marshes remain in a dynamic equilibrium with the sea level is no longer valid, and large-scale erosion or drowning of this type of intertidal wetland has been recorded since the 1970s. Well-known examples of this loss are known for the Louisiana wetlands, Jamaica Bay in New York City (Hartig et al. 2002), and the Venice Lagoon (e.g. Cola et al. 2008), although the direct link with climate change is not always evident. There is also widespread evidence that low salt marshes are encroaching on high salt marshes because of the lack of accommodation space (Stevenson and Kearney 2009). However, it is premature to forecast the end of this type of coastal wetland. For example, there are suggestions that because the groundwater in salt marshes has risen in response to higher sea levels, a greater amount of dead organic material is retained in the marshes, thus counterbalancing the impact of sea-level rise. It has also been asserted that the elevated levels of atmospheric CO_2 increase the productivity of salt-marsh vegetation, which can therefore accrete vertically more rapidly than was the case in the past (Langley et al. 2009). Nevertheless, many intertidal marshes will eventually run out of accommodation space, and in such situations salt marshes are likely to shrink or disappear.

Intertidal forested wetlands; includes mangrove swamps, nipah swamps, and tidal freshwater swamp forests

This category is dominated by mangrove forests, which are found in the tropics and at low latitudes around the world. Regionally, other types of forested wetlands are a common sight in the intertidal zone. For example, nipah palm swamps are characterized by coasts and tidal rivers in South Asia, whilst tidal swamp forests dominated by the Atlantic white cedar (*Chamaecyparis thyoides*) and red maple (*Acer rubrum*) are more common in northern America. Generally speaking, intertidal forested wetlands are well suited to adapt to the rates of sea-level rise that have been predicted under the various SRES scenarios. Because of their natural height and high growth rate of the trees, the value of forested wetlands as natural sustainable coastal barriers in times of sea-level rise has been widely recognized. Most intertidal forested wetlands have an autogenic vertical accretion rate in excess of 5 mm/year in places with a positive sediment balance. Intertidal forested wetlands, and mangrove swamps in particular, also play an important role in diminishing the impacts of extreme weather effects such as hurricanes on coasts and coastal wetlands, and dense and deep mangrove swamps can even lessen the impact of tsunamis (Adger et al. 2005; Kathiresan and Rajendra 2005; Alongi 2008).

Mangroves are an ancient group of tree species that adapted physiologically and morphologically to survive in intertidal zones. Mangroves are believed to have originated in South-east Asia in the Paleocene—the first geological epoch of the Tertiary around 60 million years ago—and the species soon spread to other tropical and subtropical regions. Plate tectonics caused a separation of the South-east Asian and American mangrove groups around 18 million years ago, and the groups subsequently developed independently from each other (Ellison 2008: 98). Mangroves display many pioneer-phase characteristics, and the Holocene distribution of mangrove swamps is closely linked to the dynamics of shoreline evolution caused by sea-level changes. The current landward extension of mangrove swamps where the influence of a higher sea level and saltwater incursion in the groundwater has caused the decline of salt-intolerant species is, in this context, unsurprising (Nicholls et al. 2007: 329). Mangrove forests have come under threat in places where significant neotectonic subsidence has occurred and where tracts of the forests have effectively drowned (Stanley and Hait 2000), or where anthropogenic-forcing mechanisms such as deforestation, damming of rivers and streams, and expansion of agriculture has reduced the resilience of mangrove swamps. It has been estimated that 35 per cent of the mangrove forests of the world have been destroyed in the last two decades of the twentieth century (Valiela, Bowen, and York 2001), with human action rather than sea-level rise identified as the cause, but the natural protection proffered by mangrove woodland is no longer available in these regions, reducing resilience to future sea-level change. All intertidal forested wetlands are relatively resilient to air and water temperature changes and to changes in salinity.

Coastal brackish/saline lagoons; brackish to saline lagoons with at least one relatively narrow connection to the sea

Coastal lagoons could be destroyed by sea-level rise if the barrier that separates the lagoons from the sea is eroded or permanently overtopped, effectively changing the lagoon into an extension of the sea itself. However, even where this is not the case, changes in the flushing regime may alter the salinity and chemistry of the lagoon, and increased inundation by seawater resulting from higher tidal range or waves will increase the salinity. Higher air and water temperatures may lead to decreased levels of dissolved oxygen and hypoxia. Increases in precipitation may have a disproportional impact on this type of coastal wetland, especially where the lagoon is shallow or where it has a relatively large terrestrial catchment. This may cause alterations in the salinity, chemistry, nutrients, and sediment transport into the lagoon. Apart from drowning, the greatest impacts of global climate change on lagoons will be on their ecology: the lower productivity of primary producers will have knock-on effects on secondary and higher producers, and whereas many lagoons are now valued for their rich ecology and biodiversity, climate change may reduce this (Anthony et al. 2009).

It is widely acknowledged that sea-level rise, warmer ambient and water temperatures, and changes in salinity will alter the nature and reduce the extent of many coastal wetlands. However, the dynamic nature of these landscapes will enable coastal wetlands to attain naturally new dynamic equilibria, and this dynamism is well documented for coastal wetlands in the Holocene. What is different in the twenty-first century is that the rate of sea-level rise may exceed the natural thresholds of some types of coastal wetlands to accrete vertically, and this would result in the drowning of many of them. What has also changed is the impact of humanity on coasts, including the influence on coastal wetlands of measures taken to offset the effects of climate change. Human action has reduced the natural resilience of coastal wetlands to the impacts of global climate change to an extent not experienced previously in the Holocene.

HUMAN ADAPTATION TO CLIMATE CHANGE AND ITS IMPACTS ON COASTAL WETLANDS

The autogenic ability of coastal wetlands to establish new dynamic equilibria in relation to sea-level change is evidenced through palaeoenvironmental studies pertaining, in particular, to the late Holocene period. The impact of human communities on the development of coastal wetlands in the past has been, generally speaking, limited or of a local nature. For example, early deforestation for agriculture in river valleys around the world accelerated the natural erosion from the valleys, increasing the sediment content of rivers and estuaries, and resulting in the raising of estuarine beds. However, case studies have shown that despite the longevity of agriculture, the most significant impact has occurred only in the last few centuries. Thus, whilst early agriculture in the Rhine basin dates back nearly 7,500 years ago, the fluvial sediments contents have increased significantly only in the last 200 years (e.g. Hoffmann 2007).

The reason for this dramatic increase of fluvial sediments lies in the intensification of agricultural practice, commencing with the Industrial Revolution and the very rapid growth in the world's population since c. AD 1750. Industrialization and population growth have also had direct impacts on the world's coasts, and this process accelerated during the twentieth century.

Around the world, few if any coasts remain wholly 'natural'. To illustrate the intensity of coastal utilization, it has been estimated that nearly a quarter of the world's population lives within 100 km of the sea and less than 100 m above current sea levels; in this relatively narrow strip, the population density is three times higher than the global average (Nicholls et al. 2007: 319). This increase has brought about intensified utilization of natural resources such as fresh water, timber, sand, gravel, fish, and shellfish. It has also increased the modification and transformation of the coast in the form of reclamations, the construction of hard defences, the drainage of coastal wetlands, and a widespread degradation of the ecosystems. Taking a broad overview of these changes, many coastal wetlands around the world are now considerably 'less natural' than they were in previous centuries; consequently, their vulnerability to the impacts of sea-level change has increased. What does this mean in the context of climate change-driven sea-level rise? Using the three 'forcing mechanisms' of the Wetlands Change Model—the ratio of sea-level rise to tidal range, sediment availability, and the presence of accommodation space—the main impacts are described below.

While the tides are driven by the gravitational forces of the moon and the sun on the water in the Earth's oceans, with the moon contributing twice the force exercised by the sun, global climate change itself is not changing the tidal range (Masselink and Hughes 2003: 67). Nevertheless, tidal ranges in specific estuaries are expected to change as a result of the human adaptation to sea-level rise. The concept of 'coastal squeeze' is relevant here. Coastal squeeze is a term that describes the loss of coastal wetlands between the sea and natural or man-made coastal defences in periods of sea-level rise (Nicholls and Branson 1998). As has already been explained, coastal wetlands accrete vertically and migrate horizontally/landwards in response to sea-level rise. However, when human communities construct embankments to protect their settlements and agricultural land, this natural process is effectively squeezed into a smaller area, and the accommodation space is reduced or removed (see also below). The impact of the construction of embankments along low-lying stretches of the coast and tidal estuaries is twofold: the tidal range increases in the absence of tidal or flood relief space, and the tidal reach extends further back up estuaries and rivers.

In the North Sea basin, where the oldest continuous embankments were constructed nearly 1,000 years ago, the historical records attest to the continued heightening of built coastal defences to prevent overtopping, and to devastating floods in the floodplains of the rivers Ems, Weser, and Elbe after the tidal reach extended further inland following the construction of continuous dykes (Van de Noort 2011b: 118). The feedback of this human adaptation to sea-level rise is positive, in that the local impact of protecting the coast through the construction of embankments increased the impact of the rising sea level on the coast. In turn the increased tidal range and reach caused accelerated erosion of intertidal coastal wetlands and the beds of rivers and estuaries, which in turn affected the ecosystems of the coastal wetlands. The same positive feedback will operate in the future. Whilst global climate change will have no direct effect on tidal ranges, locally tidal ranges and reaches will be increased and extended through the ongoing construction of continuous embankments along coasts, estuaries, and the tidal reaches of rivers, as human communities adapt to the higher sea levels caused by global climate change.

The availability of sediment is essential for coastal wetlands to adapt to rising sea levels. Without sufficient sediments, coastal wetlands are 'starved' and are unable to accrete vertically, with drowning of the wetland being the inevitable consequence. The impact of human communities adapting to climate change will have a number of positive and negative feedbacks. Across the world, the erosion of coastal wetlands is frequently linked to anthropogenic activity in the shoreline, such as sand and gravel extraction both on- and offshore, and the mining of coral (Rosenzweig et al. 2007: 92). However, it is larger-scale activities in the full river catchments that have more significant effects on the sediment budgets of coastal wetlands.

Activities increasing the sediment budgets of coastal wetlands include the combined effect of deforestation and intensification of agriculture. The construction of dams and reservoirs reduces the sediment budget. There is extensive palaeoenvironmental evidence to show that early agriculture and deforestation increased the sediment transport of main river systems. The example of the River Rhine has already been mentioned, but there are many other well-studied rivers that demonstrate how, over the last few centuries, sediment loads have increased through anthropogenic activities such as deforestation leading to widespread soil erosion and loss of sediments through the fluvial system. Well-known examples include the Huanghe/Yellow River in China which has experienced a tenfold increase in its annual sediment load in the last two millennia, linked directly to the increased intensity of agriculture (Jiongxin 2003). As a consequence of this increase, the lower reaches of river valleys and estuaries are filled with nutrient-rich sediments, altering the geomorphology and ecology.

Damming and the construction of reservoirs frequently have the opposite effect, leading to a reduction of river flow and sediment load of rivers (Nicholls et al. 2007: 319). Returning to the Huanghe/Yellow River, its sediment load reduced sevenfold in the second half of the twentieth century after the construction of two reservoirs in the upper reaches of the river and two in the middle reaches, combined with soil-conservation measures in the river floodplain (Wang et al. 2007). The impact of these measures was a negative sediment budget and as a consequence the Huanghe/Yellow River delta became erosive in nature, and the ecosystems became more fragile and susceptible to extreme weather events. Because the river's water flow was also significantly reduced, further dredging and channelization were required to keep the river navigable, causing further damage to the fragile ecosystems of the Huanghe/Yellow River delta (Zhang et al. 2009). Sediment starvation may cause the future submersion of this river delta where the tidal flats have shrunk (Fan, Huang, and Zeng 2006).

Whilst changes to accommodation space is given the lightest weighting in the Wetlands Change Model, it is in this respect that human communities have altered the future adaptive paths of coastal wetlands the most. During the Holocene, human communities have settled in the coastal zones, and in the most recent centuries this has included the concentration of population in ports, towns, and cities in the zone nearest to the sea. The driving force behind this aspect of an increasing population in the coastal zone is the emergence and development of a world economy, and it is unsurprising that many of the world's largest cities are located on the coast (Nicholls et al. 2007: 319). Associated developments—such as the draining of coastal wetlands to accommodate urban expansion or to feed the growing population, the development of docklands, and the dredging of river channels to ensure that these are navigable—have removed much of the accommodation space previously available to coastal wetlands to migrate horizontally/landwards. With less of the world's coast surviving in a natural context, the autogenic ability of coastal wetlands to respond to sea-level rise is diminishing in importance, and the desire of human communities to retain their wetlands is becoming an increasingly important issue (see Fig. 4.4).

THE IMPACT OF CLIMATE CHANGE ON COASTAL COMMUNITIES

Human communities have very good reasons to ensure that the coastal wetlands in their environment can survive into a future of global climate change and rising sea level. The socioeconomic benefits of coastal wetlands include provision of fresh water, food, fuel, and building materials, protection from the sea, a source of income in the form of tourism, and health benefits. Combining the various impacts of global climate change on the coastal zone, the IPCC's *Fourth Assessment Report* summarizes the main impacts on the coastal socioeconomic sectors as detailed below.

In the face of a growing world population, it will become increasingly difficult to meet the need for fresh water. Illustrating this, at the end of the twentieth century it was estimated that 54 per cent of accessible fresh water in rivers and groundwater was used for human consumption, and in agriculture and industrial applications. It was noted that the construction of new dams in

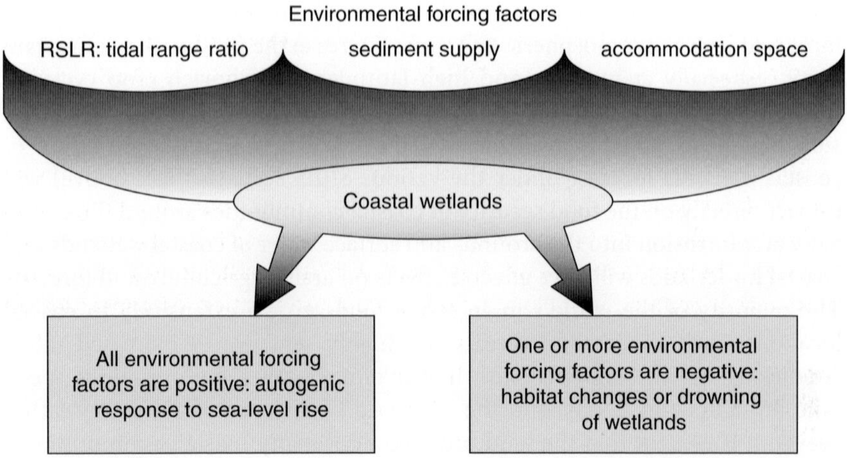

Fig. 4.4. A schematic representation of the Wetlands Change Model. Three environmental forcing factors determine the ability of coastal wetlands to respond autogenically to sea-level rise: the ratio of RSL Rise (RSLR) to the tide range, the sediment supply, and the availability of accommodation space.

rivers could raise this figure by an additional 10 per cent in the following three decades, but that the world's population was expected to grow by 45 per cent leading, inevitably, to a global-scale shortage of fresh water (Postel, Daily, and Ehrlich 1996). Coastal communities obtain their fresh water primarily from coastal wetlands, including the lower reaches of rivers and estuaries. Global climate change is expected to have a significant impact on the availability of accessible fresh water in coastal regions. Most importantly, higher sea levels and more frequent extreme weather events in the form of storms will threaten the fresh water supply to coastal human communities. This could result in saltwater intrusion into groundwater aquifers and the surface waters of coastal rivers and estuaries. Higher global temperatures will increase evaporation and evapo-transpiration, further reducing the available fresh water resource. The impact of this 'water stress' will be felt most acutely in areas where there is already a shortage of fresh water, notably on small islands and in arid and semi-arid regions. The problems will be further exacerbated in developing countries, where the effective use of available fresh water may be limited by an underdeveloped infrastructure for delivering clean fresh water to the end-users, or where existing sewage systems use the sea or estuaries as the place of discharge. The more frequent occurrence of extreme weather events will further intensify the problem of water stress, especially where the water and sewage infrastructure is not sufficiently resilient to deal with episodes of excessive precipitation or drought (Nicholls et al. 2007: 331).

Coastal wetlands play an important role in food production across the world. Global climate change is expected to have both positive and negative impacts on this role. On the positive side, increased temperatures, and increased levels of atmospheric CO_2, may increase the productivity of certain crops, especially at medium and high latitudes, and shorten crop cycles at higher latitudes (Nicholls et al. 2007: 332). However, this can only happen if sufficient fresh water is available. On the negative side, the destruction of coastal wetlands forecast under the various SRES scenarios (see above) will impact directly on the food security of coastal communities around the world. Saltwater intrusion into the ground- and surface water of coastal wetlands and coastal hinterlands will have adverse effects on arable agriculture and forestry. The majority of the world's main staple foods grown extensively in coastal locations, such as rice and cereals, are highly sensitive to increased saline conditions. When combined with higher temperatures, high humidity, periodic droughts and floods, and the increased occurrence of extreme weather events, it is evident that the food production of many coastal communities is placed under considerable abiotic stress (e.g. Wassmann et al. 2009; Wang and Frei 2011). Sea-level rise and saltwater incursion, and the impact of flood events or more extreme weather events, will be felt most strongly in deltas and lower reaches of rivers such as the Nile, the Mekong, and the Mississippi, highly productive regions that are essential for the food security of coastal and non-coastal communities alike (e.g. Wassmann et al. 2004).

Alongside agriculture, fishing has played an important part in feeding coastal communities for many millennia. Excavations on the Atlantic coast of South Africa have provided the earliest evidence for the consumption of shellfish from the deeply stratified coastal shelter Ysterfontein 1, dated in excess of 50,000 years old (Klein et al. 2004). Fishing tools and the presence of fish bones and shellfish remains in archaeozoological assemblages dated to the late/terminal Pleistocene and the earliest Holocene are known from many locations around the world (e.g. Rick, Erlandson, and Vellanoweth 2001; Van de Noort 2011b: 72–3). Archaeological and palaeoenvironmental research has shown, in some detail, how fish and shellfish populations changed with the transition from the last glacial to the current interglacial stage and through the Holocene (e.g. Tunnicliffe, O'Connell, and McQuoid 2001). There is also evidence for significant changes in fish populations during periods of relatively pronounced climate change, such as the Little Ice Age (Finney et al. 2010). Therefore, it is not doubted that global climate change will have an impact on the regional availability of fish and shellfish for human consumption.

Since the publication of the IPCC's *Fourth Assessment Report*, significant progress has been made in understanding the impact of global climate change on current fish population abundance and distribution, in particular in the North Atlantic. For example, fish living in open water have responded to the changing distribution of plankton, which has been linked directly to global warming (see above). Cod populations have expanded at the northern limits of their traditional distribution area, but cod and plaice have decreased at the southern limits (Rijnsdorp et al. 2009). However, the study of the complex interaction between climate change, its diverse impacts, and the many different species that play a role in the success or failure of fishing as a food-production activity has not as yet produced an unequivocal forecast for the future (e.g. Stige et al. 2010). What is clear, however, is that existing fishing practices will have to adapt to a rate of change that has no precedent in recent times, and this may involve changing the locations where preferred fish are caught, or changes in the human preference for some species over others.

The history of aquaculture is thought to stretch back several millennia. However, in its earliest stages, around 2,300 years ago, the practice of increasing the yield of aquatic species through human intervention was very much limited to fresh water context, such as the practice of using rice paddies for fish-rearing in China. Aquaculture in coastal and marine environments is of a slightly later date. The Etruscan method of water management of coastal lagoons in the Mediterranean captured fish early in their development. The culturing of milkfish (*Chanos chanos*) in coastal ponds in Java, dated to AD 1200–1400, is identified as the earliest example of marine aquaculture, or mariculture, in Asia (Beveridge and Little 2002: 9–15). Marine aquaculture spread to all parts of the globe, developing as a major provider of protein for human consumption, with prawn, shrimp, crabs, and lobsters especially prevalent in warmer waters, and salmon and bivalves the dominant species in colder waters (Hall et al. 2011). Aquaculture is, when compared to animal farming, ecologically 'greener' (or 'blue-er') because of the more effective conversion of foodstuff into protein for human consumption and the lower GHG production. Nevertheless, intensive aquaculture results in a range of ecological impacts, from the eutrophication or acidification of coastal waters to the use of the coastal wetlands themselves, which reduce their potential for delivering other ecosystem services. For example, in some parts of the world, mangrove forests have been removed to make way for aquaculture (especially shrimp ponds), thereby reducing resilience to coastal erosion and sea-level rise in these areas (Nicholls et al. 2007: 335).

The impact of climate change on aquaculture is, in parallel with the situation for fishing, rather difficult to predict. At higher latitudes the impact may be positive, with the vanishing ice providing new territory for expanding aquaculture, and warmer temperatures resulting in longer growing seasons and increased primary production (Barange and Perry 2009). However, the impact is likely to be negative in tropical regions and at low latitudes, where a reduced vertical mixing of the water column may reduce productivity (Brander 2007). Sea-level rise and the incursion of saltwater into fresh water or brackish zones may have a negative impact on current aquaculture production (De Silva and Soto 2009).

Coastal-specific health issues that may arise from global climate change are, mostly, directly related to exposure to higher sea levels and to the occurrence of more frequent extreme weather events. Especially when combined with peak flows in rivers following extreme weather events in the coastal hinterland, floods and more frequent storms can cause extensive flooding, loss of life, and economic damage, as is well known from the impact of the recent hurricanes Katrina and Rita. In the decade between 1992 and 2001, floods are thought to have killed almost 100,000 people worldwide, whilst affecting the lives of some 1.2 billion others directly and indirectly (McMichael, Woodruff, and Hales 2006). Apart from drowning, direct impacts include the increased risk of communicable or infectious diseases and exposure to toxic pollutants. Indirect hazards following floods include malnutrition, which is exacerbated when crops are lost, and mental-health disorders. Local floods and higher water temperatures in coastal wetland waters are likely to increase the frequency of the occurrence of cholera. Whether the twenty-first century will witness an increased frequency of flooding is, to a large extent, dependent on how coastal communities adapt to the environmental impacts of climate change.

An important function of coastal wetlands, notably of sandy beaches, is their capacity to generate income through recreation and tourism. The economic dependency of coastal communities on tourism is generally greater than that of non-coastal communities. The impact of sea-level change, and warmer air temperatures, will probably mean an increase in tourist flows to coastal resorts despite the loss of beaches, but a decline in the number of visitors to the warmest places, such as tropical islands or South-east Asia (Bigano et al. 2008). Loss of income, especially in developing countries, may have as negative feedback the utilization of coastal wetlands in other, less sustainable, manners to generate new sources of income.

As will be the case on land, coastal biodiversity will respond to global climate change, although the IPCC's *Fourth Assessment Report* offers very little in terms of what changes can confidently be expected. A general poleward migration of fish and shellfish species, for example, or a poleward extension of species' habitats, appears a very likely outcome. Some models predict the extinction of many fish species at the local level, especially in tropical areas and seas at low latitudes, and in enclosed lagoons (Cheung et al. 2009). Rising sea level is considered by some to be a direct threat to mangrove wetlands, and where there is no positive sediment budget or where accommodation space is restricted, mangrove wetlands may drown (Gilman et al. 2008). Coastal squeeze is likely to reduce the extent of intertidal and estuarine habitats, affecting primary producers and migratory birds in particular (Nicholls et al. 2007: 335). Evidence is emerging that migratory birds, who have their breeding grounds at a significant geographical distance from their non-breeding grounds, are increasingly 'mistiming' their migration (Robinson et al. 2009). However, biologists recognize the complex interaction between biotic and abiotic changes, and warn against implementing adaptations on the basis of predictions for single species.

The effect of climate change-induced sea-level rise will be most acutely felt in its impacts on settlement infrastructure, and will be greatly exacerbated by the more frequent occurrence of extreme weather events—both at sea and inland—which will increase the frequency of flooding. Urban conurbations will require new, or increasingly better, artificial coastal defences, at a very high financial cost, but one that is significantly smaller than the cost of urban flooding. For example, the 2005 Mumbai flood came at an estimated cost of 2 billion US dollars; under the SRES A2 scenario, it is estimated that the likelihood of similar floods will double by 2080, but that the cost of each flood will triple (Ranger et al. 2011). The IPCC's *Fourth Assessment Report* has estimated the average annual number of coastal flood victims by 2080 under different SRES scenarios, taking into account the impact of sea-level rise and continued migration of populations around the world to coasts, but excluding the impact of the increased frequency of extreme weather events and of the evolving coastal protection. This shows that under the most optimistic scenario, B2, an additional 3–5 million people will be the victims of floods each year. Under the pessimistic A2 scenario, the figure rises to an estimated 9–15 million people (Nicholls et al. 2007: 334).

The services provided by the ecosystems of coastal wetlands are closely interlinked, and there are anticipated and, as yet, unanticipated connections between the services that will make the situation for coastal communities worse, most of the time. And whereas improvements can be made to artificial coastal defences to protect people, especially those living in large coastal cities in developed countries, this is not an option for coastal wetlands. In one calculation, the global loss of coastal wetlands by 2100 under a scenario with a ESL rise of 0.6 m is calculated at nearly 100,000 km^2, or 10 million hectares (Nicholls et al. 2007: 339).

THE WAY FORWARD: INTEGRATED COASTAL ZONE MANAGEMENT (ICZM)

The question to be addressed next is: What is the best way to manage the coast in a time of accelerating sea-level rise? Ever since the publication of the *Second Assessment Report,* the IPCC has been promoting the concept of Integrated Coastal Zone Management (ICZM) (Bijlsma et al. 1996). The ICZM concept encapsulates an approach that can be described as: 'a continuous and dynamic process that unites governments and communities, science and management, private and public interests in preparing and implementing integrated plans for the protection and development of coastal ecosystems and resources' (Brown, Tompkins, and Adger 2002: 180). The approach recognizes the complexity involved in dealing with the dynamic coast, with changes affecting different groups and individuals in different ways, and with the sea not observing administrative boundaries. The ICZM is not a simple or single solution for the future management of a stretch of coastline, but a process that seeks to enhance the adaptive capacity of coasts and which evolves as new information and evidence become available. It bases its assessments of coastal change and of possible solutions on the best possible scientific evidence. It recognizes the interests of the many different groups in society that have a stake in the future of the coast—farmers, fishermen, residents, industry, water and energy companies, port authorities and transport providers, insurance companies, tourists and providers of tourist amenities, nature conservationists, archaeologists, and those who are responsible for coastal defences—and acknowledges the rights of these 'stakeholders' to be part of the ICZM process. This process will, at times, have to make decisions that benefit some stakeholders more than others, and at other times some significant sacrifices will have to be made, but this approach is preferable to one where single issues are dealt with unilaterally. The ICZM is described as a 'capacity-strengthening strategy', in that the sharing of information and understanding strengthens the capacity of coastal communities to find the best solutions for dealing with climate change and its impact on the coasts (Nicholls et al. 2007: 344).

In the last two decades, concepts of how to strengthen the adaptive capacity of coasts and coastal wetlands have evolved rapidly (Fig. 4.5). There is now a broad acceptance that it is better to work with natural processes, resulting in natural and sustainable protection against sea-level rise and floods, than to construct artificial dikes or to continue to nourish beaches artificially. Options for the development of 'soft' defences include the managed retreat or realignment of artificial defences, effectively providing the coastal wetlands with accommodation space to migrate horizontally/landward in response to sea-level rise (Townend and Pethick 2002). Another option is the restoration of coastal wetlands as self-sustaining ecosystems (Simenstad, Reed, and Ford 2006). These options can frequently provide sustainable flood defences and deliver the range of ecosystem services already described for coastal wetlands. Whilst this form of adaptation could reduce the impact of climate change-induced sea-level rise and more frequent storms by as much as a factor of 10 to 100, there will be places where the adaptive capacity of coastal systems cannot accommodate this change, and where artificial defences are the only alternative to abandonment of the coast (Nicholls et al. 2007). Where this is the case, problems such as the incursion of saltwater and the coastal squeeze of the coastal wetlands are likely to reduce the benefits attained from such a 'hard' solution.

Various constraints affect the functional use of the ICZM approach. These include elements that sometimes go to the heart of the political system, such as the asserted rights of private landowners and landowning corporations over

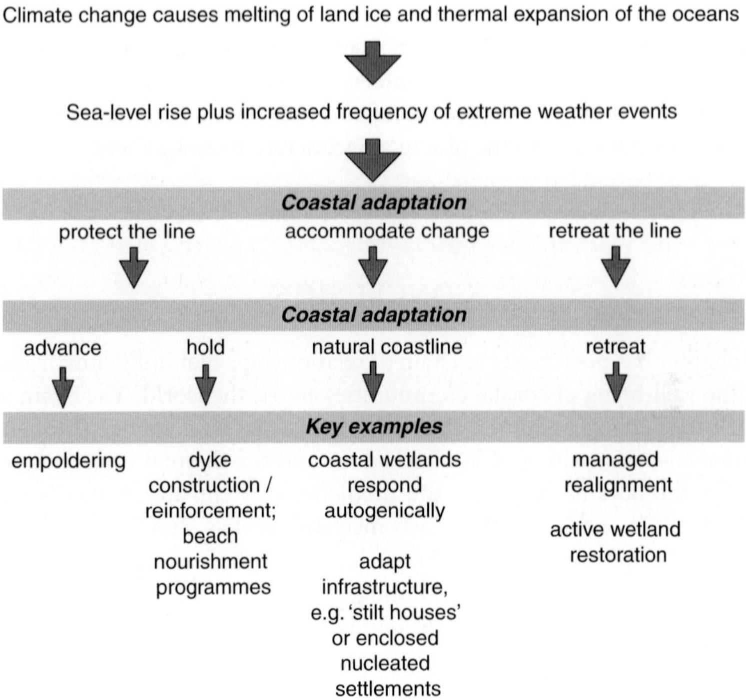

Fig. 4.5. A schematic representation of the link between climate change, sea-level rise, and coastal adaptation.

the common interest; or the role of the state, firstly in recognizing the existence of climate change-induced sea-level rise, and secondly in taking a leading role in the adaptation to climate change through the establishment of institutions, laws, and regulations, and through the funding of actions that improve the adaptive strategy of the coastal system to the impacts of climate change. The ICZM approach comes at an economic cost, which is more easily carried by developed countries than developing countries.

There are also some constraints to the ICZM approach which could be overcome. These include the lack of long-term understanding of coastal dynamics and their impacts on coastal communities, inadequate understanding of management practices and associated feedback mechanisms, and social resistance to change (Nicholls et al. 2007: 341). The latter comes, in particular, from coastal communities that are expected to make disproportionally large sacrifices for the benefit of society in general. It is, for example, unsurprising that the realignment of coastal defences, or managed retreat, for nature conservation and biodiversity reasons, is frequently strongly opposed by local communities, since 'giving in to the sea' is seen in a negative light (e.g. Myatt, Scrimshaw, and Lester 2003; Nicholls et al. 2007: 344). Climate change archaeology could provide new long-term insights into the dynamic nature of coastal change, and into the way people in the past adapted to this change. In this way, archaeology can strengthen the evidence base that underpins decision making in ICZM processes, enable learning from past successes and failures, and build the resilience of coastal communities in the face of an accelerating sea-level rise.

CONCLUSION

This chapter has presented the chain of relationships that links climate change with the well-being of coastal communities across the world. The chain starts with climate change, which is the key driver of sea-level change through two mechanisms: the melting of landlocked ice and the thermal expansion of the water in the oceans. Applying the predictions of climate change based on IPCC's SRES scenarios to these two mechanisms has shown that global sea levels will be between 0.3 and 0.6 m higher at the end of the twenty-first century than they were at the beginning of the century. Higher sea levels will have direct impacts on coastal wetlands, and this chapter has provided a number of high-level theories and formulas that can be used to calculate the impact of sea-level rise on coasts. Climate change will also increase the frequency of extreme weather events, and this is an important factor in the shaping of coasts and coastal wetlands.

The establishment of 'dynamic equilibria' for coastal wetlands is, to a large extent, determined by three forcing mechanisms: the ratio of sea-level rise to tidal range; sediment availability; and the presence of accommodation space. Sea-level change has only a limited impact on the tidal range, but its impacts on the availability of sediments and accommodation space are significant. Different types of coastal wetlands will be affected in dissimilar ways by higher sea levels. Rocky coasts and coral reefs are highly resilient to changes in the sea level, but sandy shorelines and intertidal marshes will retain their resilience only where there is a positive sediment budget, and where there is ample accommodation space. Where this is not the case, some coastal wetlands will be drowned by the rising sea. It has also been noted that changes in the ocean's water temperature and chemistry will affect coral reefs and the ecosystems in most other coastal wetlands. The importance of intertidal forested wetlands such as mangrove swamps in strengthening the resilience of coasts to sea-level rise has also been highlighted. Whilst many coastal wetlands have evolved over many millennia, an acceleration of sea-level rise, combined with human actions taken to counter the impacts of the higher sea level, will mean that sediment availability will change and that accommodation space will be significantly reduced, aggravating the impact of sea-level rise on coastal wetlands.

The next link is between coastal wetlands and the socioeconomic or ecosystem services delivered to the local communities. These include the provision of fresh water resources, an environment for farming, forestry, fishing, and aquaculture, a place for income generation in the form of tourism, and an environment that contributes to health, biodiversity, and a safe settlement infrastructure. The impact of climate change on sea levels, and the impact of sea-level rise on these services provided by coastal wetlands, is a negative one. Freshwater resources are expected to come under increased stress. Land for agriculture and forestry will be threatened by floods and saltwater incursion. The impacts on fishing and aquaculture, and on biodiversity in general, are not yet clear. In terms of human health, increased flooding poses the greatest risk, but an increased risk of infectious diseases is anticipated in a scenario with higher water temperatures and more frequent floods. Sandy beaches may shrink or disappear, but higher ambient temperatures could increase the visitor numbers to the coasts. Finally, the protection from coastal flooding offered by the coastal wetlands will be diminished when sea-level rise accelerates.

The final section of this chapter has considered the Integrated Coastal Zone Management (ICZM) concept, which has been promoted by the IPCC since 1996. The concept is now used in some developed countries, but a number of problems need to be overcome before the changing coasts can be managed in accordance with the spirit of the ICZM. These include the limited understanding of long-term coastal dynamics, the multiple impacts of changing coasts on coastal communities, and resistance to change from local communities. Exploring how people in the past adapted to their changing coast may help in overcoming these problems, and the potential for using climate change archaeology in this context will now be explored in four case studies.

5

The North Sea

INTRODUCTION

The North Sea (Fig. 5.1) is the first of the case studies, and the organization of this chapter and that of the following three have the same structure (see chapter 1). The three key questions to be answered in the case studies are:

1. How has climate, environmental, and sea-level change (and neotectonics) during the Holocene shaped the coastal wetlands?
2. What pathways were developed by coastal communities to adapt to these environmental changes and what feedbacks did these produce?
3. What can we learn from these pathways to strengthen the resilience of current coastal communities?

Much of my personal research has been undertaken in and around the North Sea. This involved taking part in fieldwork projects on the Dutch side of the North Sea coast during my student years and, subsequently, the archaeological excavations in advance of the construction of the Willemsspoortunnel in Rotterdam (Carmiggelt, Guiran, and Van Trierum 1997). On the English side of the North Sea, I directed the Humber Wetlands Project, a large-scale archaeological and palaeoenvironmental survey of coastal and terrestrial wetlands around the Humber estuary funded by English Heritage (Van de Noort and Davies 1993; Van de Noort and Ellis 1995; 1997; 1998; 1999; 2000; Ellis et al. 2001; Van de Noort 2004). In this capacity, I was one of the stakeholders in the development of the Humber Estuary Shoreline Management Plan, which was published under the title *Planning for the Rising Tide* (Environment Agency 2000). The Plan is an early example of Integrated Coastal Zone Management (ICZM) which considered the climate-change forcing of sea-level rise and its impact on the coast. The findings of the Humber Wetlands Project featured prominently throughout this Shoreline Management Plan, but the reason for including the historical environment was principally for its protection. Thus, archaeology became a 'material consideration' in the options appraisal process, with the options being: hold the line of the existing flood defences; moving the hard defences to gain benefits in terms of coastal flood

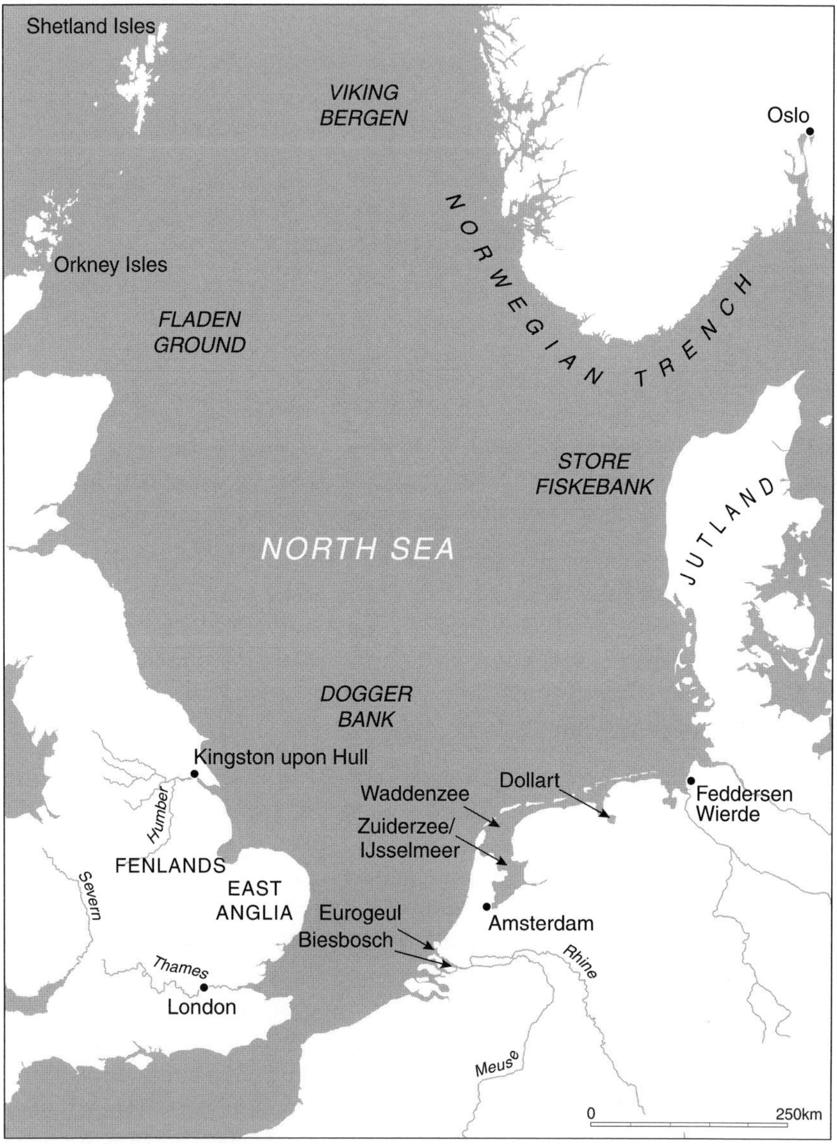

Fig. 5.1. Map of the North Sea and key sites mentioned in the text.

defence; and the creation of new intertidal habitats which would offset the loss of habitats due to sea-level rise (p. 36). In some cases, the historical environment would be protected by adopting the first option, and this was especially important for the protection of the medieval ports in the Humber estuary, including Kingston-upon-Hull. In other cases, strengthening the existing flood defences could have damaged the archaeological remains, and mitigation in the form of survey and excavations was proposed. This approach to the impact of sea-level rise on the historical environment of coastal wetlands has been replicated in many other ICZM plans in the North Sea and elsewhere. Archaeological sites and remains, and historical features, have become issues that need to be dealt with in the face of sea-level rise alongside the protection of settlements and infrastructure, agricultural lands, and ecological habitats.

However, already in the period leading up to the publication of *Planning for the Rising Tide*, there was a sense amongst the stakeholders that the historical environment could and should play a role in deciding the appropriateness of each of the flood-defence options. Noticeably, some geomorphologists argued that the use of the second option—that of moving the hard defences to gain flood-defence benefits, which was then referred to as 'managed retreat' but today is called 'managed realignment'—would be most effectively applied on the most recently reclaimed lands, and least effectively on ancient landscapes or areas that had been reclaimed from the estuary many centuries ago.

Intuitively, this made a lot of sense. In the past, coastal farmers and landowners would have possessed good knowledge of the dynamic nature of the coastal wetlands. Therefore, the earliest attempts at transformation and reclamation of these wetlands through embankment and encouraging sediment accretion was very likely to have taken place where this was relatively easy, that is, in zones with low tidal and wave energy, a positive sediment budget, and ample accommodation space. The earliest reclamations in the Humber, in the twelfth century AD, were in the inner and middle parts of the estuary. As time went on, reclamations tackled increasingly more challenging wetlands, that is, those with higher wave energy, lower sediment budgets, and restricted accommodation space, requiring more human energy and investment in the form of constructed embankments and drainage features. In the Humber, reclamations in the outer estuary were completed as late as the nineteenth century. These later reclamations had a disproportionate impact on the hydrodynamic nature of the estuary, with the high levels of wave energy deflected resulting in a significantly higher tidal range and greater tidal reach.

The argument for focusing managed retreat projects on the outer estuary, and on land recently reclaimed at the confluence of two major rivers, the Trent and the Ouse, was supported by geomorphological modelling. This, and the earliest field-based studies of managed realignment, confirmed that the tidal range and reach were reduced to the greatest extent when retreating from the most recent reclamations (Pethick 1993). An understanding of the historical development of the estuary could therefore be used as a proxy flood-defence management tool (Ellis and Van de Noort 1999). In the Humber estuary, managed realignment is now operational in two of the most recently reclaimed areas to great effect (Edwards and Winn 2006). At the same time, the more ancient reclaimed landscapes are protected from the short- and medium-term impact of sea-level change.

The contribution of the North Sea case study does not only concern the early inspiration that underpins the premise of this book. Its value in this comparative study lies primarily in the relative wealth of the countries around the North Sea, the high levels of social capital, which are exemplified in the willingness to engage with a wide range of landowners and transnational stakeholders, and the transparency in the decision-making processes. In effect, the manner in which the countries around the North Sea deal with the impact of sea-level rise closely resembles the IPCC-endorsed ICZM approach, set out previously (see chapter 4). In this way, the North Sea as a case study offers contrasts with the next three case studies, not only in terms of geomorphology and history but also economically and sociopolitically.

The first and second parts of this chapter, on the history of the North Sea and the ways in which past communities adapted to sea-level rise, are primarily based on research undertaken for my *North Sea Archaeologies* (Van de Noort 2011b). The third part, on how the socioecological resilience of current communities can be strengthened by referring to past adaptive pathways, is based on new research.

HOW CLIMATE, ENVIRONMENTAL, AND SEA-LEVEL CHANGE SHAPED THE NORTH SEA AND ITS COASTAL WETLANDS

The North Sea was created during the Mesozoic era (c. 250–65 million years ago) as part of the crustal stretching of the Atlantic Ocean (e.g. Klemperer 1988). The basin became more pronounced in the Miocene (c. 23–5 million years ago), with the process of subsidence of the basin floor commencing around 10 million years ago. At an average speed of 0.35 m per millennium, this subsidence has lowered the middle part of the North Sea floor some 3,500 m below current sea level (Cameron et al. 1992). This basin was filled over time with riverine sediments, principally from the proto-Rhine, and sediments dating to the Quaternary alone (2.5 million years ago to present) are in excess of 1,000 m in depth. Within this kilometre-thick stratum, distinct layers represent the glacial and interglacial stages. At the beginning of the Holocene (c. 11,700 years ago), this interplay between the subsidence of the basin floor and the riverine sedimentation resulted in an undulating landscape with regions of low hills—such as the Dogger Bank which formed at the mouth of the River Rhine in the early Pleistocene (Veenstra 1965)—traversed by the many rivers, and with seawater present only in the Norwegian Trench, off the south-west coast of Norway. Neotectonic movements have caused some postglacial beaches to be deposited into the Norwegian Trench (e.g. Hovland and Dukefoss 1981).

The North Sea and surrounding land masses were subject to significant glacio-isostatic adjustments during the Quaternary, and this case study is the only one where this is a major issue. The northern part of the North Sea and adjacent land masses were loaded with very deep ice sheets during late parts of the last Ice Age, or MIS 2. The furthest southward extent of the ice sheets was on a line roughly connecting the rivers Severn and Humber through central England, across the North Sea to the north of Dogger Bank, and through Jutland in Denmark, leaving only the south-west quarter of the country free of ice. The weight of this ice deformed and pressed the crust down into the underlying asthenosphere, and redistribution of the asthenosphere material lifted land in southern England and the northern Netherlands where the forebulge appeared. Following the onset of the Holocene, the isostatic dynamics were reverted. This glacio-isostatic rebound has resulted in an uplift of Scotland and northern England, Denmark with the exception of its south-west quarter, and of Scandinavia. It has also caused land levels in southern Britain and the northern Netherlands to be lowered. The rate of the uplift of the surface varies with the depth of the ice sheet that covered the land, with the regions beneath the deepest ice sheets experiencing the greatest uplift.

The onset of the warming and melting of the ice sheets some 21,000 years ago released large amounts of water back into the world's oceans, and the Eustatic Sea Level (ESL) started to rise from its lowest levels, c. 130–120 m below current levels (see chapter 4). The earliest impact of this on the North Sea was first felt in the northernmost parts of the plain. Detailed modelling of the flooding of the rest of the North Sea—from around 11,000 years ago when the ESL had risen to –55 m Ordnance Datum (OD), to around 5,500 years ago when the ESL had reached near-modern levels—was undertaken as part of the Land–Ocean Interaction Survey (LOIS; Shennan et al. 2000). These models provide an insight into the dynamic nature of the impact of sea-level change, which shows the emergence of the islands of the modern archipelagos of Shetland and Orkney around 10,000 years ago; the formation of a large embayment south of the Dogger Bank, and the extension of the English Channel northwards alongside East Anglia 8,000 years ago; the connecting of the embayment and the English Channel extension by around 7,500 years ago, cutting Britain off from the Continent; and the submersion of the higher grounds in the North Sea, such as the Dogger and Brown Banks, around 5,000 years ago (Fig. 5.2). At this point in time, the ESL had reached levels very close to modern ones, and further developments of the North Sea cannot be described at this scale.

The coastal responses to the rise in ESL in the North Sea are best characterized in two parts. In the northern part where the ice sheet had resided during the last glacial stage, the glacio-isostatic rebound resulted in a rise of the land that had outpaced the rise of the sea level by the end of the Holocene. In the southern part, beyond the reach of the ice sheets, sea-level rise resulted in the

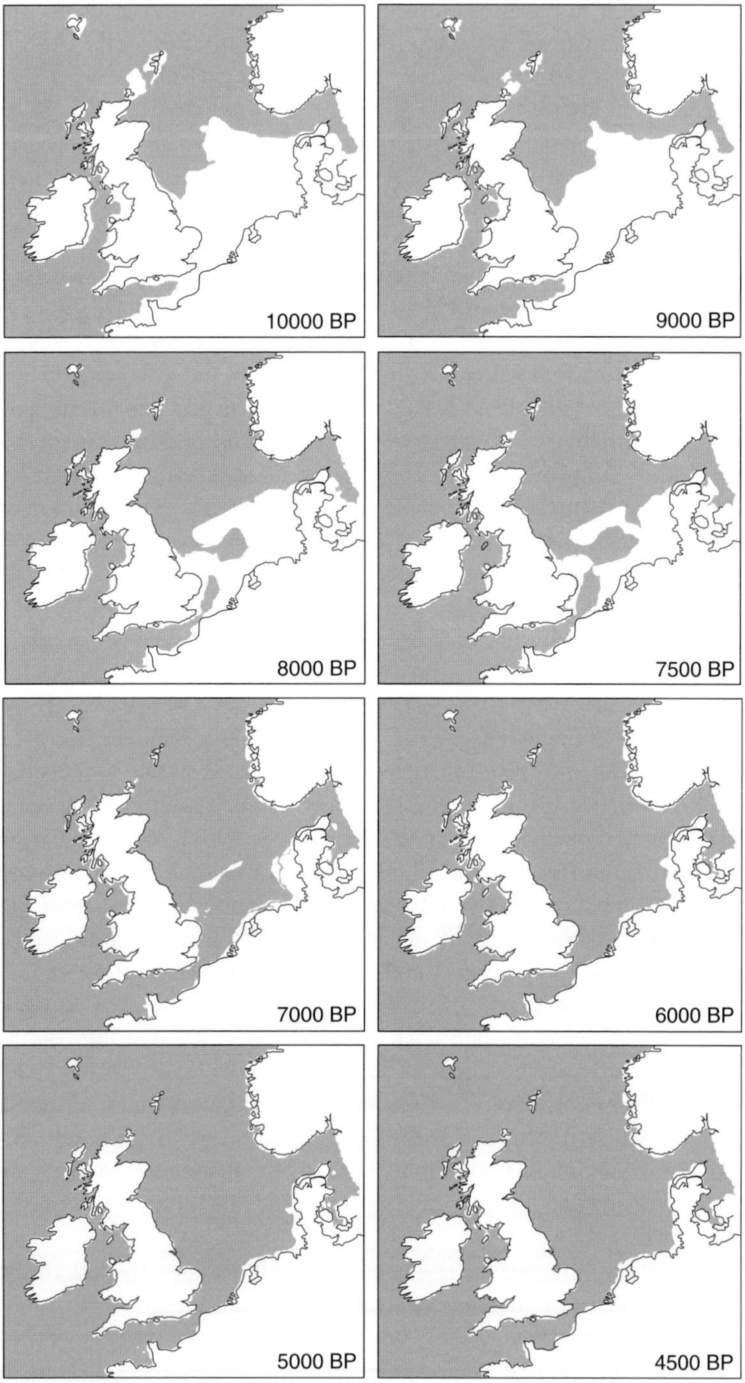

Fig. 5.2. The reconstructed emergence of the North Sea during the Holocene (after Shennan et al. 2000: 310–11). (BP = Before Present)

submergence of the coastal plains. The appearance of the coastlines of the northern part of the North Sea is one where hard and soft rocks, often elevating tens of metres above the level of the sea, dominate the seascape. The coastlines in the southern part are dominated by coastal wetlands such as salt marshes, or extensive dune systems that protect the low-lying hinterland from the sea (Van de Noort 2011b: 7–9).

Because of the glacio-isostatic dynamics, it is not possible to present a single projection of the Relative Sea Level (RSL) for the North Sea during the Holocene. Whilst we have a good understanding of the rise of the ESL, the glacio-isostatic adjustments during the Holocene include both geographical and temporal variations that can only be resolved through a significant amount of fieldwork. This fieldwork is aimed at collecting 'sea-level index points'. These are samples of organic sediments, or sediments containing organic remains, that provide proxy measures of the local height of the sea at a particular point in time.

One of the best examples of such research is presented by Ian Shennan and Ben Horton (2002) for Great Britain. They present no fewer than 52 separate RSL curves, and the differences between the curves are revealing. For example, in the south of England and south Wales, RSL rose initially very fast, from between –30 m and –25 m around 10,000 years ago, but slowed down in the more recent parts of the Holocene, to current levels at 0 m or OD. A glacio-isostatic fall of between 5 m and 10 m since the beginning of the Holocene contributed to this large rise. In north-eastern England, sea-level index points show that RSL rose from around –20 m some 8,000 years ago. This region was relatively stable in the context of glacio-isostatic dynamics, and the ESL and RSL are therefore more alike. The sea-level index points from the north-west of England, the northernmost parts of England's east coast, southern Scotland, and the Hebrides show that the highest sea levels were experienced between 5,000 and 3,000 years ago, followed by a fall in the RSL of 1 to 4 m. This is because the glacio-isostatic rebound outpaced the rise in ESL in the late Holocene. For the north-west of mainland Scotland, the curves are dramatically different. The sea-level index points show that the RSL fell here as much as 30 m in the period 15,000–10,000 years ago but subsequently was more stable. This area had the deepest ice cover during the last Ice Age, and therefore the greatest glacio-isostatic adjustment during the Holocene. Finally, the RSL curve for the Shetland Islands resembles that of north-eastern England, which can only be explained by the fact that it had little ice cover during the last Ice Age, and therefore very little glacial-isostatic adjustment in the Holocene (Shennan and Horton 2002: 521). Similar studies exist for the continental side of the North Sea, where the much deeper Scandinavian ice sheet produced a maximum glacio-isostatic uplift to date of 300 m for central Scandinavia, and a RSL fall on the North Sea and Baltic coasts in excess of 50 m (e.g. Harff et al. 2005; see also chapter 4).

For the early and middle parts of the Holocene, the rapid rise of the ESL remains the dominant force in the creation of coastal wetlands in the southern part of the North Sea, but local and regional factors become more important after c. 8,000 years ago when the rate of sea-level rise slowed down. These local and regional factors include the availability of marine and fluvial sediments, local and antecedent geomorphology such as the availability of accommodation space, and neotectonics are also important (see chapter 4). Because these factors are principally determined at the level of individual river catchments, previous conceptualizations of a supra-regional coastal evolution, such as those using the chronostratigraphic Calais and Dunkirk marine transgressive phases on the continental side of the North Sea, are no longer appropriate (Weerts et al. 2005). The modern way of studying the geology of the coast involves computer models which present three-dimensional models of the sediments, but the lithostratigraphy has been decoupled from the old chronostratigraphic framework.

In the period after c. 8,000 years ago, glacial-isostatic rebound outpaces sea-level rise in the northern part of the North Sea, with only local exceptions such as the Shetland archipelago where ice cover during the last Ice Age was thin (see above), or coastal areas of Scotland that are on the very margin of the glacio-isostatic uplift. The coasts in this part of the North Sea are dominated by hard and soft rock cliffs, often elevated above the sea by tens of metres. Former beach deposits—frequently referred to as 'raised beaches'—can be found many kilometres inland in Scotland, Norway, and Sweden (e.g. Gray 1974). Coastal wetlands that remain under tidal influence of the sea are, in general terms, restricted to the heads of estuaries, lochs, and fjords. Wetlands that develop in such locations in direct response to sea-level change are quite small, especially when compared to the extensive coastal wetlands that developed in the southern half of the North Sea. The vertical rate of sediment accretion of these small salt marshes appears to have quickened during the twentieth century, indicating that the rate of sea-level rise is accelerating again (Teasdale et al. 2011).

In the southern part of the North Sea, submerged peat deposits—'moorlog' dredged up by trawlers or, in more recent times, retrieved in cores for scientific research—attest to the presence of freshwater mires in the North Sea basin during the postglacial and early Holocene periods (Godwin 1943). The development of extensive coastal wetlands commenced only after the deceleration of the rate of sea-level rise. These coastal wetlands, mostly salt marsh, transgressed landwards and upwards with the continued rise of the sea level, often over the older freshwater peats, and the contact between the freshwater peat and salt-marsh sediments provide palaeoenvironmental material for sea-level index points. It is wholly unsurprising that the overwhelming majority of these sea-level index points date to the period after 8,000 years ago when the rate of sea-level rise decelerated (Shennan and Horton 2002: 512).

The subsequent development of these coastal wetlands is strongly influenced by local and regional factors such as the availability of marine and riverine sediments, antecedent geomorphology, and neotectonics. An important regional factor that impacted on coastal wetland development is the occurrence of extended coastal barriers in the southern half of the North Sea. The combination of a decelerated sea-level rise and tidal sediment supply produced an extensive coastal barrier along the Belgian and Dutch coast, with gaps where the major rivers met the sea. This occurred for the Belgian coast between 5,500 and 4,500 cal BC and around 4500–4000 cal BC for the Dutch coast. As a consequence of this, the back-barrier basins—effectively a series of extensive accommodation spaces created by riverine sediments—were closed off by the near-continuous coastal barrier, and extensive freshwater wetlands developed here, producing the so-called Holland Peat (Beets and Van der Spek 2000). In the last centuries BC and early centuries AD, extensive anthropogenic activities in the Holland Peat landscape, in the form of drainage for agricultural activity and through peat extraction, resulted in a lowering of the peatland surface. This led to increased volumes of tidal water entering the back-barrier basins during high tides which, in turn, caused the gaps in the coastal barriers to widen when the tides turned. In the longer term, this caused widespread floods and the expansion of coastal wetlands in the back-barrier basins (Vos 2006: 9).

Elsewhere, coastal barriers and dune systems have different histories. Thus, the coastal evolution of the Wadden Sea is characterized by the presence of a series of coastal barrier islands whose formation is dated to between 5000 and 4000 cal BC (Flemming 2002). However, this never evolved into a closed barrier, as sediment supply was less abundant than further south. The gaps between the barrier islands corresponded to the location of the antecedent river valleys. As a consequence, the extensive tidal coastal wetlands developed behind the barrier islands (Vos and Van Kesteren 2000). Responding to the continued but slower pace of sea-level rise, the barrier islands migrated slowly landwards by as much as 10 km over a 5,000-year period, with the salt marshes transgressing onto Pleistocene dry land. These landward- and upward-transgressing salt marshes impeded freshwater run-off from the older land, resulting in the formation of extensive freshwater peatlands.

Extensive dune systems exist on the coasts in the southern part of the North Sea. Their palaeoenvironmental dynamics can be observed in their stratigraphies, with aeolian sand deposits representing active dune systems, and palaeosols signifying the vegetated stabilized dune systems. It appears that there are chronological synchronies in the development of dune systems on opposite sides of the North Sea. For example, the extensive dune system on the North Sea coast of north-west Jutland developed initially around 2200 cal BC, a second phase is dated to develop from around 700 cal BC, and a third commenced around cal AD 1100, and the latter phase represents the extant dunes (Clemmensen et al. 2001). On the English Northumberland coast, dune initiation periods have been dated to 2000 BC, 800 BC, AD 500–1000, and in the Little Ice Age (Orford et al. 2000).

The sediment supply that has replenished the dune systems in the southern North Sea has been largely derived from the Pleistocene sands from the North Sea floor. However, during the Holocene the sediment supply has diminished considerably, from an estimated 41 million m^3 sand/year between 7000 and 4000 cal BC, to 27 million m^3 sand/year in the following two millennia, to around 7 million m^3 sand/year today, whilst the tidal transport direction of sand from the North Sea has evolved from an onshore one before 5000 cal BC to an along-shore one today (Van der Molen and Van Dijck 2000). In other words, the principal source of dune sand is diminishing rapidly, and the underlying sustainability of the dune systems in this area is therefore affected. Without implementing management systems that fix the sand and prevent large-scale erosion, the future of the dune systems that protect most of the Dutch and Danish and parts of the English coast is in question, and this type of coastal wetlands is under serious threat.

What is the future impact of the rise of the ESL on the coastal wetlands in the North Sea? As before, the answer to this has to be presented in two parts. Where the glacio-isostatic adjustment continues to outpace sea level in the northern part of the North Sea, the impact of an accelerating sea-level rise will be limited. It was already noted that the relatively small wetlands at the heads of estuaries, lochs, and fjords have experienced a quickening of the rate of vertical sediment accretion. These wetlands will continue to migrate landward and upward as long as sufficient accommodation space is available (Teasdale et al. 2011). Certain regions in the northern part of the North Sea where the rate of glacio-isostatic lift is very small—such as the Shetland archipelago, those parts of Scotland's coast where the ice cover during the last Ice Age was limited in depth, and near the furthest southern extent of the ice in northern England and southern Scandinavia—may experience a rise in their RSL for the first time in many millennia (De La Vega-Leinert and Nicholls 2008). Nevertheless, the impact of this will be more localized than further south, requiring local coastal flood-defence solutions.

In the southern part of the North Sea, RSL rise may reach 1 m during the twenty-first century. Because the population density is much higher, and infrastructure is much more developed, in the southern than in the northern part of the North Sea, accommodation space for the transgression of coastal wetlands is very limited, and a rise in the RSL of up to 1 m will have far-reaching impacts on the coastal wetlands, many of which are expected to be drowned to greater or lesser extents by the rising sea (Fig. 5.3). This will have far-reaching consequences for the services delivered by the coastal wetlands and the lives of local communities.

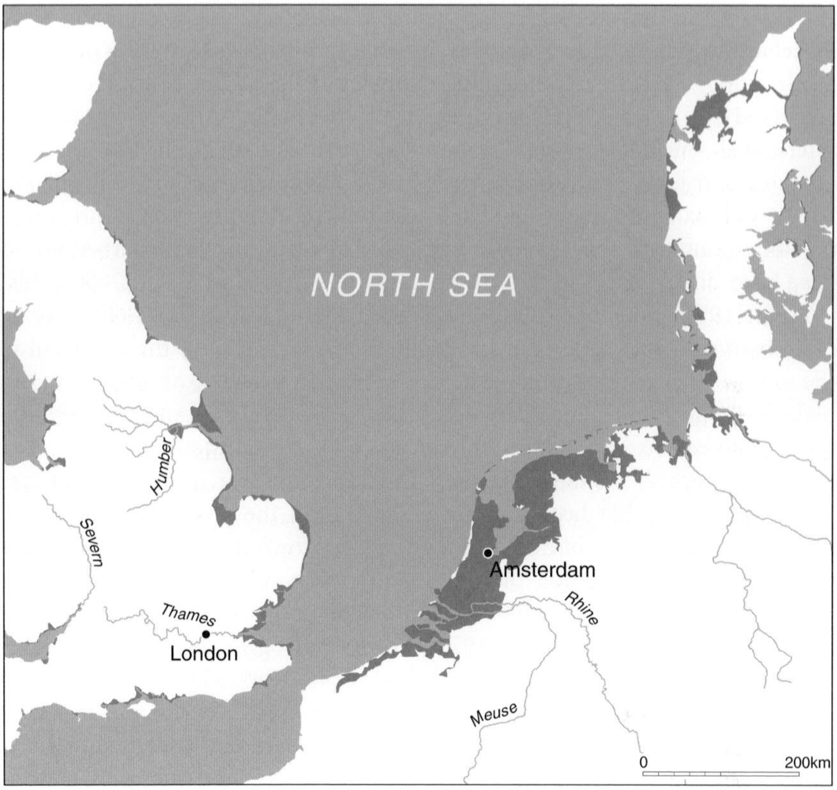

Fig. 5.3. Projected marine transgression of the North Sea (dark-grey tint) based on a rise of the RSL of 1 m and excluding the impact of constructed coastal defences (based on the model presented in Weiss, Overpeck, and Strauss 2011).

PAST ADAPTIVE PATHWAYS TO CLIMATE, ENVIRONMENTAL, AND SEA-LEVEL CHANGE

There is ample evidence for human presence on the North Sea plain before it was submerged during the Holocene. Research into this can be traced back to Clement Reid's (1913) study *Submerged Forests*, but the first unequivocal evidence for human presence on the North Sea plain before its submergence came in the form of the discovery in 1931 of a Mesolithic antler barbed point by the fishing trawler *Colinda*, from the Leman and Ower Bank off Norfolk. The barbed point was attributed to the Maglemosian Culture which was, by that time, already known from the opposite terrestrial sides of the North Sea (Clark 1936), and later radiocarbon-dated to 11,950–11,300 cal BC (Housley 1991). The realization that, in the early Mesolithic, hunter-gatherers were able to travel across and live on the North Sea plain was subsequently reinforced with discoveries from other higher grounds, notably Viking Bergen and Store Fiskebank which produced a series of stone implements (Louwe Kooijmans 1971; Long, Wickham-Jones, and Ruckley 1986). The faunal evidence recovered by trawlers for this warmer period—the Bølling-Allerød/Windermere interstadial—includes a range of mammals such as mammoth, elk, red deer, aurochs, and horse, but also walrus (Van Kolfschoten and Van Essen 2004: 78–9), and many of these would have been hunted. From the Europoort region off Rotterdam, no fewer than 500 fragments of antler and bone barbed points have been discovered, and these were dated to between 9450 and 7050 cal BC (Verhart 1995), and three perforated antler axes, dated to 7180–6820 cal BC (Glimmerveen, Mol, and Van der Plicht 2006). Dredging of the Eurogeul in the same area produced direct evidence of human presence on the North Sea plain in the form of human remains, which provided date ranges of between 9660–9220 cal BC and 7600–7100 cal BC (Glimerveen et al. 2004: 50). The faunal evidence shows that mammals, including red deer, aurochs, and horse were hunted but that boar appears to have been the favourite target (Post 2000). A further group of finds come from the Brown Bank, and this includes axes, adzes, and a shaft-hole pick, all made from bone and dated by inference to the sea-level curve for the southern bight of the North Sea to around 8000 cal BC (Louwe Kooijmans 1971). From the Dogger Bank comes the last Mesolithic artefact, dated to c. 6050 cal BC (Coles 1999: 57).

There can be no doubt that climate change and sea-level rise affected these early inhabitants of the North Sea plain. The temperature rose quickly in the early Holocene, changing the vegetation and fauna, and the animal bone evidence suggests that different species were hunted with the disappearance of what was, in earlier times, favourite game. During the Mesolithic, the North Sea filled with water, and the key adaptive pathway adopted by the people that lived here was to move to higher grounds. In view of the low population density in this archaeological period, this may not have been a major problem but, even then, archaeological research has observed that this was a time of significant changes in the economic basis of hunter-gatherer fishers, underpinning new social structures. This is most explicitly shown in Denmark where the late Mesolithic Kongemose and Mesolithic-Neolithic Ertebølle sites frequently include stationary fishweirs and middens (*køkkenmøddinger*) on the coast (e.g. Fisher 1995). The middens included the remains of shellfish alongside several fish species dominated by eel, sea mammals, and seabirds. This shift from the focus on large mammals as a key source of food to a more broad-spectrum economy has also been observed elsewhere around the North Sea, and concerns both marine and terrestrial foods. The concept of the 'broad-spectrum revolution' was initially applied to the emergent Neolithic in the Levant around 9000 cal BC by Kent Flannery (1969; see also chapter 8). He linked this to population growth, with increased demographic pressure demanding the raising of the carrying capacity of the environment, something that was achieved through the diversification of the food sources. The flooding of the North Sea would have led to an increase of the population on the dry margins, and as such sea-level change is likely to have been a factor in the adaptive pathway that led to the adoption of a more broad-spectrum economy in the late Mesolithic around the North Sea. Finally, the connection by land between Britain and continental Europe was severed around 6000 cal BC, and technological developments became increasingly more varied.

It had been assumed that the North Sea plain had been completely abandoned by the end of the Mesolithic due to the submergence of the sea, but Bryony Coles (1998) has raised the question whether this is correct, or whether people could have lived on the highest grounds of the Dogger and Brown banks into the early Neolithic. There is some archaeological evidence pertaining to the Neolithic from the North Sea: three polished stone axes from the Brown Bank (Louwe Kooijmans 1971; Maarleveld 1984), and two small polished tuff axes from the Dogger Bank (Van de Noort 2011b: 59). Sea-level modelling of the North Sea (e.g. Shennan and Andrews 2000) suggests that the tops of the Brown and Dogger banks would have been submerged a considerable time before the early Neolithic. Others have argued that the Dogger Bank was flooded, even earlier, by the Storrega Slide tsunami of c. 8200 cal BC (Weninger et al. 2008). It is therefore unsurprising that some archaeologists have argued that the polished stone axes were lost by early seafarers (e.g. Louwe Kooijmans 1985: 14). Alternative explanations are that the tops of the Brown and Dogger banks survived as tidal islands or in the social memory of people whose ancestors inhabited 'Doggerland', and that the polished axes represent deliberate depositions. If this is indeed the case, then we have here an early example of an adaptive pathway to sea-level rise that uses the deliberate, structured, or votive deposition of objects as a way of dealing with the loss of ancestral land.

The practice of structured deposition represents a familiar way of dealing with (rapid) environmental change in the past. In effect, the dynamic landscape is socialized through a range of ritualized activities that seek, at the same time, both to offer explanations of why environmental change is occurring and to help communities redefine their sense of place in the changing environment. In the case of the expanding North Sea, this focused on the attribution of other-than-human agency to the sea itself. The concept that the sea was itself an active agent—either directly or indirectly through an otherworldly proxy such as ancestors or Neptune, Aegir, or the Christian God—or was the abode of mythical creatures that created storms and caused ships to sink can be found in many cultures (Van de Noort 2011b; Mack 2011).

On the margins of the North Sea, this practice of structured deposition is widespread for the Neolithic and is found in a range of contexts where the rising sea level changed the landscape. Whilst polished axes play a pivotal role in the early stages of the Neolithic, over time a greater range of artefacts alongside human remains is utilized in this practice. To name a few examples, polished axes of Neolithic date have been found on the margins of the wetlands in the Humber estuary in the UK, where sea-level rise and the resulting impeded drainage of fresh water from the rivers and streams resulted in the landward extension of alder carrs (Van de Noort 2004: 95). In the Thames estuary, exotic axes of stone imported from Scandinavia found their way into the expanding floodplain (Edmonds 1995: 133), alongside maceheads and pottery which were deliberately placed on the water's edge. In Denmark, early Neolithic structured depositions include whole stone axes but also funnel beakers and animal skulls, spatially closely associated with human internments (Koch 1998). In the Bronze Age, structured deposition on the edge of the transgressive wetlands involves large number of bronze objects, especially weapons, and such deposits are known from Denmark, the Netherlands, and England. In short, the structured deposition of a range of artefacts and human remains represents an adaptive pathway that enabled communities to socialize their changing landscape, helped in the coming to terms with the transgression of wetlands onto their ancestral lands, and in defining a new sense of place.

Defining this new sense of place also involved the construction of new types of monuments on the edge of the North Sea. Two such monuments from the English Fenlands are 'Seahenge' and Flag Fen. The Bronze Age timber circle widely known as Seahenge is an enclosure of split oak timbers with a diameter of c. 7 m, and with an inverted oak tree trunk at its centre (Fig 5.4). Dated by dendrochronology to 2049 BC, the function of this site remains something of an enigma, but the excavators have argued that Seahenge derived its significance from the space it occupies, that is, the liminal area between sea and land (Brennand and Taylor 2003). The Flag Fen site includes a complex of timber alignments constructed between the first half of the thirteenth century and second half of the tenth century BC. The alignment functioned in part as a walkway to a platform that was used for ritual performances, in part as a symbolic defence against the sea which transgressed over previously valued agricultural land. Excavator Francis Pryor (2001: 431) describes this as a symbolic weir or dam against the steadily rising waters of the North Sea.

During the Bronze Age, we note the first archaeological evidence of the utilization of coastal wetlands, and in particular salt marshes, by human communities as an adaptive pathway to sea-level rise. Salt marshes are excellent grazing grounds for cattle and sheep, not only providing fresh fodder outside the normal growing seasons for 'dryland' pasture but also preventing foot rot in livestock (Rippon 2000). The archaeological evidence comes in the form of structures that gave access across the intertidal creeks to salt-marsh areas. For example, a number of wooden trackways comprising hurdles made of hazel—with other species such as alder, willow, and poplar also used—have been found in the intertidal zone of the Humber estuary; these trackways date

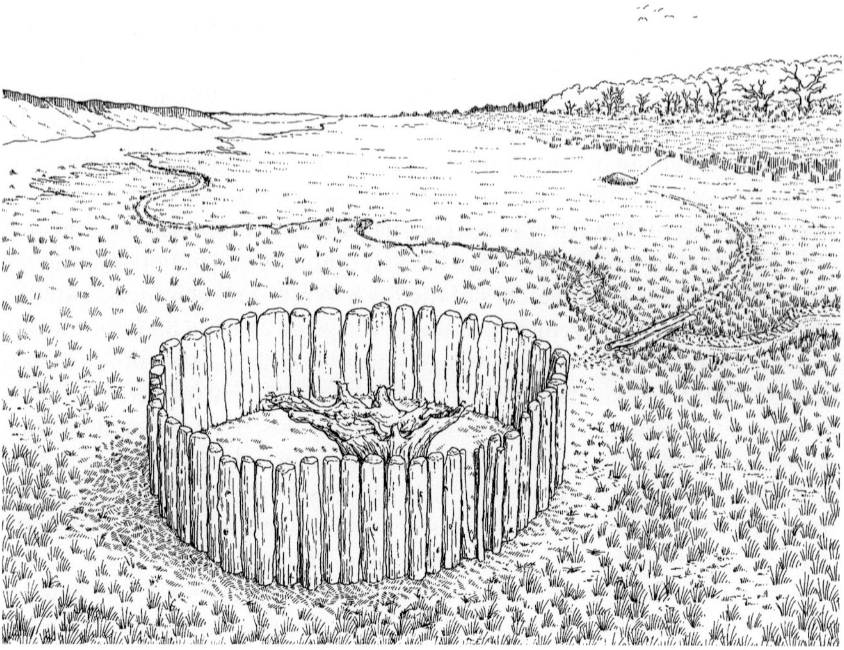

Fig. 5.4. Reconstruction drawing of 'Seahenge'. Original drawing by David Dobson (in Brennand and Taylor 2003: 67 © The Prehistoric Society).

to the middle and later parts of the Bronze Age (c. 1500–800 cal BC; Van de Noort 2004: 53–7). Because these trackways are often aligned parallel to the shore, it seems unlikely that their primary purpose was to provide access to the estuary itself. Instead, the trackways would have provided cattle and sheep access to areas of salt marsh that would otherwise have been unreachable. An experimental project involving Dexter cattle and replica hurdles has confirmed this potential use of wooden trackways (Bidgood 2002). Elsewhere in the North Sea, Bronze Age trackways in the Thames estuary (Meddens 1996) may have performed similar functions.

In the Iron Age, the archaeological evidence for the utilization of salt marshes is more extensive. Salt marshes continued to be grazed by cattle and sheep, frequently from settlements that were based on higher and dryer grounds and that were linked to the coastal wetlands by means of droveways, but increasingly from settlements that were based within the coastal wetlands themselves. Thus, in the Assendelfder Polder in the Netherlands, detailed analysis of the dung from 'site Q' revealed the existence of a settlement with a predominantly pastoral focus. It has been inferred that the salt marshes and levees in this coastal wetland landscape were utilized for cattle grazing, and that sheep grazed the peatlands (Brandt and Van der Leeuw 1987).

In the Wadden Sea, a particular effective adaptive pathway was followed. From c. 500 cal BC to AD 1000, artificially heightened islands known as terps (Dutch: *terpen* and German: *Wierden*, *Wurten*, or *Warften*) were inhabited in the intertidal landscape. Extensive excavations of some of these terps have revealed a number of wealthy settlements, based predominantly on stock-breeding and the use of the salt marshes as the principal grazing grounds. The Feddersen Wierde in the German state of Niedersachsen is the most extensively excavated of this type of site (Haarnagel 1979). It concerns a village that was established in the second half of the first century BC on a shore ridge in the salt marshes in the lower reaches of the River Weser, and continued until the fifth century AD. At its zenith, this was a village of 25 large houses arranged concentrically around a central 'village square' (Fig. 5.5). Many of the houses excavated show in plan how the living quarters and the stables were integrated in a single structure. This would have offered shelter to the cattle during particularly cold spells, but it also enabled the dung to be collected effectively and this was used in manure to enrich the soils of the terps, improving the productivity of the small arable fields above the summer high-tide line. This practice, and the sediments laid down during winter tides, enabled the terp to grow in height and extent during its habitation. The success of this practice can be seen at Feddersen Wierde, where all large houses have their own granary. Roman imports such as Samian ware, glass, pearls, and basalt stones were present in this settlement 280 km north-east from the nearest Roman *limes*—most likely exchanged for meat products. The archaeological evidence reinforces the sense that the terps represent a highly successful economic utilization of coastal wetlands, despite Pliny the Elder's disparaging remarks in his *Natural History* (XVI: 2–4) that the Chauci who lived here eked out a meagre existence from the fish caught at low tide.

The longevity of the terps as a phenomenon—about 1,500 years but with an interruption in the fifth century AD—lies in the fact that the artificial islands did not disrupt the natural processes that had shaped the salt marshes in the first place, and accommodated the natural upward and landward transgression of the salt marshes in times of sea-level rise. The terps did not alter the tidal range, nor alter significantly the sediment budget, nor reduce or remove the accommodation space for these coastal wetlands to respond to the changes in the sea level. After c. AD 1000, a number of terps had so-called ring-dikes added, especially in the eastern and north-eastern parts of the North Sea (Meier 2004). These prevented summer tides from flooding the lower flanks of the terps, which provided land for arable agriculture, but winter floods were not prevented, thus ensuring the continuous build-up of sediments. The construction of terps represents a sustainable adaptive pathway to sea-level change not found in more recent periods in the North Sea.

The limited palaeoenvironmental evidence available for sea-level change in the Roman period suggests that the impact of the sea on the lands around the

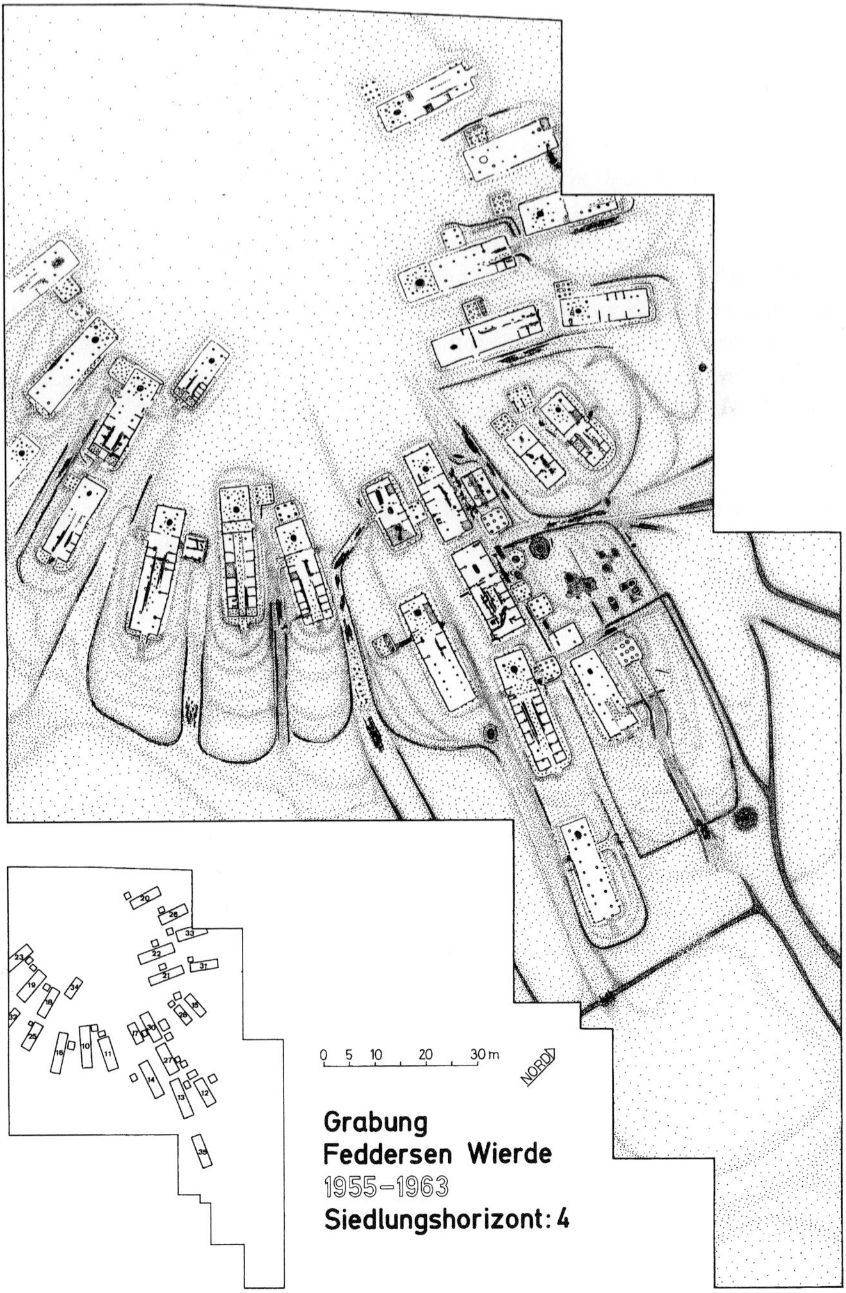

Fig. 5.5. Feddersen Wierde Phase 4 (from Haarnagel 1979 © Niedersächsischen Institut für Küstenforshung).

North Sea was rather limited. There are several potential explanations that can be presented here. One argument is that the Roman period coincided with a regional 'climatic optimum' as observed in a retreat of some glaciers in the northern hemisphere, Californian dendrochronology, and elevated levels of ^{14}C (Denton and Karlén 1973). Whilst archaeologists have little doubt that the period between c. 200 BC and AD 400 was one of relative warmth in Europe and the Mediterranean, because this was a regional climatic optimum that did not reflect the global climate, it did not have a significant impact on sea levels. Another argument is that during the Roman period, the frequency of impactful storm events was much lower than in preceding or succeeding periods. It is certainly true that Roman-period settlements and field systems have been found at low elevations around the North Sea, which would have been regularly flooded in earlier and later periods (e.g. Van de Noort 2004: 118–22). However, it remains unclear to what extent this can be attributed to changes in the sea level or has a more local explanation. For example, changes in the sediment load of rivers—following intensification of arable agriculture in river valleys following the expansion of the Roman Empire northwards—could have resulted in the build-up of sandbanks and sand barriers on the coast, providing suitable land for settlements (e.g. Buckland and Sadler 1985). A third argument is that coastal settlements in the Roman period did not have to be sustainable in the long term, as these served an urban market economy, and that the effective exploitation of coastal wetlands during a few flood-free decades was, in effect, a profitable business. The modification of coastal wetlands during the Roman period on both sides of the North Sea, for example the English Fenlands (Jackson and Potter 1996; Crowson et al. 2000), may be explained in this way. This form of coastal wetland utilization in the Roman period included the first albeit small and discontinuous dikes, as those near Wijnaldum, Donjun, and Vlaardingen on the Dutch coast, and at Walraversijde on the Belgian coast (Bazelmans, Gerrets, and Vos 1988; De Ridder 1999; Durnez 2006).

There is very little archaeological evidence for adaptive pathways for the period after the fifth century AD outside the Wadden Sea region. There is evidence of the flooding and drowning of some Roman settlements and roads on both sides of the North Sea, but the link to sea-level change is, at best, tentative. Some of the settlements could have been flooded because of a rise in the RSL, but such submergence could have been caused by local environmental phenomena such as storms, the erosion of barriers following anthropogenic alterations to the coastline, as has been argued for Holland (see above), or changes in the sediment budgets following widespread reforestation of previous agricultural land. Alternatively, embankments and ditches providing local flood protection for farms and settlements in the coastal wetlands were no longer maintained following the collapse of the Roman market economy.

This situation changes in the early ninth century AD. This point in time signals a new adaptive pathway to how coastal communities engage with the sea: the construction of dikes. The process is not synchronous on opposite sides of the North Sea, but urbanization appears to have created new markets, especially for livestock, which made the utilization of coastal wetlands more economically viable than was the case in the preceding centuries. The earliest post-Roman evidence for the construction of sea dikes comes from the North Sea coast of England. In the early ninth century, historical charters refer to the Thames in north Kent being embanked, and excavations and palaeoenvironmental analysis have revealed the existence of a sea bank at several locations in Norfolk, around AD 850 (Hall and Coles 1994: 127; Rippon 2000: 168; Crowson, Lane, and Reeve 2000: 225). Continuous dikes fully embanked the Fenlands of East Anglia by the tenth century. On the opposite side of the North Sea, the earliest dike building is thought to date to the ninth or tenth century in Holland, the eleventh and twelfth centuries in Zeeland, and the eleventh century in Flanders (Rippon 2000: 185). In the Wadden Sea, the raised roads that linked to terps became prototype dikes, dating back to the tenth and eleventh centuries, whilst in the eastern part of the terps area individual settlements were surrounded by so-called ring-dikes which in the thirteenth century were connected to form continuous sea walls (Rippon 2000: 185; Behre 2003: 49). Many of the earliest dikes were not built to keep the worst of the winter storms at bay, but sought to increase the productivity of the embanked land during the summer months. Over time, the economic benefits of year-round protection became an increasingly desired goal, and by the fourteenth century the majority of the coastal, estuarine, and riverine wetlands around the southern North Sea basin that did not benefit from the natural protection of dunes had been embanked, and the lower-lying back-marshes were increasingly put to agricultural use (Rippon 2000: 208–11; Tol and Langen 2000: 361).

The construction of continuous dikes was an adaptive pathway with far-reaching consequences. Dikes and embankments alter the three controlling factors of coastal wetlands (see chapter 4), that is, they change the tidal range, the availability of sediments, and the accommodation space for coastal wetlands. Dike construction resulted in the increase of the tidal reach—that is, the extent of tidal energy into riverine systems—and the tidal range—the height of the water at high tide—which required not only that the coastal sea defences were made higher and stronger but that riverine defences needed to be built further and further upstream. This process was already underway by the end of the fourteenth century. In terms of changes to the sediment regimes, because land behind the dikes was no longer flooded, no new material was deposited here as had previously been the case. Where the land surface behind the dikes subsided because of soil wastage—the combined effect of more intensive agriculture, dewatering, and peat wastage—high tide was frequently higher than the land itself. The reduction or removal of accommodation space meant that coastal, estuarine, and riverine wetlands—the type of environment that could have dissipated the increasing wave and tidal energy—diminished in extent. The result was something of a vicious circle of extending, strengthening, and increasing the height of the artificial sea and riverine defences to reduce the occurrence of floods, leading to increases of the tidal range and reach, and restriction of the deposition of sediment on the subsiding land and further reduction of the accommodation space available to the wetlands, which became the reason for further enhancements to the embankments at a later date.

The need for ever higher dikes in a period during which the sea level barely rose was one primarily determined by the interaction of socioeconomic and environmental factors. The exclusion of the sea from the land brought with it increased produce and profits in the short term, but because of the ever-growing investment in infrastructure behind the dikes in pursuit of short-term benefit and the environmental impact of dike construction, this was not a sustainable practice in the long term. It was also not a practice without its failures. Floods have been recorded for much of the second millennium AD. Floods direct from the sea had the greatest impact, and the Dollart Bay on the modern border of the Netherlands and Germany, and the Biesbosch in the Netherlands, are examples of two extensive wetlands created by sea surges, in the thirteenth and fifteenth centuries respectively. Floods on the major rivers were especially frequent in the period between AD 1350 and 1750, and historically recorded floods along the rivers Rhine and Meuse were nearly annual events (Fig. 5.6; Tol and Langen 2000: 368). There is little doubt that many of these floods are the consequence of the failure of local authorities to collaborate effectively, and the social and political situation in the Netherlands in these centuries was not conducive for the development of high social capital amongst the people responsible for the prevention of floods. The centralization of authority at the very end of the eighteenth and in the nineteenth centuries produced more effective collaboration and solutions, and the number of recorded floods dropped significantly during this time (Tol and Langen 2000: 368).

Despite state-formation processes and significantly improved engineering of the system of embankments and drainage, the risk of floods during storm surges did not disappear in the twentieth century. For example, the floods in January 1916 killed 21 people in the Netherlands, and the political response included the decision in favour of the closing off of the Zuiderzee (Van Koningsveld et al. 2008: 373–4); the 'Watersnoodramp' of 1953, during which 1,836 people were killed—alongside 307 in East Anglia, 29 in Flanders, and over 220 at sea—led to the political decision in favour of the closing off of the many sea-mouths in Zeeland, known as the Deltawerken (Van Koningsveld et al. 2008: 373–4; Van der Ven 2003; Fig. 5.7). This pattern seems to have

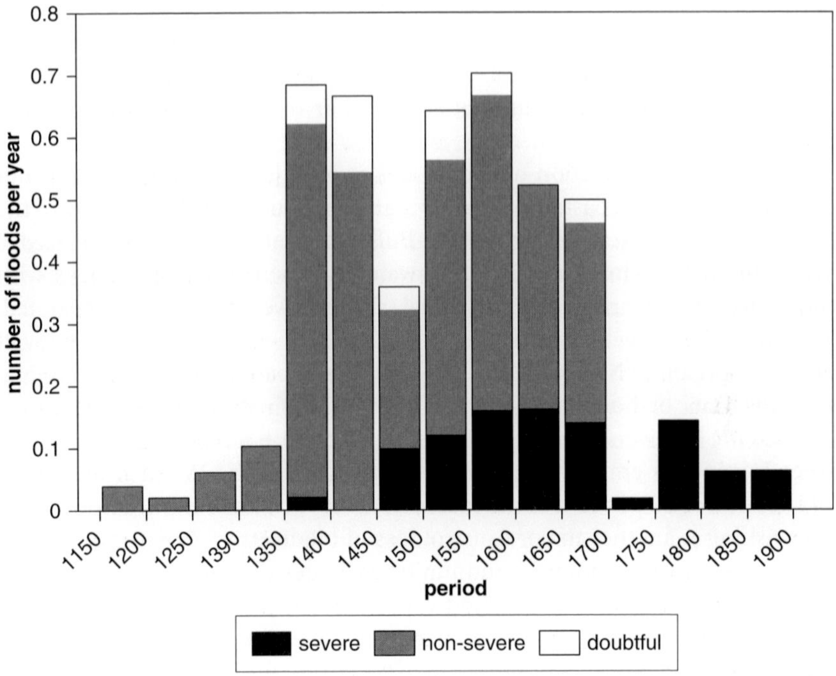

Fig. 5.6. Historically recorded floods along the rivers Rhine and Meuse, 1150–1850 (based on Tol and Langen 2000: 368).

Fig. 5.7. The flooded village of ‘s-Gravendeel in Zeeland, the day after the 1953 Watersnoodramp (© ANP; photo by R. Winterberger).

been repeated at other times: the social, economic, and, above all, political motivations to improve the state of the flood defences emerge only after dikes are breached or nearly breached. The recent floods in the Dutch rivers in 1993 and 1995, for example, brought about a major rethink on the purpose and design of riverine defences.

Alongside the protection of people and economic assets, the protection of 'nature', and with it coastal wetlands, became an issue in the last quarter of the twentieth century (e.g. Zwart 2003). Building continuous dikes replaced a continuum of marine, brackish, freshwater, and terrestrial ecosystems with one where the marine and terrestrial spheres were physically separated, and where the brackish and freshwater habitats were frequently removed altogether (Van de Noort 2011b: 118–19). The impact of this was a dramatic simplification, or homogenization, of the food web structure which in turn reduced the biodiversity of all habitats. For example, it has been calculated that since continuous embankments were constructed in the Wadden Sea, some 144 species, or 20 per cent of the macrobiota, have become extinct, noting that over-fishing is another major cause of loss of biodiversity (Lotze et al. 2005).

The European Economic Community, and later the European Union, has attempted to regulate the fishing industry in the North Sea since 1970 through the Common Fisheries Policy, with mixed success as scientific advice was diluted in the political processes (e.g. Daw and Gray 2005). The European Union Birds and Habitats directives, originally adopted in 1979 and 1992 respectively, have provided in practice a formal basis for the nation states around the North Sea to protect intertidal 'biotopes' or ecosystems. The European Union Water Framework Directive of 2000 has formalized and extended this environmental and ecological protection to whole rivers, estuaries, and coasts, focused on the quality of water. These policies and directives are early examples of transnational agreements on the shared use of the sea, and potentially provide a basis for transnational adaptation to the impacts of climate change.

The need to rethink the hard-defences strategy and develop more sustainable solutions that acknowledge and utilize the ability of coastal wetlands to play a role in protecting land from floods, especially in a time of sea-level rise, has resulted in the widespread adoption of the practice of managed realignment. There are currently some 60 managed realignment projects around the North Sea and adjacent estuaries, including the two examples in the Humber estuary introduced at the beginning of this chapter (http://www.abpmer.net/omreg/). The majority are specifically designed to compensate for the loss of intertidal wetland habitats as a consequence of specific developments, such as harbour facilities; a smaller number are specifically aimed at reducing the impact of sea-level change. In terms of cost–benefit analysis, managed realignment projects within the context of habitat protection have been shown to have a positive value. At its current small scale, managed realignment has also been shown to be cost effective in terms of provision of protection against floods at peak tides (Turner et al. 2007).

The depletion of the natural sources of sand for the replenishment of the dunes on the North Sea coasts of the Netherlands, Denmark, and England has already been highlighted (see above). During the later Middle Ages and the early modern period, dune systems were valued as these accommodated large rabbit populations, with the animals initially cherished for their meat and later for their fur. Warreners looked after the rabbits, protecting them from predators and controlling their numbers (e.g. Beekman 2007). Rabbit populations control the dune vegetation, effectively preventing extensive afforestation of the dunes, but also increasing the biodiversity by bringing calcium-rich sands to the surface when burrowing. The collapse of rabbit populations in dunes, sometimes after a period when their numbers exploded, can therefore result in short-term fixed dunes dominated by limited numbers of plant species, which in turn can undermine the long-term dynamics and sustainability of this ecosystem (Van Wijk et al. 2006). The control of rabbit populations by warreners, as happened in the later Middle Ages and early modern period, contributed to the dynamic stability of coastal dune systems.

When dunes erode in the modern era, one option is to construct hard defences to prevent flooding of the low-lying hinterlands. However, since the mid-1980s, an alternative option has been adopted—especially in the Dutch coast—that of beach nourishment using sands from dredging projects (Pilarczyk, Van Overeem, and Bakker 1986). This alternative is not without its problems, and the need for the right grain size of sediment and the ongoing loss of sand through both cross-shore and longshore erosion processes are recognized. Nevertheless, beach nourishment appears to provide good value for money, especially where the material used has been obtained from dredging for the purpose of maintaining deep-water navigation channels, whilst maintaining the coastal beaches and their value for tourism, but with as yet unknown ecological consequences.

STRENGTHENING THE RESILIENCE OF COASTAL COMMUNITIES

As noted in the opening statements of this chapter, the countries around the North Sea have pioneered the ICZM approach advocated by the IPCC, and many 'best practice' examples of how coastal communities should deal with the impact of sea-level change have been tried and tested here. These coastal communities have a long history and unrivalled expertise in the construction of coastal defences, such as the Deltaworks in the Netherlands and the Thames

Barrier in the UK. There are many studies addressing issues such as the strengthening of coastal defences by deploying a variety of innovative technologies, using beach nourishment to offset the loss of sands in the coastal dunes, and how managed retreat can help in attaining European directives on habitats.

Nevertheless, the rise in sea level will have a number of important consequences. The increase in the risk of floods will not be linear—i.e. in line with the changes of the RSL alone—but it will be exacerbated by the increased frequency of extreme weather events. In the southern North Sea basin, much of the population, heavy industry, and key energy producers are situated close to the sea or along estuaries, directly or indirectly for reasons of ease of access to global markets, and the cost of a single flood will be potentially very high indeed. Capital investments in coastal defences, especially in the Netherlands, Germany, and in parts of the UK such as the Thames estuary, will need to increase significantly if current infrastructure is to be protected from sea-level change (e.g. De la Vega-Leinert and Nicholls 2008: 349). Higher sea levels will also mean increased problems in river discharge, and floods that are a combination of high tides and peak river flows will become more common. Again, this will require capital investments in new dikes and defences. Other aspects of the infrastructure that may need adapting include port and harbour facilities, whilst a significant number of modern and ancient towns may experience floods more frequently in the twenty-first century.

The consensus amongst coastal communities around the North Sea is that the engineering expertise exists to deal with the impacts of sea-level rise (e.g. Sterr 2008; Van Koningsveld et al. 2000; De la Vega-Leinert and Nicholls 2008: 349; Correljé, Broekhans, and Roos 2010). However, higher sea levels and the strengthening of coastal defences will increase further the squeezing of coastal and riverine wetlands. This will result in the amplification of tidal range and reach in estuaries around the North Sea, requiring further coastal defence works and affecting the availability of freshwater resources for the urban conurbations and agricultural irrigation. Agriculture in the coastal zone will also be affected by saline intrusion into terrestrial groundwater, reducing the productivity of the land. Sea-level rise may affect the sustainability of the beaches around the North Sea, a major source of income through tourism for coastal communities, although higher ambient summer temperatures may attract increasing numbers of tourists to the beaches. Further squeezing of the intertidal zone will lead to the drowning of salt marshes and intertidal mudflats and sand flats. This change in ecosystem will alter the biodiversity of the ecosystems, with changes in migratory bird behaviour a very noticeable aspect, further affecting the recreational and tourism potential of the coastal wetlands. The changes to the intertidal flora and fauna will also affect local fisheries and aquaculture, although the links with the impact of sea-level rise and the squeezing of coastal wetlands is not well understood.

There is now a growing appreciation, at least amongst those directly involved in its study and management if not the general public, that sustainable coastal management means increasingly working with the dynamics of sea and sediment or, as one study proclaims, 'Working with nature implies a gradual paradigm shift moving from a perspective of water (nature) as an enemy toward a perspective of water (nature) as an ally' (Van Koningsveld et al. 2008: 376). On the Dutch side of the North Sea, much more so than is the case in Britain, this is reflected in the detailed study of the Holocene geomorphological history of the coasts, as a basis for understanding the character of this enemy/ally.

However, on both sides of the sea, the adaptive pathways of how people in the past dealt with the dynamic sea have not played a role of particular significance in deciding how to deal with sea-level rise in the future. In the last 12,000 years or so, the coastal communities employed a range of adaptive methods to the rising sea level: migration, structured deposition, the utilization of the intertidal salt marshes, and the construction of terps and dikes. In what way can the study of these adaptive pathways help build the socio-ecological resilience of modern-day coastal communities?

Migrating away from the coast as an adaptive pathway to a climate change-fuelled sea-level rise is an unlikely scenario for the twenty-first century in the North Sea basin. This was the most obvious adaptive technique in prehistoric and early historic periods, when population densities were low and land ownership was not legally determined or recognized. It is conceivable that in specific locales the interest of the majority is served by the removal of human habitation, for example where managed realignment has a far-reaching impact on the sea and the coast. However, it seems very unlikely that large-scale migration will be considered part of the solution. The very high value of settlements and infrastructure under threat of sea-level change, including the capital cities of London and Amsterdam, makes anything other than a 'hold the line' approach economically and sociopolitically unacceptable. The recognition that migration was an important adaptive pathway to climate change in the past in this region, but that this is no longer open to modern societies, provides us with a warning that we should be careful in using examples of how people in the past adapted to climate change, because the contexts have often changed beyond comparison.

The use of structured depositions and the construction of new monuments as, or as part of, adaptive pathways to climate change, provide an insight into how societies come culturally and emotionally to terms with the loss of familiar and socialized landscapes. Others have already argued that 'the landscape is the main focus for social memory, with both myth and history inscribed on the landscape' (Gosden and Lock 1998: 5), and the loss of land and the dramatic changing of the familiar landscape leads to an inevitable loss of sense of place. In the Neolithic and the Bronze Age in the North Sea basin, this loss was counterbalanced through the creation of new practices that sought to socialize the landscape in direct relationship to the transgressing North Sea. This form of adaptation, and at this level of humanity, is rarely considered within climate change science or by the researcher working with the IPCC. Nevertheless, modern communities will need to discover their own ways of re-socializing the coastal landscapes, and redefining their sense of place, in a rapidly changing coastal world.

One can imagine a modern equivalent to the structured depositions of the Neolithic and Bronze Age, for example in Antony Gormley's installation *Another Place*. This artwork comprises 100 cast-iron figures in the image of the artist, and these have been displayed around the North Sea basin—at Cuxhaven in Germany, Stavanger in Norway, and De Panne in Belgium—before being transferred to the Mersey overlooking the Irish Sea (Fig. 5.8). Gormley's thinking behind the installation concerns the imagination of far-away places, but the location of the statues on intertidal beaches, where the sea 'plays' with them, has greatly added to their appeal, significance, and place in modern memories and stories (McEwan and Gormley 2006). In a way, this art installation has re-established the connection between coastal communities and a living or dynamic sea, and it has helped to turn the perception of the sea from the enemy of old into a modern-day ally.

The earliest evidence for the utilization of intertidal salt marshes in the North Sea basin dates to the middle Bronze Age, around 1500 BC. The value of salt marshes in livestock breeding has already been rehearsed, and in a modern

Fig. 5.8. Reconnecting coastal communities with their coast and the sea: one of the cast-iron figures of Antony Gormley's *Another Place* installation (© Ron Davies).

context products such as 'salt-marsh lamb' can be sold at a premium price because of the distinctive taste of the meat. Nevertheless, very few of the intertidal salt marshes in the North Sea basin are utilized for food production, and this is especially the case where new salt marshes are created through managed retreat projects. As noted earlier, the majority of managed retreat projects serve to uphold the European Habitat and Birds directives, and this focus on nature conservation is usually to the exclusion of people and their need for food production. The adaptive pathways from the past suggest that a more holistic approach to intertidal salt-marsh creation provides sustainable solutions for both habitat protection and food security.

In the Bronze Age the salt marshes were used by farmers who were themselves based on higher and dryer land, but the Iron Age farms in the Wadden Sea were based within the coastal wetlands to optimize the use of this valuable resource for food production. The farms on the terps in the Iron Age, Roman period, and early Middle Ages were large and their inhabitants were prosperous. The terps-based coastal communities represent an adaptive pathway in the context of sea-level rise where the interplay between the sea and the sediment budget was not disturbed, and where accommodation space—where young salt marsh could expand onto—was considered the most valuable of all landscapes. Coastal communities in the twenty-first century could reconsider using the salt marshes as grazing grounds for sheep and cattle on a scale that has not been considered to date. This would require that higher-than-springtide refuges are constructed for the livestock to retreat to, alongside short trackways across tidal creeks for ease of access during low tides. In areas of managed realignment, where existing embankments are demolished to allow the sea to transgress on former land, sections of old dikes could be preserved and usefully perform the function of refuges at a minimal costs. For this to become possible, the current segregation of habitat restoration and economic utilization of intertidal habitats would need to be set aside. Research into this has already shown that sustainable grazing of salt marshes—either one head of cattle or two sheep with their lambs per hectare—increases biodiversity (Bakker et al. 1993: 92). One could also imagine that farms and small settlements could be constructed on these refuges. It will not appeal to everyone, but the relative isolation of new terps would offer a very different lifestyle, and one that would be attractive to tourists and birdwatchers who seek a (temporary) retreat from the modern world. The very long history of living on terps in a salt marsh would add an authentic sense of place to the experience.

In the medieval period, a number of terps were provided with ring-dikes, which excluded floods during the summer period but did not seek to prevent floods during the winter months. Experimental archaeology has shown that some cereal crops can be successfully grown in such environments—and indeed on the ridges of unprotected salt marsh—as long as saltwater does not flood the growing crops (Van Zeist et al. 1976). The use of summer and winter dikes is the norm in many river floodplains but, building on this adaptive pathway from the past, it could also be considered for coastal wetlands and zones. It would reduce the tidal range and reach in estuaries during the winter months, when high peak flows in rivers are more likely to cause floods, but leaving productive agricultural land for the summer months.

As noted earlier, the construction of continuous dikes commenced in the tenth century AD, and this particular adaptive pathway has continued to be employed by the coastal communities in the southern part of the North Sea. In the description of the impact of sea-bank construction presented above, some of the negative feedbacks have been highlighted—increased tidal range and reach, coastal squeeze of wetlands, the need to build ever higher and stronger embankments—but dike construction should also be recognized for strengthening the social capital and social identity of coastal communities. Building dikes was a community effort. In some places, and at certain times, ecclesiastical and lay institutions and authorities took a leading role in the construction of sea banks (Rippon 2000; Van de Noort 2011b: 115–23), but local communities had a strong sense of ownership of the dikes, not least because of their dependency on these for subsistence and personal safety. The need to maintain the dikes and drainage systems behind the dikes, and the essential collaboration required at local and regional scales, have been credited with instigating a range of local institutions that have served to strengthen both the social capital and identity of coastal communities (e.g. Van der Vleuten and Disco 2004). For example, the first democratic organizations in the Netherlands in the thirteenth century (the so-called 'Waterschappen') have been credited with contributing to the collaboration and ultimate unification of the distinct political units that were to become the Netherlands (Tol and Langen 2000: 360–5). This early sense of social capital should not be overstated, and there are ample early examples of 'over-diking', where dikes are made higher than adjacent dikes so that, during storms, a neighbour's land is flooded first, relieving the pressure on one's own land. Nevertheless, the high value of the social capital amongst nations, institutions, and communities around the North Sea, with regard to the need to protect the human populations and their economic interest and the need to protect ecologically important habitats, strengthens the resilience of the coastal communities in a time of accelerated sea-level rise.

Choosing between increased capital investments in strengthening the coastal defences to 'hold the line', or adopting solutions that seek to work with the dynamic sea—either through managed realignment or beach nourishment—is essentially a matter of political decision making, albeit constrained by financial resources. In the southern North Sea basin, where the impact of RSL change in the twenty-first century is most noticeable, the countries with very long coastlines—Denmark, the Netherlands, and Britain—have adopted policies that promote the latter approach as the preferred option, whilst the hold-the-line option is more selectively applied to towns and areas of high economic value (De la Vega-Leinert and Nicholls 2008; Fenger et al. 2008; Van Koningsveld et al. 2008). Germany, which has a relatively short coastline and one that is of key economic importance, has a preference for the strengthening of hard defences of the North Sea coast (Sterr 2008).

Deciding to work with the natural dynamics of the sea requires high levels of social capital, mainly because these options inevitably adversely affect the interests of some stakeholder groups. For example, moving the flood defences inland as part of a managed realignment project usually means taking agricultural land out of production, with the livelihood of farmers affected, but this will benefit other stakeholders, such as residents whose properties are protected because of the resulting reduction of the tidal range and reach. High levels of social capital can overcome the tendency to find piecemeal solutions to the impacts of sea-level rise—reflecting the interests of specific groups or individual landowners—and aim for the 'greater good'.

CONCLUSION

RSL rise in the twenty-first century will affect primarily the coasts in the southern part of the North Sea basin; in the northern part, the glacio-isostatic adjustment will see the land continue to outpace the rise of the sea level. The rise in sea level could be as much as 1 m by the end of the century. Whilst this rise may appear relatively small, it represents a significant change in that the sea level has been stable for several millennia. During this stable period, coastal communities have established themselves in large numbers on the coast, and their resilience to the impacts of climate change on the sea level will be severely tested.

In the last 10,000 years, coastal communities have adapted to the rising sea in different ways, including migration away from what were once contemporary coastlines; structured depositions played a role in the re-establishment of the social relationships between coastal communities and the sea at times when familiar landscapes were flooded: economic utilization of the salt-marsh environments, which commenced in the Bronze Age and peaked with the construction of terps between c. 500 BC and AD 1000; and the construction of hard flood defences from c. AD 1000. Current adaptive pathways of coastal communities include the construction of more, higher, and stronger hard defences to protect urban settlements and areas with industrial infrastructure, combined with the release of relatively lower-value land for managed realignment and continuous—and probably ever-increasing—nourishment of beaches with dredged sand. On an archaeological timescale, these are not sustainable adaptive pathways, but it is acknowledged that they represent the application of the 'best practice' ICZM and will—politically, economically, and probably socially—be adequate for the twenty-first century. The archaeological study of how people in the past adapted to rising sea levels can strengthen the resilience of coastal communities around the North Sea in a number of ways, which can be summarized thematically as follows:

First, it shows that sea-level change in the North Sea basin is not a phenomenon unique to the modern era and that long-term understandings of sea-level change, coastal morphological and ecological change, and changes in the way coastal communities lived with the sea are required to inform the future management of the coastal zone. The more educated coastal communities are on these matters, the greater the likelihood that future adaptive pathways will be sustainable in nature. The confidence in governmental institutions to protect coastal communities from the sea is quite remarkable, given the recent and not-so-recent history of floods and dike breaches, and the sharing of a greater knowledge base should form the bedrock for the sustainable management of coastal zones and wetlands. The example mentioned at the beginning of this chapter, how archaeological and historical knowledge of the reclamation history of the Humber estuary could have been used as a source of information for future management solutions, illustrates this point.

Second, a greater understanding of past successes and failures in dealing with coastal change, and of the risks involved with investing in settlements and infrastructure behind the dikes, is also likely to help build social capital. If there is one 'lesson' to be learnt from the past, it is that no one institution or individual working on their own, be that a government in charge of flood defence or a farmer who works on the coast, can produce sustainable and workable solutions in this context (e.g. Correljé, Broekhans, and Roos 2010: 12). The social capital in the North Sea basin extends, exceptionally, across national borders. It has, for example, been used for the implementation of an ecologically and geomorphologically sustainable management of the Wadden Sea and, under the umbrella of the European Union, for the control of fishing of species that were threatened by extinction.

Third, in the past, people around the North Sea reconstituted their relationship with familiar landscapes that were drowned through affirmative actions, such as the construction of new types of monuments and the structured depositions of valuable objects. These actions played a role in reformulating the sense of place of coastal communities. In the current century, this will also be required as coastal landscapes are set to change dramatically, ranging from ever higher dikes that will visually separate coastal communities from the sea to the disappearance of coastal wetlands and their bird populations. Modern coastal communities will have to find ways in which they can reformulate their relationship with the sea and their sense of place with the new landscapes, and the examples from the past may offer some guidance.

Fourth and finally, the coastal wetlands around the North Sea were in the past highly valued economic landscapes. Today, very few coastal landscapes are used for grazing sheep and cattle, despite the fact that their produce can be sold at premium prices. Where managed realignment projects seek to create new intertidal environments for the purpose of nature conservation, for example where guided by the European Union Bird and Habitat directives, people and livestock are frequently not part of the new landscape. Whilst this may, at first sight, appear logical, research has clearly shown that where salt marshes are grazed the biodiversity is increased. The reintroduction of the grazing of young salt marshes should therefore be encouraged. Integrating the very recent concerns with nature conservation within a holistic approach that recognizes the adaptive pathways from the past would make managed realignment more sustainable in terms of maintaining biodiversity and supporting food security, and gain more support from the various stakeholders within the coastal communities.

6

The Sundarbans

INTRODUCTION

The Sundarbans are one of the largest halophytic wetlands in the world (Fig. 6.1). In geographical terms, they are the delta formed by the rivers Ganges, Brahmaputra, and Meghna on the Bay of Bengal. This delta is defined by the River Hooghly to the west, and the River Meghna in the east; in the north by the maximum extent of the tidal influence, and in the south by the open sea of the Bay of Bengal. In the modern political world, the Sundarbans can be found in the South 24 Parganas district of West Bengal, India, and the Bakarganj and Khulna districts of Bangladesh. The total extent of the Sundarbans before the seventeenth century was c. 17,000 km^2, of which about one-third is located in modern India and two-thirds in Bangladesh. The name 'Sundarbans' probably comes from the Bengali words *sunder* and *ban*, meaning 'beautiful forest', or, alternatively, is derived from the mangrove Sundari tree (*Heritiera fomes*). The name is of a relatively recent date (Sarkar 2009: 11). These coastal wetlands are, famously, the home of the Royal Bengal tiger, but the ecosystem also supports or supported leopards and panthers, estuarine crocodiles, buffaloes, rhinoceros, chital deer, Gangetic dolphins, and many species of snakes, shellfish, and birds (Mandal 2004: 32–61).

The coastal wetlands of the Sundarbans are among the most enigmatic in the world. Ancient and modern literature has used them as the backdrop for many legends and tales, although they rarely stay in the background of these stories and the dynamic interaction of sea and sediment frequently becomes the main theme, as in Amitav Ghosh's (2004) *The Hungry Tide*. The palaeoenvironmental history and archaeology of the Sundarbans, which provide the basis for this chapter, have been researched to a high standard. There is an urgent case, however, for more extensive archaeological fieldwork and for the publication of past excavations. I visited the Indian Sundarbans in February 2012, undertaking fieldwork with the aim of contextualizing the archaeological sites in their environmental setting and to study the impact of sea-level rise on coastal communities. The Indian Sundarbans were selected for fieldwork because the rapid accumulation of sedimentation ceased here some 2,500 years ago (see

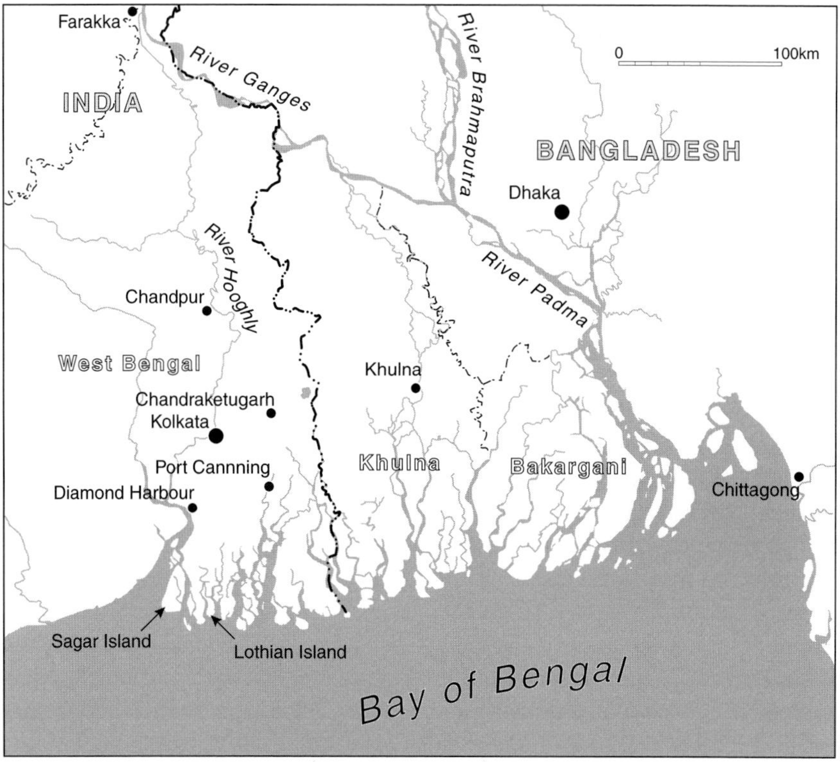

Fig. 6.1. Map of the Sundarbans and key sites mentioned in the text.

below), leaving a historical landscape with plentiful features of a significant age and of considerable archaeological interest. The sediment accretion in the Bangladeshi Sundarbans is of a more recent date, concealing historical landscapes that are over 200 years old, and this part of the Sundarbans was not visited. In the writing of this chapter, Professor Dilip Chakrabarti, a long-standing specialist in the archaeology of India and Bangladesh, acted as a critical friend.

The key reasons for selecting the Sundarbans as one of the case studies are fourfold. First, intertidal mangrove forests are recognized for their outstanding ability to respond 'naturally' or autogenically to sea-level rise. Since the Sundarbans are the largest block of halophytic mangrove swamp in the world, their selection as a case study is fully justified on this basis alone. Moreover, this region has been subjected to neotectonic movements that add a layer of 'real-world' complexity to the ability of the local mangrove swamps to adapt to sea-level rise. In effect, the eastern parts of the Sundarbans are sinking, producing locally a higher rate of Relative Sea Level (RSL) rise, which can be presented as a proxy for what happens to this type of coastal wetland under the worst-case SRES scenario.

Second, the Sundarbans present us with a rich archaeological heritage and a living culture that is closely connected to the development of the coastal wetlands. Archaeological evidence is available for people living in, or on the edge of, the Sundarbans for nearly 2,500 years, and directed palaeoenvironmental research in recent decades provides the environmental setting for these findings. Equally, the living culture of the Sundarbans is closely connected to the environment, and this has been recorded in recent decades to a very high standard (e.g. Jalais 2010).

Third, the very rapid population increase in both India and Bangladesh during the second half of the twentieth century and first decades of the twenty-first century places very significant pressures on the coastal wetlands and the coastal communities of the Sundarbans. The populations of these two countries have quadrupled since the partition in 1947, and the thirst for new agricultural land is very acute. This places tremendous pressure on the accommodation space of the coastal wetlands. The availability of such space is essential for the successful adaptation by intertidal mangrove forests to the rising sea level. It is likely that the full impact of climate change and sea-level rise will be experienced here earlier than in most other coastal wetlands around the world.

Fourth, the political situation in the region is one of strained relationships between the various stakeholders and there are no signs of the development of an Integrated Coastal Zone Management (ICZM) plan. This is visible on two different levels. At a national level, transnational relationships are characterized as much by enmity as by cooperation. This is a reflection of the recent history of Bangladesh, following the partition of 1947 when Bangladesh became East Pakistan. During the war of liberation in 1971 the Indian Army joined forces with Bangladeshi fighters against Pakistan. An independent Islamic Bangladeshi state was subsequently declared. Apart from the religious tensions, current political issues that adversely affect the relations between India and Bangladesh include the rights of ethnic Indians in the many 'enclaves' in Bangladesh and those of the Bangladeshi living in enclaves in India, along with the uncontrolled migration of Bangladesh refugees to India, and the close relationship between Bangladesh and China. Furthermore, the Farakka Barrage—constructed between 1961 and 1975—diverts fresh water from the Ganges into West Bengal with far-reaching consequences for the downstream ecosystems, and this continues to be a source of much controversy between the two countries.

The key institution available for transnational cooperation, the South Asian Association for Regional Cooperation (SAARC), has had only limited success in dealing with the issues of sharing the limited freshwater resources or constructing coastal defences (Ray 2010). Within India and Bangladesh, the relationships between the various communities are also strained, and the institutions and structures that seek to bring these different groups together remain relatively poorly developed. The main oppositions in the Sundarbans are between the groups of farmers and fishermen on the one hand and on the other the nature conservationists, represented on the ground by the various forestry departments and under the scrutiny of international NGOs. Other oppositions exist between the traditional fishermen and those practising aquaculture—which is largely focused on the export of tiger prawns and has extensive adverse impacts on the health of the riverine and intertidal ecosystems. In short, the Sundarbans provide a case study in which the rights of different groups with an interest in the coastal wetlands are relatively poorly organized and integrated.

This chapter will present, first, the geological and palaeogeographic history of the Sundarbans and, second, the archaeology and history of the Sundarbans. The third part, on how the socioecological resilience of current communities can be strengthened by researching past adaptive pathways, is based on recent fieldwork.

HOW CLIMATE, ENVIRONMENT, AND SEA-LEVEL CHANGE SHAPED THE SUNDARBANS

In the geological history of the Bay of Bengal, the separation of the Indian tectonic plate from the Antarctic tectonic plate in the early Cretaceous era, around 120 million years ago, is of critical significance (Rao, Krishna, and Sar 1997). The Indian plate drifted subsequently north-westwards, colliding with and subducting beneath the Eurasian tectonic plate from around 55–45 million years ago, which slowed down the rate of drifting. It is estimated that the total shortening of the plates following the collision lies in the order of 2,000 km, with the Indian plate rotating by 45° anticlockwise (Le Pichon, Fournier, and Jolivet 1992: 1,096). The subduction and folding of the edges of the plates following the collision created the Himalayas.

The River Ganges and its many tributaries drain much of the water from the Himalayan glaciers. However, its many 'sources' include not only glaciers in the Himalayas, such as the Gangotri, but also a range of rivers from the slopes of the Aravalli mountains to the west and the Vindhya range to the south. Major tributaries of the Ganges in its upper and middle reaches include the rivers Chambal, Yamuna, Ghaghara, Gandak, and Son. With a length in excess of 2,500 km, the Ganges drains an area of around 1 million km^2. Reaching the India–Bangladesh border, the Ganges splits into the Hooghly channel which continues south past Kolkata to the Bay of Bengal, and the Padma channel which continues east into Bangladesh where it joins the River Brahmaputra—which has its source in the Himalayas as the

Tibetan River Yarlung Tsangpo—and the River Meghna, which drains the Dhaka region of Bangladesh. The rivers form a broad delta with a very large number of channels defined by the River Hooghly on the west and the River Meghna on the east. These channels create a landscape of hundreds of islands. The Sundarbans are the delta that is under direct influence of the tides.

The Sundarbans are, effectively, a tide-dominated sediment deposition centre at the mouths of the rivers Ganges, Brahmaputra, and Meghna, and their many tributaries. These rivers formerly transported downstream in excess of 1 billion tons of alluvial sediments each year, and these three rivers combined had the greatest river-sediment discharge in the world (Milliman and Meade 1983: 2). Most of these sediments were deposited in the Bay of Bengal where they formed a large subaqueous delta comprising an upper, middle, and lower fan, with the latter extending as far south as Sri Lanka. A major marine canyon—the 'Swatch of no Ground' or the Ganga Trough—played an important role in the transport of the sedimentation into the Bay of Bengal and beyond the edge of the continental shelf. However, about one-third of the sediments were deposited within the Sundarbans.

Coring through the late Pleistocene and Holocene sediments of these coastal wetlands has established a clear differentiation between the harder and often concreted clays and sands, dating to the Pleistocene and representing a terrestrial landscape with riverine deposits, and the much softer sediments, frequently intercalated with peat, which represent the tidal or deltaic sedimentation regime of the Holocene (Stanley and Hait 2000). The coring programme located the top of eroded Pleistocene deposits in excess of 50 m below the current surface of the land in some locations, and radiocarbon dates from samples of these deposits usually date to before 9000 cal BC. Marine-transgressive sediments found in a number of cores between the riverine-terrestrial land surface and the fully tidal landscape—and frequently containing mangrove pollen—were dated to the period before 7500 cal BC. Radiocarbon dating of the base of the tidal deposits found in the cores revealed dates in the range of 7500 to 5300 cal BC. However, there was significant variation in the thickness of the tidal landscape deposits. As a general rule, the accumulation of these sediments was lower in the north and north-west, usually less than 30 m, and greater in the south-east and south of the Sundarbans, often over 50 m (Stanley and Hait 2000: 33).

The outline relationship between sea-level dynamics and coastal wetland development is clear. Around 10,000 years ago, the Eustatic Sea Level (ESL) was around 50 m below current levels, and the earliest marine transgressions onto the edge of the continental shelf date from around this time, when they formed the delta-front platform south of the current Sundarbans. Between 9000 and 7500 cal BC, the marine influence transgressed onto the alluvial plain. A deltaic wetland was formed from c. 7500 cal BC, the oldest parts being in the south of the Sundarbans with the more northern areas being influenced from c. 5300 cal BC. The three factors that determined this interaction were: sediment accumulation, which raised the land by an annual average of up to 7 mm for the southern part of the Sundarbans during the Holocene; land subsidence at an average rate of 5 mm/year for the southern Sundarbans, caused by the sediment loading of the continental-shelf edge and neotectonic activity; and a rising ESL in the order of 2 mm/year. However, intercalated peat layers representing submerged forests, such as those found at Port Canning and at Khulna, and shell beds buried beneath the deep sediments (Mandal 2004: 2–3; 17–18), show that the landscape development was not linear or uniform, and that periods of marine transgression were interspersed with periods during which freshwater and brackish wetlands developed. Few of these peat layers are independently dated, and we should assume that they represent a dynamic landscape development, rather than Sundarbans-wide periods of marine transgressions and regressions.

More detailed analysis of sedimentary sequences has shown a very direct link between climate change and the sedimentary regime of the River Ganges, on a range of timescales (Goodbred 2003). It shows that during the last glacial stage, the lack of rainfall in the region limited glacier activity, and very little sedimentation was dispersed through the river system during this time. This changed in the postglacial period, and the increased precipitation, concentrated in the monsoon months, caused deep erosion of the river channels and floodplains in the upper reaches of the Ganges, leading to the rapid increase in the growth of the sediment fans in the Bay of Bengal. Subsequent Holocene sea-level rise caused a landward marine transgression, and a partial shift of the deposition centre from the Bay of Bengal to the coastline on the continental shelf, creating the current delta. It has been estimated that during the rapid climatic amelioration of the postglacial and Early Holocene periods, the sediment load of the Ganges was 2.3 times greater than its current load (Goodbred and Kuehl 2000). Over the last 7,000 years, with temperatures stabilizing, the sediment load has diminished. During this period the infilling of the river channels in the upper reaches of the Ganges system, previously subject to erosion, has contributed to the emerging dynamic balance described earlier, between sediment accumulation, land subsidence caused by sediment loading, and sea-level change.

The absence of land ice in the last glacial stage in the Bay of Bengal means that no account has to be taken of any glacio-isostatic adjustments. The ongoing north-west movement of the Indian plate, however, caused a range of neotectonic adjustments throughout the Late Glacial and the Holocene periods. These neotectonic adjustments or earthquakes have created a landscape characterized by uplifted terraces and subsiding basins. These features have caused changes in the drainage pattern of the major rivers (Kuehl et al. 2005). One of the most important of the neotectonic features that have played a role in the development of the Sundarbans is a hinge zone—here, the boundary between two tectonic blocks dating back to the Eocene, and a related Tertiary morphologic lineament running NE–SW across the Sundarbans—which tilted the eastern part of the region downwards. This has caused the rivers Ganges and Brahmaputra to find new courses to the east of their previous channels. Before the twelfth century AD, the Hooghly on the western edge of the Sundarbans was the main outflow channel for the Ganges. By the sixteenth century, the majority of the flow was through the Padma channel eastwards towards the Brahmaputra (Chakrabarti 2001: 127). By the middle of the nineteenth century, the main channels of the Ganges/Padma and Brahmaputra had joined the various channels of the River Meghna near Chandpur in Bangladesh, towards the eastern limit of the delta (Mandal 2004: 23; Chakrabarti 1992: 4). This easterly redirection of freshwater flow has caused saline intrusion to extend increasingly northwards and westwards in recent centuries.

The geological hinge zone provided the Sundarbans during much of the Holocene with extensive accommodation space in what is now the Bangladeshi part of the Sundarbans (Stanley and Hait 2000: 33–6). In other words, the sediment deposition centre of the Ganges—and to a lesser extent of the Brahmaputra—shifted during the late Holocene, from the western to the eastern parts of the Sundarbans. The West Bengal part of the Sundarbans is much older, but receives decreasing amounts of fresh water and sediments, whereas the Bangladeshi Sundarbans, of more recent date, receive increasing volumes of water and sediments from the Ganges–Brahmaputra–Meghna rivers. This insight into the long-term geological history is confirmed by coring of the upper sediments of the Sundarbans. Examination of the upper 7 m of sediments has indicated that progradation, or the extension of the Sundarbans southwards into the Bay of Bengal, occurred in distinct phases. The westernmost parts of the Sundarbans progradated between 4000 and 1000 cal BC, the central part between 2500 cal BC and cal AD 1750, and the easternmost parts extending since cal AD 1750 only. This eastward shift of the extension of islands and peninsulas into the Bay reflects, principally, the eastward shifts in the main channels of the Ganges (Allison et al. 2003).

The earliest presence of mangrove pollen in the sediments dates to the time of the creation of the delta, between 9000 and 7500 cal BC (see above), and this reinforces our understanding of the pioneering nature of mangroves on tidal and tropical coasts. The coastal wetlands of the Sundarbans are dominated by the halophytic *Heritiera fomes*, referred to locally as the Sundari or Sundri tree. As with other halophytic mangrove trees, the Sundari's habitat is in the intertidal zone between mean sea level and high-tide level. It has played a key role in the trapping of sediments to produce the deep stratigraphies described above. The Sundari tree prefers regular freshwater flushing, and is therefore noticeably much larger and more luxuriant in the eastern Sundarbans where the greatest flow of fresh water occurs. The mangrove forests are interspersed with numerous dendritic tidal channels, many of them formed by

Fig. 6.2. A view of the Indian Sundarbans. The Sundari trees grow on either side of the tidal channels and their air roots play a pivotal role in the trapping of fine sediment. This enables mangrove swamps in their natural setting to respond autogenically to sea-level rise.

the frequent storms and cyclones that affect the coastal morphology. Mudbanks occur within the wider channels (Fig. 6.2).

Landward of the mangrove swamps, freshwater and brackish swamp forests developed, reflecting the lower levels of salinity here. Dominant tree species for this part of the swamp forest include a number of mangrove trees and shrubs such as *Heritiera minor*, *Xylocarpus molluccensis*, and *Bruguiera conjugate*, for which there are no common English names, as well as non-mangrove species such as *Pandanus tectorius* or Thatch Screwpipe, and the *Hibiscus tiliaceau* or Sea Hibiscus along the banks. Seaward of the mangrove swamps are sandy beaches.

There is evidence that the Sundarbans extended further northwards in the past. For example, during excavations undertaken as part of the Metro Railway Project Work in Kolkata two peat bands were encountered at different depths: these were dated to around 5950 cal BC and 800 cal BC respectively. The depth of the sequences of alluvial sediments in the floodplain of this channel of the Ganga, known as the Bhagirathi, is in itself not surprising, but the abundant presence of mangrove pollen in the peat indicates that the freshwater and brackish swamps extended further north and west in prehistory (Barui and Chanda 1992). More recently, two cores from the Rajarhat area in the North 24 Paraganas, c. 20 km north-east of Kolkata, were investigated through pollen analysis. The older of the two samples, with a midpoint radiocarbon date of 5500 cal BC, contained the pollen of halophytic plants, dominated by the mangrove *Heritiera fomes*. This reflected a vegetation pattern similar to the current composition of the Sundarbans. The younger of the two samples, with a midpoint of 820 cal BC, suggests vegetation in which *Heritiera* was less dominant; this is attributable to fluctuations in the freshwater supply, and to the existence of a less dense mangrove forest (Barui 2011). The results of this younger sample could reflect local hydrological conditions or possibly the impact of the first farmers on the region (see below). The more recent historical geography of the Sundarbans suggests that no significant changes affected the swamps in the second half of the second millennium AD, and it has been argued that the list of the mahals in the rent roll of 1582 implies a northernmost boundary for the forest very similar to that of the late nineteenth century (Hunter 1875: 380–1; cited by Chakrabarti 2001: 130).

The Sundarbans are, famously, home to the Royal Bengal tiger whose reputation as a man-killer is well known. During the twentieth century, an average of 20 people lost their lives to the Bengal Tigers of the Bangladeshi Sundarbans each year (Islam, Alam, and Islam 2007). Besides the tiger, the Sundarbans are home to a range of mammals that have adapted to living in the delta, including wild pig, Chital (spotted deer), wild buffalo, and rhinoceros, alongside smaller mammals of the Muridae family including various types of rats and mice. The largest reptile in the Sundarbans is the Estuarine crocodile, with the smaller Marsh crocodile only just surviving in the upper reaches of many rivers. Unsurprisingly, the nutrient-rich rivers and estuaries of the Sundarbans are naturally stocked with a wide variety of fish species, mostly of the carp and dogfish families, although sharks are frequent visitors to the Sundarbans' rivers. The Gangetic dolphin inhabits the freshwater parts of the system. Amongst the shellfish, species of crab and shrimp are abundant in the mudbanks and along the edges of the islands (Mandal 2004: 32–60).

In the context of the Wetland Change Model (McFadden, Spencer, and Nicholls 2007; see above), the resilience of this coastal wetland in dealing with sea-level rise is dependent upon three aspects: tidal range, sediment budget, and accommodation space (see chapter 4). The tidal range in the Sundarbans lies between 3.5 and 5m in the estuaries (Chaudhuri and Choudhury 1994), and this relatively large range provides an active and dynamic mechanism by which the Sundarbans delta can autogenically respond to the anticipated rise in the sea levels.

What will happen with the sediment budget under the various climate change scenarios is much less clear. This is principally a function of the difficulties of simulating monsoons within General Circulation Models (GCMs). But it seems likely that the South-east Asian monsoon will become stronger and more intense. This would cause the River Ganges, which receives 80 per cent of its waters and transports 95 per cent of its sediments during the monsoon months, to produce (under natural circumstances) a higher positive sediment budget for the delta, thus theoretically strengthening the resilience of the eastern part of the Sundarbans (Goodbred 2003). The Brahmaputra, however, is more dependent on ice meltwater from the Himalayas: its water supply may decrease by as much as 20 per cent, leading to a lower sediment load being transported by this river to the Bay of Bengal, although under certain SRES scenarios this may be offset by increased precipitation rates (Immerzeel, Van Beek, and Bierkens 2010). Nevertheless, the sediment load carried by the Ganges has reduced in recent times, and this has had a direct impact on the Sundarbans. Using satellite imagery, the net loss of coastal wetland between 1989 and 2010 has been calculated at 170 km^2, or an average loss of 4.6 km^2 per year (Rahman, Dragoni, and El-Masri 2011). The research found that the erosion varied both spatially and temporally, and that the rate of sediment accretion was decreasing. The latter is unlikely to be caused by climate change; rather this is the direct impact of the construction of dams, notably the Farakka Barrage which became operational in 1975, and which is the only probable cause of a 30 per cent decrease in the sediment load of the Ganges/Padma in Bangladesh between 1974 and 1979 (Tanzeema and Faisal 2001). A reduction in the sediment budget of this magnitude evidently weakens the resilience of the Sundarbans to respond to sea-level rise (Fig. 6.3).

The greatest impact on the resilience of the Sundarbans in responding to sea-level change is that of the reduced, and rapidly reducing, availability of

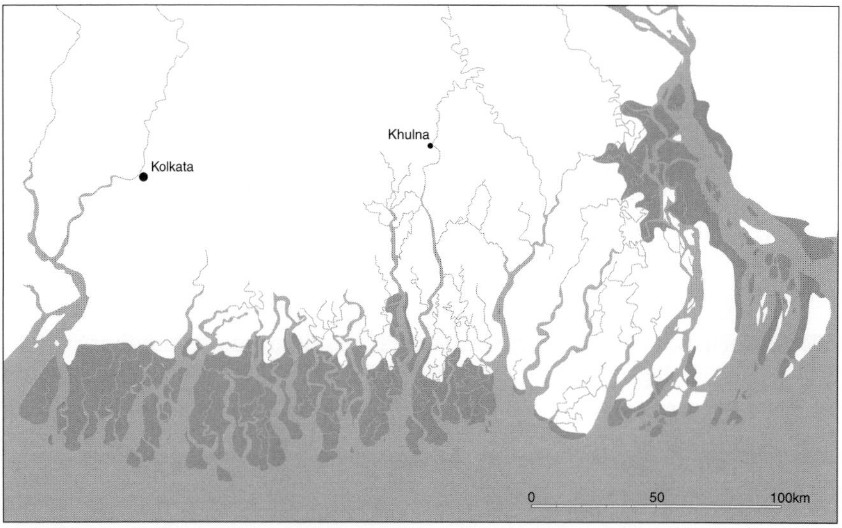

Fig. 6.3. Projected marine transgression of the Bay of Bengal on the Sundarbans (dark-grey tint) based on a rise of the RSL of 1 m and excluding the impact of constructed coastal defences (based on the model presented in Weiss, Overpeck, and Strauss 2011).

Fig. 6.4. Another view of the Sundarbans. The construction of a bund on the edge of Bali Island has removed the accommodation space for the mangrove swamps, which are no longer able to trap fine sediments. Widespread erosion is the outcome.

accommodation space. During the last three centuries, about two-thirds of the Sundarbans' mangrove woodlands have been cut to make way for agricultural land, and very little remains of the freshwater swamp forest landwards of the mangrove swamps (Richards and Flint 1990; see below). In a scenario of rising sea levels, the mangrove swamps—which play a pivotal role in raising the lie of the land and in protecting it from the impacts of storms and cyclones—will be increasingly squeezed between the rising sea level and the advancing deforestation taking place to accommodate the need for agricultural land and space for aquaculture (Fig. 6.4).

PAST ADAPTIVE PATHWAYS TO CLIMATE, ENVIRONMENTAL, AND SEA-LEVEL CHANGE

The geographical development of the Sundarbans during the last 10,000 years defines, to a very large extent, what we know of the archaeology of the region. This manifests itself in two ways. First, much of the early prehistoric evidence has been buried beneath many metres of the sediments accreted in the coastal wetlands during the Holocene, and evidence of the activity of hunter-gatherers-fishers, especially, will only be uncovered in very deep excavations. This bias can be addressed by considering the archaeological evidence from the surrounding higher and dryer grounds, and extrapolating this onto the Sundarbans. Second, the shifting of the deposition centre of the Ganges from west to east during the late Holocene is an important phenomenon. As a direct consequence of this, large-scale sediment accumulation ceased c. 2,500 years ago in the Indian Sundarbans (Allison et al. 2003; see above). This means that the historical landscape contains many archaeological features that have their origin in the second half of the last millennium BC. This historical landscape is, to a certain extent, representative of human activity elsewhere in the Sundarbans, where younger sediments have masked archaeological features.

Palaeolithic material is not known from the Sundarbans. The closest evidence for human presence of this date comes from the Lalmai Hills—the southern extension of the Mainamati Hills—some 70 km north-east of Chandpur on the River Meghna in Bangladesh. From 11 findspots come a large number of handaxes, scrapers, points, and blades from retouched fossil wood, representing activity during both the lower and upper Palaeolithic (Chakrabarti 1992: 34–42). On the Indian side of the Sundarbans, Palaeolithic sites are known from the higher grounds in Orissa to the south-west, and in Bihar to the north-west (Chakrabarti 2010: 71–2). Aceramic Mesolithic material is equally absent from the Sundarbans, and the nearest evidence for human activity during the first half of the Holocene comes from Birbhanpur on the banks of the Damodar River, around 100 km north-west of Kolkata. Excavations by the Archaeological Survey of India in the 1950s found concentrations of cores, and microlithic non-geometric flakes and tools of locally available materials including quartz, chert, and fossil wood. In situ evidence included ten holes or pits, probably dug for the posts of a dwelling. This site is dated to the early or middle Mesolithic on the basis of an absence of geometric tools and pottery (Lal 1958). It is highly unlikely that the natural resources of the Sundarbans, especially fish and shellfish, were not recognized as an important source of food during the Mesolithic by communities living in or on the edge of the Sundarbans. Any archaeological evidence relating to this time would have been buried beneath the sediments of the Ganges and Brahmaputra, and this is an instance where the absence of evidence should not be taken as evidence of absence.

The introduction of agriculture into South Asia has been the subject of a recent comprehensive review of the published literature by Dorian Fuller (2006). Pollen cores from across India include peaks of microscopic charcoal in the mid-Holocene, with dates ranging from 7000 to 6000 cal BC in the north-west of India to 3500 cal BC in peninsular India. This implies that forests were cleared through the use of fire, presumably for the purpose of encouraging new plant growth as a means of attracting wild game, but also, possibly, to provide opportunities for the development of agricultural activities. Early agricultural activity in the Near East influenced South-west Asia, and the earliest evidence for the domestication of plants and animals in Pakistan and the Indus Valley dates back to around 7000 cal BC. The Ganges is likely to have been a focus for early agriculture, which may have developed independently from agriculture in the Indus Valley. The Gangetic Plain is the natural habitat of a number of wild progenitors. Wild rice was probably harvested here as early as the seventh millennium BC, and domesticated rice from the third millennium BC. The limited evidence available hints at the earliest domestication of livestock around 2200 cal BC for the Bihar region on the eastern Gangetic Plain, some 500 km north-east of Kolkata. The influence from the west, in the form of domesticated varieties of wheat and barley, is also dated to this period. Some crops that originated in Africa, such as hyacinth bean, cowpea, and sorghum, were introduced to the Gangetic Plain in the early second millennium BC.

Around 2500 cal BC, the first permanent villages were constructed in the Gangetic Plain. By 2000 cal BC sedentary villages of farmers growing rice, millet, and pulses, and with domestic livestock of cattle, sheep, and goats, became commonplace. There is ample evidence for continuity of habitation into the Chalcolithic, in the mid-second millennium BC. Alongside the agriculturalists on the Gangetic Plain, a 'ceramic Mesolithic' persisted in adjacent regions. This was characterized by a subsistence based on hunting-gathering-fishing activities, including the gathering of wild rice, but with cord-impressed and rusticated ceramic wares being used (Chakrabarti 2010: 116). Whilst there is no archaeological evidence for a ceramic Mesolithic in the Sundarbans, it seems probable that any human activity in the region could be characterized as such, with an emphasis on fishing, hunting, and gathering of shellfish. The evidence from the coastal mound sites of the Orissa region on the Bay of Bengal south-west of the Sundarbans—with stratigraphies dating from the second through to the first millennium BC—shows that fishing was an important part of the economy. This is implied by the presence of harpoon and other projectile points, and by the location of the mounds alongside rivers in the coastal wetlands. The material from the mounds in the Orissa region shows that domesticated animals, and domesticated plants including rice and pulses, were also consumed at these sites (Fuller 2006: 47).

The origin of domesticated rice remains a hotly disputed topic. Opposing hypotheses include the single-origin hypothesis for the domestication of *Oryza sativa* from wild progenitors in the Yangtze Valley in China, and a multiple-origin hypothesis that identifies several regions from which rice originated, including the Yangtze and Gangetic valleys, producing the two main varieties *Oryza sinica* or *japonica* and *Oryza indica* from different wild progenitors. Wild rice was used at the site of Lehuradewa in Uttar Pradesh in the seventh millennium cal BC, as yet the earliest evidence in South Asia for this (Fuller 2006: 41). At nearby Koldihawa, rice was found in a context that included wattle-and-daub houses, microliths, and polished stone celts or axes, and handmade pottery—cord-marked and incised ware, plain red ware with ochrous slip, and crude black-and-red ware—with radiocarbon dates only slightly later than those at Lehuradewa (Chakrabarti 2010: 205–7). However, unambiguous evidence for the production and consumption of domesticated rice in India—from Uttar Pradesh, Bihar, and Orissa—dates in all cases to the early second millennium BC. The evidence includes charred grains from cultural deposits, the use of rice and rice husks for the tempering of ceramics and as imprints in pottery.

The oldest archaeological evidence for the use of domesticated rice in the Sundarbans dates to the later part of the second millennium BC and falls in the Chalcolithic, or Copper Age. The evidence comes in the form of rice and rice-husk impressions in pottery from Amri, in the southern parts of modern Kolkata, then within the tidal mangrove landscape (Vishnu-Mittre 1989). Whereas the rice variant cannot be ascertained from the impressions, traditionally the variety of rice most commonly used in the Sundarbans is *boro*. It grows as a submerged crop and, once sown, needs no transplanting. Its cultivation on mudflats near the big rivers is an age-old practice, utilizing the dynamic coastal wetlands without embanking the mudflats (Chakrabarti 1992: 21). Salt-tolerant varieties of rice grown in paddies, including Matla and Hamilton, are also known to have been grown in the Sundarbans in the past (WWF-India 2011: 11). The use of these varieties of rice represents evidence of coastal communities adapting their subsistence strategies to living in a dynamic tidal landscape.

Chalcolithic sites in West Bengal are identifiable by the presence of various variations of black-and-red wares, alongside the presence of copper and very occasionally iron artefacts, and terracotta female figurines. About 70 sites of this period are known from West Bengal, but none has to date been recorded from the Sundarbans (Chakrabarti 2010: 241–3).

The earliest archaeological evidence for proto-urban settlements in the Sundarbans dates back to the pre-Mauryan period. These settlements represent another adaptive pathway within a tidal region where dynamic sedimentation and sea-level rise formed the landscape. The famous site of Chandraketugarh, c. 35 km to the east of Kolkata and at the northern margins of the contemporary mangrove swamps (see above) is one of the sites containing Northern Black Polished ware (NBP). The presence of NBP is an important archaeological marker for the beginning of the early historic period. NBP ware has its origins as early as the eighth century BC, but is thought to be found from around 500 BC in the lower Ganges floodplain (Chakrabarti 2010: 353–4): this has been confirmed by independent radiocarbon dating. Chandraketugarh was a large fortified settlement, thought to be over 25 ha in extent on the basis of the mud ramparts that still survive as earthworks, and which stand in places to a height of 6 m. Evidence has been found for activity dating to the pre-Mauryan era; the excavations, limited to a few trenches, revealed remains of a wooden housing complex dating to the second century BC. The settlement spread beyond the ramparts in the first century AD. In the Gupta period (AD 320–520), a *sarvatabhadra*-type temple complex constructed from burnt bricks was added to the northern part of the site; the complex included a large square sanctum *cella*. The end-date for activity on this site is from the Pala-Sena periods (eighth to thirteenth centuries AD). The material culture retrieved from Chandraketugarh is very varied, and includes cast and punch-marked copper coins, silver and gold coins, objects of ivory, bone, steatite, glass, semi-precious stones, and a very large number of terracotta figurines and plaques depicting a wide range of subjects from the religious to the erotic. Some of the punch-marked coins have a 'ship motif'. The material culture from Period III, from the first century BC to the third century AD, includes pottery and amphorae from the Mediterranean and replicas thereof (Chakrabarti 2001: 136–9). The fortified settlement at Chandraketugarh was located alongside the River Vidyadhari, then a tributary of the Hooghly/Bhagirathi. A series of palaeochannels can be seen to the west of the settlement in aerial photographs and satellite images (Archaeological Survey of India 2000–1). This proto-urban settlement was, undoubtedly, engaged in trade along the Ganges, both upstream and downstream, especially during the period 350 BC–AD 500. The settlement itself is likely to have been a centre for political and religious activity, and a focus for a large number of artisans engaged, most notably, in the production of moulded terracotta figurines.

In Bangladesh, about 100 km north of the present edge of the Sundarbans, lies the early historic site of Wari-Bateshwar. This was another Mauryan proto-urban settlement, with material culture that includes NBP ware dating back to the fifth century BC. It is located near the confluence of the rivers Meghna and Brahmaputra. The presence of very large numbers of punch-marked coins, alongside rouletted ware, which has its origin in the Mediterranean, and glass beads of South-east Asian origin, suggests that this was a major transhipment port, or emporium, linking the Bay of Bengal with the Ganges Plain (Jahan 2010). Several authors have ascertained that this is Ptolemy's Sounagora. In Hinduism, this place is considered sacred (Chakrabarti 2001: 127).

Apart from Chandraketugarh, a number of other urban or proto-urban Mauryan sites are known from the 24 Parganas district. These include Atghara (terracottas, rouletted ware, cast copper coins, and grey pottery), Bhangankhali (on the River Matta, which yielded Sunga-Kushan pottery), Harinarayanpur near Diamond Harbour, where the pottery included Mauryan NBP wares (Archaeological Survey of India 1956–7), and possibly seven other sites (Chakrabarti 2001: 162). The material culture from six of these includes Mediterranean artefacts, including rouletted ware and amphorae. In 2007, the Archaeological Survey of India revealed to the press information on recent discoveries of quartz and garnet beads dating from the fifth to third centuries BC, locally referred to as *pratnaputi*. These had been found with terracotta figurines and at temple sites, burials, and monuments in the Sundarbans, notably in the floodplain of the Adi Ganga, from the southern outskirts of Kolkata to its modern confluence with the Hooghly between Diamond Harbour and Kulpi. The beads are thought to have been used for exchange, and were marked with the figures of gods and goddesses, and their symbolic representations. Excavations beneath the police station in Jaynager in 2004–5, also located near the course of the Adi Ganga (see below), revealed settlement traces and material culture from the second century BC. The excavations have to date not been published.

The case of Chandraketugarh and the other proto-urban walled sites in the Sundarbans present an adaptive solution for living in a tidal landscape without adversely affecting the ability of the mangrove swamps to respond to sea-level changes over the centuries. The area surrounded by the mud walls was not only a place to live, but would have been large enough for keeping livestock and growing root crops and vegetables in gardens. Thus, the mud walls would not only have had a defensive and sociopolitical function, but would also have protected people, animals, and foodstuff from floods during the monsoon months. Rice, the principal staple food, could have been grown outside the walls in paddies or on the sandbanks alongside the river. This represents an adaptation to hydrogeological changes and sea-level rise which, as the archaeological evidence clearly shows, produced a long-term sustainable solution for living in a flood-prone landscape.

Several researchers have considered the ancient landscape of the Adi Ganga, or old or original Ganges south of Kolkata. The Hooghly, the current channel from Kolkata to Sagar Island—the Sundarbans' most westerly island on the Bay of Bengal—is relatively recent in origin, gaining in significance only from the fifteenth century AD onwards; it is not surprising that this channel holds no sacredness for local communities. The older channel, the Adi Ganga, was the focus of settlements and temples dating back to the pre-Mauryan–Mauryan periods, as evidenced by the presence of Mauryan NBP ware or the distinctive and very large Mauryan burnt-clay bricks (measuring up to 0.49 x 0.38 m and 0.07 m thick). Great ritual significance continues to be attributed to this palaeochannel, or *antah-salila*—buried river—and to the place where the Adi Ganga formerly reached the sea near Sagar Island, where it remains the focus of an annual pilgrimage, and where there is a temple (Chakrabarti 2001: 127; 135). The course of the Adi Ganga ran from the Kalighat outskirts of modern Kolkata, taking a south-south-easterly direction to Rajpur, Baruipur, and Dakshin Barasat, after which the main channel turned south-west to join what is now the Mariganga, which separates Sagar Island from Lothian Island.

The historical landscape of the Adi Ganga can be analysed from aerial photographs and satellite images, and a number of features were visited during the fieldwork. These include the meandering channels of the Adi Ganga, and the dendritic creek patterns of the tidal waters. In this flatland, the channels appear regularly spaced and of a near-linear design, which led some early scholars to suggest that these were hand-dug canals (e.g. Willcocks 1930; quoted in Chakrabarti 2001: 128–9). However, with the advance of aerial photography, this drainage system has shown itself to be wholly natural in origin. The settlements are predominantly based on the levees or banks of the old rivers, and the older roads all follow the sinuous paths of the palaeorivers. This is an obvious place for people to live, for several reasons that are as valid in the present as they were in the past. The lower-lying palaeochannels continue to provide drainage for the settlements, which reduces the risks of flooding during the monsoon. The riverbanks are further elevated because the coarser sediments settle down here and create natural levees or banks, whereas finer sediments remain suspended in floodwaters at lower-energy levels, and settle on the floodplain behind these riverbanks. Finally, the palaeochannels remain important features in the landscape. Where water still flows, fishing and aquaculture may complement agricultural activities, and rivers provide easy access to other settlements. Where water has ceased to flow, the lower ground is well-suited to ponds and tanks, and the nutrient-rich beds of the palaeochannels can be used for rice cultivation or horticulture. Present-day coastal communities have adapted to the challenges posed by this landscape by selecting settlement locations similar to those chosen in the past.

Research undertaken in the Indian Sundarbans shows that the antiquity of this historical landscape reaches back into the Mauryan period, possibly into the fifth century BC. The recent and not-so-recent discoveries of settlement sites of that period alongside the Adi Ganga imply a significant degree of continuity in the locations in the landscape where people choose to live. This continuity is most clearly shown in the long-term use of temples and ritual places and through the offering of *pratnaputi*, or beads, at a number of ancient temples. For example, excavations of the large temple mound at Atghara produced Mauryan NBP pottery as some of the earliest datable evidence, and a Jaina terracotta figurine of the tenth–twelfth centuries AD representative of the latest datable evidence. It is also evident that the settlement of Atghara was inhabited over a long period, as shown by the earthwork remains of tanks, wells, and houses (Chakrabarti 2001: 142). The Adi Ganga was the focus for the Mauryan proto-urban settlements discussed earlier, and remained the focus for later settlements and temples as well. A collection of Gupta terracotta figures has been discovered at Dakṣhin Bishnupur on the banks of the Adi Ganga, and at nearby Mathbari two Buddhist stupas (mounds covering relics or objects, or commemorating events, relating to Buddha or his prominent followers) were constructed in the tenth or eleventh centuries AD. The larger of the two stupas measures 6 m in height and 150 m in circumference. Finally, the Hindu Tripureshwari temple at Chhatrabhog, on the left bank of the Adi Ganga, is modern but has an ancient antecedent, possibly predating the fourteenth century AD (Chakrabarti 2001: 143–4). The historical landscape of the Indian Sundarbans contains a number of other eye-catching elements that are ancient but still in use, notably the rectangular tanks for retaining fresh water, with the very largest of these considered to be in excess of 1,000 years old.

In the Faridpur district of Bangladesh, the large walled city of Kotalipara is situated on the River Ghagar on the northern edge of the Sundarbans. The site measures 4 km by 4 km and is clearly visible on aerial and satellite photographs. On the basis of a few finds, it has been dated to the sixth century AD, and is associated with the Vanga Kingdom. Kotalipara was known as Chandravarmankot, as recorded on copperplate inscriptions. A second walled town existed on the opposite side of the river. No excavations have taken place here, and the land inside the walls is now largely waterlogged. It is thought that the current low-lying position of Kotalipara is a result either of neotectonics since the foundation of the town, or of hydrological changes in the Ganges–Brahmaputra delta; the town is believed to have been built originally on slightly higher ground as an important transhipment port (Chakrabarti 1992: 158).

As was the case of the Mauryan walled settlement of Chandraketugarh and the other proto-urban settlements in the Indian Sundarbans, the extraordinary extent enclosed by the mud wall of Kotalipara—about the same size as Rome in the heyday of the Republic—would not only have provided a defensive location for a community, but would also have protected gardens and livestock from flooding during the monsoons, with enough land available to accommodate rice paddies as well. The ability of the surrounding wetlands to respond to any changes in sea level was not compromised. If it had not been for the fall of the land following earthquakes between the twelfth and sixteenth centuries, or the impact of the eastward shift of the main channel of the Ganges, Kotalipara would have remained habitable for many centuries.

A number of archaeological sites are known for the Indian Sundarbans in the later first and early second millennia AD. The temple of Jatar Deul, situated east of Raidighi across the River Mani or Moti, is the most evocative of these and the oldest standing structure in the Sundarbans (Fig. 6.5). The temple probably dates to the tenth century AD. A copperplate found nearby in the nineteenth century, dating to AD 975, recorded in Sanskrit that the building was erected by Raja Jayantachandra, but the plate is no longer available. The standing remains are of a brick-built tower 30 m in height and with a 9.75 m by 9.75 m ground plan. The structure would have towered above the mangrove forests that dominated the landscape at the time. It stands on one side of a brick-built terrace or podium, c. 3 m high, which would have provided safety from any floods. Nearby finds from the River Mani include a terracotta vessel dated to the seventh or eighth century AD, and a black stone figure of a soldier probably dating to the eleventh century. Several other temples are known from the Sundarbans, including the remains of a large temple complex at Mandirtala on Sagar Island. The *ratha*-type temple is dated to the Pala period, but artefacts from the area date as far back as the early centuries AD, and the antiquity of this complex may well be some time early in the first millennium

Fig. 6.5. The Shiva temple of Jatar Deul (c. 40 km south-east of Port Canning) on its 'podium'. The temple originated in the tenth century AD and is still in use. The monument symbolizes the successful adaptation of early communities living in the coastal wetlands of the Sundarbans.

AD (Chakrabarti 2001: 144–7). At Rakshaskhali, deep in the Sundarbans, a copperplate inscription dated to the twelfth century was found (Chakrabarti 2001: 158).

The Mogul invasion in the twelfth century is said to have caused some disruption to the communities in the Sundarbans, but the archaeological monuments and temples dated to this and subsequent centuries show that no wholesale depopulation occurred. By the early sixteenth century, the

Portuguese historian João de Barros was describing the region in his *Década da Ásia* as a populated country. During the reign of Maharaja Pratapaditya, who temporarily established an independent Hindu state in Bengal in the second half of the sixteenth century, the communities in the Sundarbans appeared to have continued to thrive, but this started to change in the seventeenth century. A combination of adverse storm events and the presence of Portuguese pirates caused a degree of depopulation, and a subsequent expansion of the mangrove forests and swamps to within a few miles of Kolkata (Mandal 2004: 62–3). However, the Sundarbans remained inhabited, and the depiction of the region as a wilderness devoid of human presence has been described as a colonial perspective which ignores the large-scale salt production at Sandwip—just east of the River Meghna—and the many clearances near the Bay of Bengal attributed to the Maghs, who settled here after the fall of their independent kingdom of Arakan in the late eighteenth century (Sarkar 2009: 16–24).

The reclamation of the northern part of the Sundarbans' forests had already commenced when the East India Company seized control of Mughal Bengal—following the victory at the Battle of Buxar in 1764—but the deforestation and resettlement accelerated under the Company's control. From 1770 onwards, Bengali peasants could claim tenure in return for bringing the Sundarbans' wastelands under cultivation, but the majority of land was held by the large landowners, or *zamindars*. The new form of reclamation was accompanied by the introduction of new varieties of rice, notably *aman* rice which was sown during the monsoon months when salinity levels were at their lowest and which needed subsequent transplanting. Coconut, date, mango, and jackfruit were also introduced into the region. The settlers brought new ways of protecting themselves and their crops from flooding, such as raised beds. The construction of large-scale embankments in the coastal wetlands started in the late eighteenth century. Fishing was important to the settlers, providing a ready source of food during the first difficult years, whilst aquaculture included the growing of certain fish and prawn species within the flooded rice paddies. Those parts of the mangrove swamp that were too saline for reclamation were stripped for fuel and building timber (Richards and Flint 1990; Mandal 2004: 62–82). Boat building, salt production, and the collection of beeswax and honey were undertaken by specific castes. Boat building and salt production continued into modern times, but the clear separation between the castes, and the formal links between caste and occupation, started to break down as a result of land reforms and redistributions. The dynamic nature of the Sundarbans, and the erosion and accretion of the islands are thought to have contributed to the need of coastal communities to adapt their modes of subsistence, regardless of caste divisions (WWF-India 2011: 13).

During the late eighteenth, the nineteenth, and the twentieth centuries, the reclamation of the mangrove swamps continued under different forms of government, with an increasing emphasis on enabling peasants drawn mainly from the West Bengal region north of the Sundarbans—initially mainly low-caste Hindus but later Muslims and Hindus from other castes as well—to own their own land (Mandal 2004: 76). A village community society developed, comprising largely self-sufficient households with some degree of shared landholding, even though the land itself was frequently owned by larger landowners such as the *zamindars*. After the independence of India, the West Bengal Land Reform Act and subsequent legislation enabled the landless to own their own homestead in the Indian Sundarbans, with a maximum extent of just over 1 acre (Jalais 2010: 62). The speed of reclamation is mirrored in the rapid increase in the population in the South 24 Parganas district in West Bengal and the Backerganj and Khula districts of Bangladesh: between 1880 and 1980, the total population increased from 5.5 to 25 million (Richards and Flint 1990: 25). There were attempts to protect the mangrove forests, initially in the late nineteenth century with the aim of preserving a sustainable source of timber, and later for reasons of nature conservation. Although these attempts were not always successful, they ultimately resulted in the protection of significant parts of the Sundarbans under the auspices of the state and national governments. More recent attempts to protect the remaining mangrove forests have come from external institutions whose key concern is the protection of the habitat of the Royal Bengal tiger, and for whom the presence of coastal communities is 'seen as a hindrance' to the development of a natural Sundarbans (Jalais 2010: 9). The balance between the need for land for food production and the protection of the Sundarbans as a nature reserve continues to be an area of discourse between the many stakeholders.

The reclamation of the Sundarbans' islands was in full flow by the end of the eighteenth century, and this phase of habitation has produced the historical landscape that has remained largely unchanged into the twenty-first century for much of the reclaimed parts of the Sundarbans. This landscape, as observed on Bali Island and elsewhere in the Indian Sundarbans, comprises individual homesteads, or *bhite*, with houses either constructed on the inside of the clay embankment, or *bãdh*, or on a 'podium' of clay, of c. 1 m height, both positions providing some protection from floods (Fig. 6.6). The walls of the houses are made of wattle and daub with thatched roofs of rice straw. Many houses have an outhouse for the storage of foodstuffs. A potted basil bush—considered sacred in Hindu religion—and small shrines are co-located on the elevated land. Each house has its own freshwater pond, for washing, cooking, cleaning cooking utensils, irrigation of the garden, and providing water for domestic animals. The natural woodlands have been largely cleared, but stands of palm trees around the freshwater pond provide shade and timber for building. The low-lying land surrounding the houses is used, overwhelmingly, for paddy rice cultivation, with most fields less than an acre in size. The

Fig. 6.6. A homestead on Bali Island. The most important structures have been built on a 'podium' of clay, an adaptive solution that strengthens the resilience of the family.

rice straw is piled up high in bundles, sometimes in the shape of houses, and this is transported on cycle-vans to roadsides and settlements where there is a market for rice thatch. Goats graze freely on the stubble of the harvested paddies. Cows are tethered and their dung is shaped into 'patties' and dried in the sun, providing the main source of fuel for cooking fires. The traditional village economy in the Indian Sundarbans is one of predominant self-subsistence at village level. Surpluses that are transported to local markets include rice thatch, timber, and fish.

The *bãdh* is between 3 and 4 m in height, and at its base 5 to 6 m wide. It is made of clay and sediments dug from the foreshore on the riverside of the *bãdh*, and the square pits dug for this purpose can be seen everywhere where flood banks exist. The banks are occasionally covered with bricks set in a herringbone-pattern, which protect them from erosion caused by people, animals, and rain. These brick paths are seen, by the local communities, as a symbol of progress and status (Jalais 2010: 35). The lie of the land inside the *bãdh* is frequently between 2 and 3 m below the high-tide level of the deltaic waters. The digging of clay pits in the intertidal muds is a contributing factor to the absence of a protective mangrove cover, and very little foreshore vegetation exists: the flood banks are therefore completely exposed to waves and the impact of cyclones. Villages across the Indian Sundarbans suffered badly from cyclones in 2007 and again in 2009, with many banks being breached, resulting in extensive flooding and loss of human and animal lives and crops. Living near the *bãdh* is more dangerous than living towards the centre of the islands, as storms, breaches, tigers, and crocodiles will have the greatest impact on those who live on the edges. Those who can afford to live near the centre of the island near the local school are considered, and consider themselves, to have attained a higher status in the village.

Fishing takes place both on open water and in the drainage channels on the island. Fishing in open water involves stationary nets or fishtraps set from boats parallel to the river close to the intertidal shore, and left overnight with the traps emptied early in the morning. Line-fishing in open water is restricted to catching crabs. Fishing in the drainage channels involves, principally, nets thrown from the bank and pulled up within seconds (Fig. 6.7), but occasional fishtraps across the drainage channels have been observed. Traditionally, aquaculture involved the capture of juvenile carp fish with nets and the raising of these on the rice paddies. The physical evidence for aquaculture in the modern landscape comes principally in the form of land-use change of agricultural land and forestry into aquaculture farms. Tiger prawns (*Penaeus monodon*) can bring the successful aquaculture farmers wealth beyond the prospects of ordinary farmers (Chopra, Kapuria, and Kumar 2009: 132–69). The introduction of nylon nets is widely considered a threat to the fish stock. In particular, the practice of catching tiger-prawn hatchlings using these nylon nets near the edge of the tidal shore is considered unsustainable by the established fishermen and

Fig. 6.7. Net-fishing from the edge of Bali Island.

conservation organizations, as the young of 52 other species, for which there is no commercial interest, are caught and discarded in this practice.

Extensive tracts of mangrove forests are now restricted to the protected nature reserves, and these bring some tourists to the Sundarbans. The absence of flood banks ensures the continued interplay between water, sediments, and mangroves. Processes of erosion and accretion can be seen throughout the Indian Sundarbans, and there is no apparent negative sediment budget. The presence of extensive mudflats at low tide in the river channels corroborates this observation. However, the *Heritiera fomes* woodland is severely depleted, and the individual trees are not as lush as their counterparts in the Bangladeshi Sundarbans. This is a direct consequence of the low volume of fresh water transported downstream by the western channels of the Ganges and the absence of regular freshwater flushing of the mangrove swamps. The depletion of the mangrove swamps diminishes the ability of the un-embanked islands to respond to a rising sea level.

The stories in *punthi* literature—dating from the seventeenth to nineteenth centuries—from the lower Bengal delta reflect directly the division between the reclaimed and the unreclaimed parts of the Sundarbans. *Punthi* literature principally concerns the conflict between people and nature, and in this dialectic relationship the tigers represent untamed nature or wild(er)ness. The main characters in *punthi* literature are not the regular Hindu or Muslim deities, but 'lower' deities or saints of direct relevance to the woodcutters, honey gatherers, boat builders, and farmers. The prose was meant to be read as mantras, protecting those who entered the forests by appealing for protection from the resident deities. The most important deity in the Sundarbans is the people's protector Bonbibi ('good woman'), a female deity with a Muslim origin but venerated by Hindus and Muslims alike (Fig. 6.8). Bonbibi not only rules the reclaimed parts of the Sundarbans, but also protects the people of the Sundarbans when they enter the forest, the domain of Dokkhin Rai, the lord of the tigers. Bonbibi won the right of ruling the reclaimed lands when her brother Shah Jongoli defeated Dokkhin Rai in a fight, and her power over him in protecting woodcutters and honey gatherers and all who enter the forest is traced back to this event.

A common theme in the *punthi* literature is the importance of coexistence and a shared system of beliefs among the Hindus and Muslims, something that is maybe not unique, but is certainly of great significance on the Indian subcontinent (Sarkar 2009: 30–54). The belief in Bonbibi remains strong in the modern community in the Sundarbans, especially amongst those who have to enter the forest for their work and who are in danger of encountering tigers. The sharing of deities provides the coastal communities with a potential basis for developing social capital across the two main religious groups; but the reality is somewhat different, and the communities remain divided across their (real or perceived) caste origin and occupation (Jalais 2010: 70–5).

Fig. 6.8. A shrine to Bonbibi at Saznekhali. Bonbibi is on the right, with her brother Shah Jongoli and the tiger behind her. Dokkhin Rai stands to the left.

According to Annu Jalais's (2010) account of the modern woodcutters, honey collectors, and forest fishers of the Sundarbans, the belief in Bonbibi continues to instil a deep-rooted respect for the forest and its non-human inhabitants, especially the tigers. One can only take from the forest, which is after all the storehouse of Bonbibi, what one really needs. It is the custom that the produce of the forest is shared equally amongst the group of workers who enter the forest, and when 'enough' has been collected or caught, one should leave the forest. This sharing of food is not restricted to humans only, and 'what ties humans and non-humans [including tigers] in a symbolic web of kinship is the common forest and the shared food and environment it provides' (Jalais 2010: 75). Even though wood, honey, and fish can be sold for high prices at the local market, one should never grow rich from the proceeds of the forest. To seek to profit from the forest is to seek the revenge of Dokkhin Rai. Equally, the collection of tiger-prawn hatchlings should be undertaken in a similar vein, but where prawn collectors seek to grow rich they incur the wrath of Bonbibi and provoke attacks by crocodiles.

In summary then, evidence of early prehistoric activity of hunter-gatherers-fishermen may exist beneath the deep stratigraphy of sediments, but actual archaeological evidence from the Sundarbans has not been found. It is apparent that the Indian Sundarbans were inhabited from at least the Mauryan or pre-Mauryan period onwards, with the oldest proto-urban settlements dating to the fifth century BC; it seems probable that similar habitation existed in the Bangladeshi Sundarbans, but any archaeological evidence of this remains concealed beneath the thick sediments. The most important proto-urban settlements were located on the major channels of the Ganges, allowing for trade and exchange up and down this river and into the Bay of Bengal, linking the Sundarbans to the Mediterranean and South-east Asia. Archaeological evidence for continuity of habitation of many of these settlements, and continuity of ritual practices at specific locales and temples, is available for the last centuries BC and the first and second millennia AD. Historically, periods of unrest and invasions are recorded—for example, following the decline of the Gupta in the sixth and seventh centuries, and in the seventeenth century when Portuguese pirates enslaved many of the Sundarbans' inhabitants—and these would have caused periodic disturbance in habitation and a recovery and expansion of the mangrove swamps. However, these lacunas are not reflected in the archaeological settlement record (Chakrabarti 1992: 70). The reclamation of the Sundarbans commenced in the second half of the eighteenth century, became centrally organized during the colonial rule of Bengal, and accelerated after the independence of India and former East Pakistan.

STRENGTHENING THE RESILIENCE OF COASTAL COMMUNITIES

There is no tradition of transnational ICZM in the Bay of Bengal involving the widest range of stakeholders, as advocated by the IPCC as good practice, although such a management plan is being developed for West Bengal (WWF-India 2011: 24). Of particular concern is the lack of cooperation over the use of fresh water in the region. The implementation of policies that serve national rather than transnational objectives adversely affects the ability of coastal communities to adapt to sea-level rise. The example of the Farakka Barrage—which is considered a key cause for the net loss of some 170 km^2 of mangrove swamps between 1973 and 2010 (Rahman, Dragoni and El-Masri 2011; see above)—is telling: during my time in the Sundarbans, the chief minister of West Bengal accused the prime minister of India of secretly diverting more fresh water into Bangladesh, in order to gain political favour there as part of attempts to build transnational bridges, whilst leaving West Bengal without sufficient water for agriculture, domestic and industrial use, and adversely affecting fishers (*Times of India*, 22 February 2012). Despite the existence of the South Asian Association for Regional Cooperation, governments pursue national policies of 'freshwater security' ahead of transnational policies for coastal sustainability (Ray 2011: 5–65). Disputes about fresh water are one of the main causes of political antagonism between nations: the relationships between India and Bangladesh deteriorate whenever too little or too much water is diverted at the Farakka Barrage (Ray 2011: 143–8).

The impact of climate change on the sea level, and on the frequency of extreme weather events such as cyclones, is already noticeable in the Sundarbans. The recorded rise of the RSL in the later parts of the twentieth century in the Bay of Bengal has been greater than the global figures: between 1977 and 1998, sea level rose here annually by 4.0–7.8 mm against global average figures of 1.8 mm/year for the period 1961–2003 and 3.1 mm/year for the period 1993–2003 (SAARC Meteorological Research Center 2003; Bindoff et al. 2007). Several islands in the Ganges–Brahmaputra–Meghna delta have been submerged beneath the rising sea, and the inhabited island of Ghoramara is partially submerged and likely to disappear in the forthcoming decades (Ray 2011: 71). This higher rate of RSL is the product of ESL rise combined with the ongoing descent of the edge of the sediment-loaded tectonic plate and the relative dearth of sediments from the Ganges–Brahmaputra–Meghna river system reaching the Sundarbans. The predicted increased frequency of extreme weather events is, in addition, already causing problems to the coastal communities. Recent research involving modelling predicts that the Sundarbans will disappear in the next 60 years, or that the Sundarbans will drown if sea level rises by as little as 28 cm (WWF-India 2010; Loucks et al. 2010). Sea-level rise will have devastating effects for the Sundarbans, as expressed so evocatively in Tushar Kanjilal's (2000) treatise *Who Killed the Sundarbans?*

In the WWF-India (2010) report *Sundarbans: Future Imperfect. Climate Adaptation Report*, the eyewitness accounts from residents identify a number of environmental changes that are at odds with the traditional understanding of the environment. Alongside the rise in sea level and the increased occurrence and intensity of extreme weather events such as cyclones, these include an accelerated erosion of the islands by the sea and rivers, changes in monsoon rainfall patterns and changes in fishing patterns. The evidence provided in this report supports the changes observed by the eyewitnesses, but the time frame for this—from 1990 to 2008—is simply too short to attribute these unequivocally to climate change. The report does, however, highlight the urgent need to build the resilience of the coastal communities.

Whilst a series of designations from the late nineteenth century onwards have sought to protect parts of the intertidal mangrove forests of the Sundarbans—and the Sundarbans National Park in India and the Sundarbans in Bangladesh were declared World Heritage Sites in 1973 and 1977 respectively—formal policies in India and Bangladesh have been slow to accept the existence of climate change in the twenty-first century and the likely impact this will have on the coastal wetlands. The assertion made by the Bangladeshi Centre for

Environment and Geographic Information Services in 2010 that the country's coastline was not under threat, even if the sea level rose by 1 m (Ray 2011: 10–11), exemplifies policies of denial of the impacts of climate change on the Sundarbans. Practical action, such as the repair of embankments organized by the Irrigation Department of the West Bengal state, are of a piecemeal nature and lack a catchment-wide or long-term plan (WWF-India 2010: 13).

Recognizing that the national governments of India and Bangladesh, or their respective sub-national administrations, have not yet grasped the situation facing the Sundarbans in the context of climate change, WWF-India published a vision for the Indian part of the Sundarbans in 2011 under the title *Indian Sundarbans Delta—A Vision*. In broad outlines, it envisions that in Phase I, the Indian part of the Sundarbans will become a single Biosphere District—replacing the North and South 24 Parganas districts—which will work closely with lower-level administrative units and other stakeholders in the realization of the vision, and which has as its first objective the restriction of land acquisition in the Sundarbans by outsiders. A 'Green Line' is drawn through the reclaimed land surrounding the reserves where the mangrove forests survive. During Phase II, facilities and physical infrastructure—offices, housing, civic amenities, industrial zones with buildings for secondary and tertiary production activities—should be developed around existing towns on the margins of the Biosphere District outside the Green Line. In Phase III, individuals and families living within the Green Line should be encouraged to move to the newly expanded towns, providing the space for the regeneration of the tidal mangrove forests in Phase IV, which WWF-India hopes to complete by 2050 (WWF-India 2011).

The adaptive pathways proposed in WWF-India's 2011 vision for the Sundarbans are primarily aimed at delivering the greatest benefit for threatened animal species, in particular the Royal Bengal tiger, rather than supporting the coastal communities in their adaptation to the impacts of climate change. This plan—and the attitudes underpinning similar visions for the Sundarbans that place the rights of the tiger above those of local communities—belongs to a convention held by many outsiders that the Sundarbans is a 'beautiful and exotic garden' that has been despoiled by its human inhabitants (e.g. Kanjilal 2000). These outsiders frequently live at a great distance from the Sundarbans, and through influential NGOs such as WWF exert pressure on national and regional governments to protect the wildlife, often at the expense of local communities and individuals. Annu Jalais (2010: 176–202) refers to this phenomenon as the case of the 'cosmopolitan tiger'.

In developing a sustainable plan for the future, the WWF-India vision does not seek to draw on adaptive pathways followed by the inhabitants of the Sundarbans in the past. What, then, has the study of the past to offer the current communities in the Sundarbans?

In assessing the likelihood that the mangrove swamps will be restored as a habitat for a range of threatened species, including the Royal Bengal tiger, following the retreat of human activity, the WWF-India plan could refer to the significant reduction of population living in the Sundarbans in the seventeenth and eighteenth centuries. As shown in palaeoenvironmental and historical research, this relatively short period of abandonment resulted in the rapid expansion of the mangrove forest—as far as the outskirts of Kolkata. However, the WWF-India plan does not reflect on the longer-term perspectives provided by the research presented in this chapter on the eastward shift of the freshwater discharge of the Ganges as a consequence of (neo)geological changes. This shift has a twofold impact on the coastal wetlands of the Sundarbans. The first impact is the dearth of sediments brought down by the Ganges into the tidal Sundarbans. These sediments have, effectively, formed the Sundarbans during the Holocene; a negative sediment budget will prevent the coastal wetlands from autogenically responding to sea-level rise. It is not surprising that the western islands are more prone to erosion than their counterparts in the Bangladeshi Sundarbans. The second impact is the absence of freshwater 'flushing' of the ecosystem during the monsoon months, essential for the growth and health of mangrove trees. The diverging nature of the mangrove swamps on either side of the Sundarbans, with the Sundari trees in the east described as lush and luxuriant, but those in the west characterized as stumped and diminutive, is a direct reflection of the effect of freshwater flushing. Attempts to restore the mangrove swamps to their former 'natural' glory in the Indian part of the Sundarbans will, therefore, depend on the 'unnatural' diversion of Ganges fresh water at the Farakka Barrage into West Bengal.

The adaptive pathways from the past show two distinct phases of living in the Sundarbans. In prehistoric and early historic times through to the sixteenth century, the critical natural dynamics between the rivers, waves, tides, and sediments remained essentially untouched by the human presence. This meant that, as long as the sediment budget was positive, the mangrove swamps reacted naturally to sea-level rise and the lie of the land adjusted to the height of the sea level. Following the reclamation in the late seventeenth and subsequent centuries, these natural dynamics were halted as a result of deforestation, the cultivation of the land, and the construction of the embankments, or *bãdh*. Periodic floods could no longer increase the height of the land, and mangrove swamps could no longer autogenically react to sea-level rise because their accommodation space had been taken away. This has led to the drowning or flooding of some of the islands and widespread erosion of the islands' natural banks. Natural changes, such as the eastward shifting of the main channels of the River Ganges, have exacerbated the problems a great deal: in the eastern half of the Sundarbans, the lack of freshwater flushing reduces the resilience of the mangrove ecosystem in responding to sea-level rise, whilst in the western half, floods during peak flow periods—especially dangerous for the rice harvest—have become more frequent. The adaptive pathways adopted in the period before the seventeenth century are therefore more likely to produce inspiration for sustainable solutions for the future. The archaeological and historical record provides two such adaptive pathways that have relevance in this context.

The first adaptive pathway is the construction of the walled proto-urban settlements, such as Chandraketugarh and Kotalipara. These have been described by archaeologists as transhipment ports, connecting the Ganges Plain with the Bay of Bengal and beyond, and as places where artisans and craftsmen lived and worked, such as the sculptors of Chandraketugarh who made terracotta figurines. However, the mud walls provided protection not only from enemies or raiders, but also from floods during the monsoon season and during extreme weather events. These walled settlements were of very great size and the protected areas were much larger than was required for accommodating the houses of the population. The settlements would easily have accommodated gardens and livestock, and possibly even some rice paddies. Importantly, these proto-urban settlements did not affect the natural dynamics of the water and the land in the wider landscape, and mangrove swamps continued to adjust autogenically to changes in the sea level. As such, these walled settlements represent an adaptive pathway that was well suited to the landscape and accommodated the impact of climate change and sea-level rise over very long periods in a sustainable manner.

Are walled nucleated settlements on the islands in the Sundarbans an adaptive pathway that could be adopted in the modern world? The current use of the landscape, with homesteads scattered across it but concentrated on the inside of the embankment, is not sustainable because it does not allow for the autogenic adaptation of the mangrove swamps or the accretion of sediments in a time of sea-level rise. The inevitable result will be increased occurrences of floods and erosion, the overtopping of the *bãdh* and the drowning of whole islands. A mud-walled nucleated settlement located near the centre of the island would protect the communities and their most important possessions, including harvests, from the flood dangers. The embankments could be adapted to allow accommodation space for the mangrove swamps. A possible strategy is to adopt a system of 'summer and winter bunds', replacing the single *bãdh* surrounding each island. Thus, the islands would have two sets of embankments, a lower one on the outside and a higher one on the inside. This would create a margin where the rivers would regularly flood the land and increase the lie of the land through sediment accretion, and an inner core for rice paddies that would be flood-free for the larger parts of the year. Occasional floods in the inner core would be restricted to the period from July to September, when the rivers are swollen by the Asiatic monsoon and when the cyclones in the Bay of Bengal are at their most common (Longshore 2000: 44). It has to be recognized that such changes would require a significant shift in modern cultural perceptions in India and Bangladesh, which places the homestead at the very core of the post-independence movement to provide land for the landless; but the existence of a local precedent from the past may help to provide modern walled settlements with a sense of place.

Constructing mud-walled settlements could also help in increasing cooperation and collaboration within coastal communities. During the reclamation of the Sundarbans, the *zamindars* organized and funded the construction and maintenance of the embankments that surrounded most inhabited islands, but this role has now passed to the Irrigation Office of the state of West Bengal in the Indian Sundarbans. The technical expertise for the maintenance of embankments is not well developed in the area, and the resources to maintain the embankments are also limited (WWF-India 2011: 25). Local communities have no role in the repair of the *bădh*, and when these are breached the burden of repair falls on the homesteads nearest the breach. Maintaining a mud wall surrounding a nucleated settlement would be in the interest of the whole community; it could foster a sense of mutual dependency and of the need for collaboration, and form a basis for the strengthening of communities' social capital and resilience.

The second adaptive pathway from the past is the use of rice varieties that can be grown on lands that have relatively high levels of salinity. In the Sundarbans, the use of the salt-tolerant *boro* variety is of a very long-standing tradition, and its use in the past on sandbanks outside the bunds represents a specific example of how past adaptive pathways can be used in a dynamic coastal wetland. The use of salt-tolerant rice varieties does not need to be restricted to the wetlands outside the embankments, and in a scenario where the bunds are reduced in size to allow for the accretion of sediments on existing islands, the increased use of salt-tolerant rice varieties would be a prerequisite to making a more dynamic interaction between land and sea acceptable to coastal communities.

Finally, it has to be said that the construction of mud-walled nucleated settlements towards the centre of the islands which allow the mangrove swamps to re-establish their role as natural barriers, and the increased use of salt-tolerant rice varieties by coastal communities do not deliver the objectives sought by WWF-India for the Sundarbans. These objectives are primarily focused on the creation of an extensive habitat for threatened non-human species, rather than building the resilience of human coastal communities. The WWF-India vision essentially seeks the creation of a nature reserve with minimal human presence, and the development of townships outside the core of the Sundarbans is seen as the means to achieve this. However, as this study of the archaeology of the Sundarbans has shown, people have lived here for many millennia side by side with the now threatened species. That an urgent need is perceived to remove the human communities from the Sundarbans is a reflection of a modern-day conservationism paradigm. The aim should not be the removal of coastal communities from the Sundarbans in order to create space for nature, but rather to strengthen the resilience of coastal communities and to enable people to live alongside endangered species. They could adopt the attitude of the woodcutters, honey collectors, and forest fishers who only enter the forest to take what they urgently need in times of hunger, and aim for the coexistence of humans and non-humans under the watchful eye of Bonbibi.

CONCLUSION

The RSL rise in the twenty-first century will affect the whole of the Sundarbans more so than most other coastal wetlands as the edge of the continental shelf, deeply loaded with Holocene sediments from the Ganges–Brahmaputra–Meghna rivers, continues to fall. The greatest RSL rise will be experienced in the Indian part of the Sundarbans, where there is a negative sediment budget following the eastward migration of the main channel of the River Ganges. Records over the last decades suggest that the rate of RSL rise for the Sundarbans is at least double and possibly quadruple the rate of rise for the ESL.

Archaeological evidence for human activity in the Holocene is practically absent before c. 500 BC, but it is probable that the Sundarbans were utilized for their rich resources of fish and shellfish from at least the Mesolithic. In the last 2,500 years, coastal communities have adapted to living in this tidal landscape through the development of very large mud-walled settlements, which not only protected people from attack but also provided security from floods. The enclosed space would have accommodated houses alongside gardens for horticulture and livestock. Importantly, despite the very large expanse of these walled settlements, the dynamic interaction between sea and land remained effectively unaltered, and coastal wetlands such as the dominant mangrove swamps could autogenically adapt to changes in the level of the sea. The long-term sustainability of this form of living in coastal wetlands is reflected in the archaeological evidence from the Adi Ganga. Rice cultivation using salt-tolerant varieties made it possible to farm effectively in the tide country, without upsetting the balance between changes in the sea level and sediment accretion. The resources of the forests were utilized, especially wood, honey, fish, and shellfish, but never to the extent that the ecosystems were fundamentally damaged.

The current adaptive pathways of coastal communities in the Sundarbans have their origin in the late seventeenth century when reclamations began to take place, involving the construction of embankments or bunds at the margins of the islands and the wholesale conversion of the mangrove forests into agricultural land. This process has removed the accommodation space for the mangrove swamps, and the natural protection provided by this coastal wetland against sea-level rise and cyclones has been significantly diminished. On any timescale, these are not sustainable adaptive pathways, and it is more than likely that serious floods and the drowning of inhabited islands will happen well before the end of the twenty-first century. The archaeological study of how people in the past adapted to rising sea levels can be used to strengthen the resilience of coastal communities in the Sundarbans in a number of ways, which can be summarized thematically as follows:

First, an understanding of the long-term interaction between sea-level change, coastal morphological and ecological change, and changes in the way coastal communities lived with the sea, appears to be missing from institutions and communities that could play a role in the future management of the Sundarbans. Local communities and institutions such as the government of West Bengal or WWF-India have not fully recognized the geological changes that will inevitably affect the future of the Sundarbans. The creation of a biosphere and extended nature reserve in the Indian Sundarbans can only be achieved by diverting significant amounts of fresh water and sediments in the Ganges at the Farakka Barrage, and this will affect the sustainability of the Bangladeshi Sundarbans. More education will be essential if future adaptive pathways are to deliver a sustainable way of life.

Second, and following directly from the observation above, the future sustainability of the Sundarbans depends critically on transnational cooperation and on collaboration between local communities and institutions which operate on a range of levels, from the district level to the global NGOs. Future transnational collaboration on the Sundarbans depends critically on the sharing of fresh water from the Ganges. The basis for building social capital requires an acknowledgement of the legitimacy of the large range of stakeholders and the authenticity of their claims to the Sundarbans. This has to include the acknowledgement that—as the archaeology of the region shows beyond any doubt—people have been part of the Sundarbans for at least 2,500 years, coexisting with threatened species including the Royal Bengal tiger.

Third, in the past, people in the Sundarbans reconstituted their relationship with the socialized but dynamic landscape through affirmative actions. For the last three centuries, this is expressed in the *punthi* literature which demonstrates a philosophy of sustainable coexistence of coastal communities and the natural environment, including endangered species such as the Royal Bengal tiger. However, the archaeological evidence presented in this chapter shows that living sustainably in the Sundarbans goes back to c. 2500 cal BC. This was, in part, achieved through two sets of affirmative actions. The first set concerns the establishment of concepts of continuity in a landscape that was dynamic and continually changing. The continued sacred nature of the Adi Ganga—even though this is no longer a channel of the River Ganges—and the long-lasting significance of temples and rituals that are present in, and connected to, the landscape provide existing examples of this. The second set of affirmative actions in the past concerns the construction of the mud-walled towns. These not only provided protection in times of war, flood, and storms, but also created communities with higher levels of social capital and therefore greater resilience to rapid environmental change, whilst allowing for the autogenic adaptation of the mangrove swamps to higher sea levels. Moving from the dispersed settlements with the *bãdh* enclosing complete islands to a model with nucleated settlements and two sets of embankments would also create space for increased biodiversity. In the past, these affirmative actions played a role in reformulating the sense of place of coastal communities, and these may serve as examples for modern coastal communities.

Fourth and finally, the coastal wetlands of the Sundarbans were in the past highly valued economic landscapes and were utilized not only for their natural products such as timber, fish, and honey, but functioned also as natural rice paddies. Today, *boro* and other salt-tolerant rice varieties have been replaced by salt-intolerant varieties that have higher yields. Coastal communities in the Sundarbans who want to develop sustainable rice agriculture will need to embrace these salt-tolerant varieties. By doing so, the occasional floods that are essential to adjusting the lie of the land in times of sea-level rise would become part of a sustainable rice agriculture, leading to greater food security whilst maintaining biodiversity, and gaining more support from the various stakeholders within the coastal communities.

7

The Wetlands of Florida's Gulf Coast

INTRODUCTION

The state of Florida is flanked on the east by the Atlantic Ocean and on the west by the Gulf of Mexico (Fig. 7.1). It has a coastline of over 2,000 km in length. Florida sits on a tectonically distinct but stable unit, known as the Florida Plateau. The geological origin of the Plateau is predominantly one of marine sediments, and this has resulted in a flat landscape with few hills: Florida's highest point is Britton Hill near the border with the state of Alabama, at just 105 m above sea level. Two-thirds of the Plateau is submerged, a direct consequence of rising sea levels since the Last Glacial Maximum, leaving about one-third of the Florida Plateau above sea level. Florida has no sea cliffs that provide a natural protection against rising sea levels, and the different coastal wetlands found along both the Atlantic and Gulf coasts are vitally important in protecting coastal communities from the rising sea, and from sea surges caused by the frequent storms and hurricanes.

Since the middle of the nineteenth century, the state of Florida has experienced a rapid rise in its population—from fewer than 100,000 in the 1850s to over 18 million in 2010—and this growth has been accommodated by intensive urban development, especially along the coast. In the process, many of the natural coastal wetlands that provided protection against storms and sea-level rise have been damaged to the point that their ability to respond to future sea-level rise has been compromised. The importance of the coastal wetlands has been recognized for many decades, and this study is able to draw from the proceeds of a rich research culture in the fields of sea-level and palaeoenvironmental change. Similarly, and contrary to the notion amongst 'mainstream' archaeologists specializing in the American South-west that Florida was something of a backwater, archaeological research into Florida's coastal communities has produced many high-quality publications in recent decades.

The central and southern wetlands of Florida's Gulf Coast were selected for fieldwork during April–May 2012. A large number of different wetlands were visited—from Crystal Bay in the north to the Everglades in the south—with the aim of gaining contextual understanding of their development, their

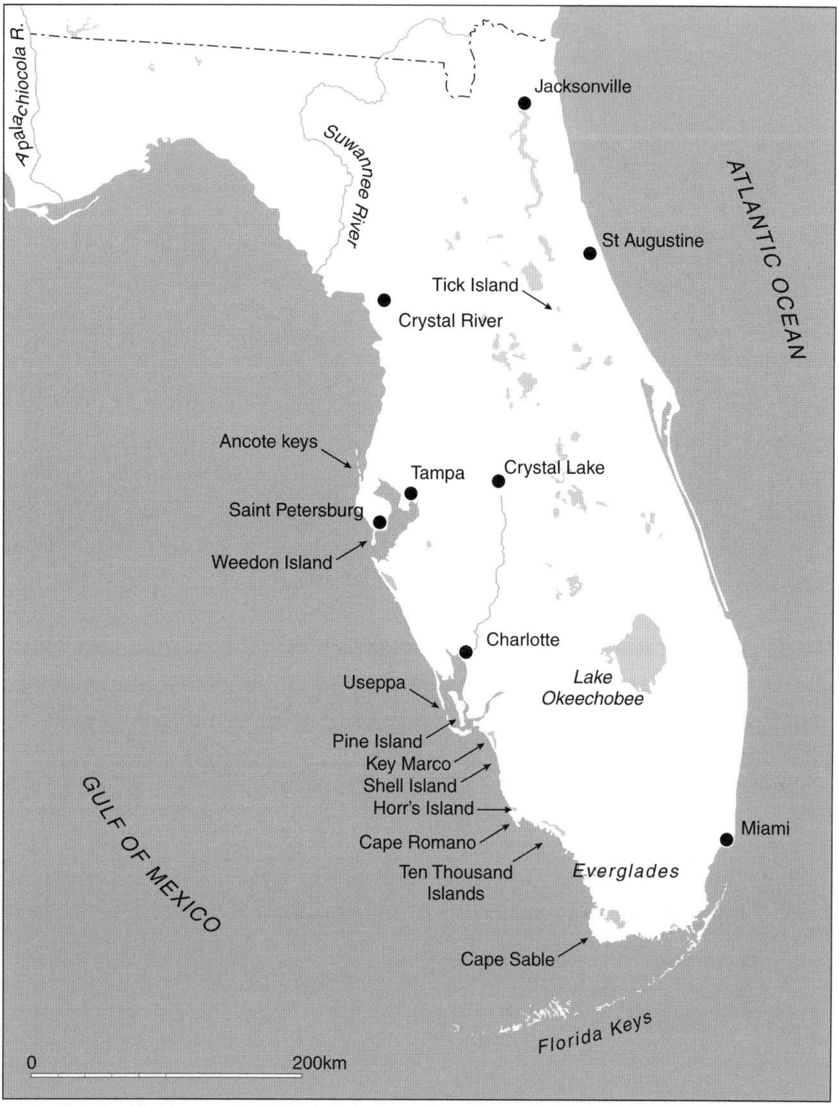

Fig. 7.1. Map of Florida and key sites mentioned in the text.

archaeology, and future prospects. Generous support was provided by a large number of colleagues who are acknowledged in the preface. Dr Nancy White and Dr Brent Weisman of the University of South Florida, both with over 30 years of experience in the archaeology of Florida's wetlands, acted as critical friends.

The reasons for selecting Florida's Gulf Coast as one of the case studies were fourfold. First, the tidal and wave energy levels in the Gulf of Mexico are low when compared to the other case studies, and this limits the ability of coastal wetlands to respond autogenically to future sea-level rise. Different types of wetlands are affected by this in different ways, and the diversity of the coastal wetlands along Florida's Gulf Coast—which includes mangrove swamps, barrier islands, and oyster reefs—offers an intriguing set of intellectual issues for which no single answer is the right one.

Second, as already noted, Florida's archaeology has been studied to a very high level and provides many examples of living sustainably on the coast. From c. 7000 cal BC onwards, indigenous coastal communities were specializing in the utilization of fish and shellfish in the estuarine, intertidal, and marine environments of the Gulf Coast. When the Spanish encountered sedentary tributary chiefdoms of the Calusa Indians in south Florida early in the sixteenth century AD, they noted a continued reliance on the resources from the sea. The 'maritime hypothesis' on the primacy of coastal sedentism and comparative richness of resources in the development of long-standing early complex societies remains a key theme in current archaeological research and debates.

Third, as can be seen elsewhere in the world, the coast of Florida has undergone a rapid urbanization with increasing population growth in the twentieth and early twenty-first centuries, and this development has removed accommodation space for many of the coastal wetlands. For Florida, a state that relies economically on income from tourism inextricably linked to the coast—with over 70 million visitors in 2004—population growth is also threatening to take away its main source of income.

Fourth, the political situation in Florida offers an interesting contrast to the other case studies. In Florida, nearly all stretches of coast are in the hands of individuals or private organizations, but the responsibility for coastal protection lies with the US government and the state of Florida. This creates difficulties when attempts are made to develop Integrated Coastal Zone Management plans (ICZMs), as governments and private landowners work on different timescales, the former ostensibly seeking long-term sustainability, the latter short-term financial gain from coastal developments. Add to this situation the elected politicians who, frequently, have very different opinions about the existence of climate change caused by the burning of fossil fuels, the impacts on sea levels, and the responsibility of government to tackle climate change-related problems, and Florida provides a case study of a wealthy nation that remains poorly organized and integrated when it comes to adapting to climate change.

It would be amiss not to acknowledge that the potential of Florida's archaeology to offer a different understanding of the ways in which climate change, and especially sea-level rise, impacts on coastal communities has already been identified. In his paper 'Why Florida archaeology matters', Brent Weisman (2003: 217) points explicitly at this, noting that 'we may look to technology for our shield [in times of climate and sea-level change], but in so doing must realize that the ancient people in their way did the same'.

This chapter presents, firstly, the geological and palaeogeographic history of Florida's Gulf Coast, considering in some detail the ongoing debate on mid and late Holocene sea-level high-stands (periods of relatively higher sea levels observed in geomorphological formations); and, secondly, the archaeology and history of the Floridian coastal communities. It concludes with a consideration of the current state of the coastal wetlands and an investigation into how the resilience of coastal communities can be strengthened using research into adaptive pathways from the past.

HOW CLIMATE, ENVIRONMENTAL, AND SEA-LEVEL CHANGE SHAPED THE FLORIDA WETLANDS

Geologically, the Florida Plateau has its origins in the Ordovician period of the Palaeozoic Era (with a mid-date around 450 million years ago), and has a deep base of volcanic rock and marine sediments. In terms of plate tectonics, the Florida Plateau is in a relatively stable position; no earthquakes have been recorded here in the last centuries, with the exception of a few minor seismic shocks which caused limited damage and no loss of life (Smith and Lord 1997: 25). The emergent geology of Florida is of a younger age, and consists of a range of marine sedimentary formations dominated by limestones and shales, clays, and sandstones deposited from the Eocene onwards (commencing c. 55 million years ago). These sediments originate principally from the North America land mass. The shallow-marine origin of the Eocene and younger deposits has resulted in horizontal strata with only minor dips and folds (Cooke 1945).

Florida was never covered by land ice during the cold stages in the Quaternary, and therefore no glacio-isostatic adjustment of the land mass took place. However, two opposing processes exert pressure on the Florida Plateau. The first, causing uplift, is the process of dissolving carbonate in Florida's limestone geology, or what is known as 'karstification'. Calculations based on measured carbonate loss for the whole of Florida suggest that the maximum uplift during the Quaternary could have been as much as 3.5 mm/century, totalling 58 m over the last 1.6 million years (Willett 2005). The second process, potentially causing isostatic fall, is caused by the build-up of sediments on lower parts of the Florida Plateau during the warm stages of the Quaternary. However, this isostatic response for Florida was at a much smaller scale than the one that has been observed in the Mississippi delta part of the Gulf of Mexico (e.g. Törnqvist et al. 2004).

The flat and generally low-lying nature of Florida has ensured that the sea has had a major influence on the development of the landscape (Fig. 7.2). For example, peninsular Florida's central ridge system was formed on the coast

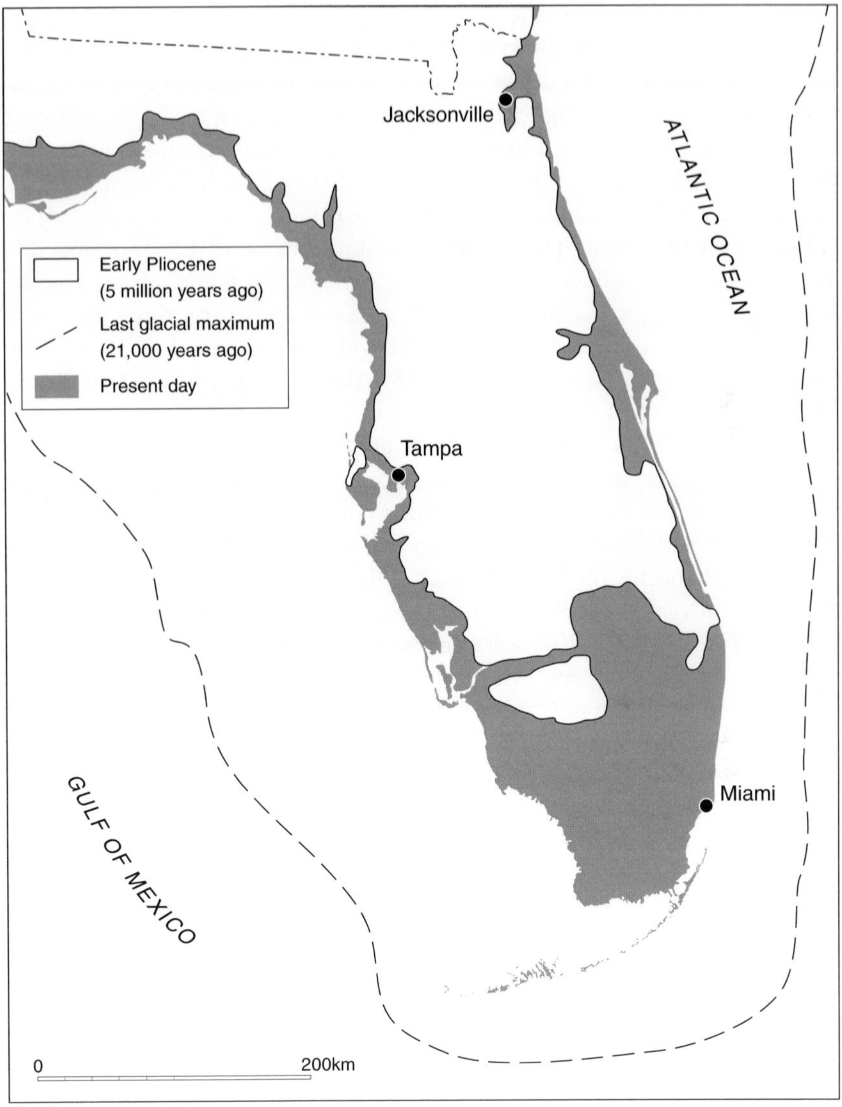

Fig. 7.2. The impact of ESL change on the Florida Plateau: extent of the land mass of Florida in the Early Pliocene (c. 5 million years ago), at the Last Glacial Maximum (c. 21,000 years ago), and today.

where wave action reworked older deposits. Terraces, reflecting higher sea levels in the pre-Quaternary past, are recognizable in the landscape of Florida as abandoned shorelines. The c. 80-m shoreline has been dated to the Upper Miocene (7.5 to 5.3 million years ago), and the 30-m shoreline to the Pliocene (5.3 to 3.4 million years ago). Lower-lying relict shorelines in the landscape of

Florida are of more recent date, including the 18-m shoreline which is thought to date to either the late Pliocene or the earlier Pleistocene (Alt and Brooks 1965). It should be noted, however, that the isostatic uplift caused by karstification may explain the higher than expected elevations of the Upper Miocene and Pliocene shorelines in Florida, which do not readily compare with relict shorelines of similar dates in other regions that are stable in terms of plate tectonics (e.g. Adams, Opdyke, and Jaeger 2010).

During the Quaternary, the changing Eustatic Sea Level (ESL) continued to have a significant influence on the formation of the landscape of Florida. During glacial stages, the land mass of Florida was up to three times the size of the modern state of Florida, exposing the wide continental shelf on the Gulf Coast and the much narrower shelf on the Atlantic Coast. At the Last Glacial Maximum when the ESL was some 120 m below current levels, the entire area of the Florida Plateau was above sea level, and the south-west corner of Florida was some 270 km further west of its current south-west extremity at Cape Sable. The Florida Keys were part of this land mass (Widmer 1988; Griffin 2002: 18). Because of the relatively high rate of the rise of the ESL in the early part of the Holocene, the lower terraces of the Florida Plateau were flooded by the rapidly advancing sea. The type of coastal wetlands that developed in the period before c. 4000 cal BC were generally short-lived, as they were rapidly inundated by the rising sea. The remains of these coastal wetlands survive, to varying extents, on the now submerged parts of the Florida Plateau. With the deceleration of the rate of sea-level rise in the later Holocene, a range of new types of coastal wetlands developed.

On the Atlantic Coast, the barrier system, tidal inlets and flats, and intertidal marshes have developed largely over the last 6,000 years alone, although about 20 per cent of the Atlantic Coast of Florida—especially that of the Anastasia Formation of cemented molluscan grainstone—is made up of earlier, Pleistocene deposits. The remaining 80 per cent consists of barrier islands of quartz-sand of Holocene date, and these barriers are frequently established on top of the Anastasia Formation. The Atlantic Coast barrier system has benefited from a predominantly north–south longshore sediment transport, which has brought large amounts of sediment down the North American coast (Davis 1997: 158).

Florida's Gulf Coast is best described by dividing it into four sectors. The most southerly sector is dominated by mangrove forests and swamps, combined with long oyster and sabellarid-worm reefs. This stretch of coast, from Cape Stable to the south to Cape Romano near Naples to the north, includes the Ten Thousand Islands, a landscape of mangrove islands and tidal channels formed in the near absence of any sediment from the land. Florida's mangrove islands and swamps developed only in the last 3,000 years, under conditions of a very slow sea-level rise; this provided the opportunity for the establishment of oyster reefs, which led to topographic highs becoming available for the pioneering mangroves. The seaward progradation of this type of coastal wetland over the last 3 millennia is estimated to be up to 8 km around Cape Sable. Three species of mangrove exist in Florida: the red mangrove (*Rhizophora mangle* L.), the white mangrove (*Laguncularia racemosa* Gaernt), and the black mangrove (*Avicennia germinans* L.). Mature mangroves appear to be zoned, with the red mangrove occupying the seaward end, and the black mangrove the landward end of the coastal mangrove forest (Fig. 7.3). Research has shown that this is the result of competition between the different mangrove species. Competition between the shade-tolerant red mangrove and the shade-intolerant white mangrove, for example, favours neither in the early pioneering stages of mangrove forest development; but as the forests mature, and a shaded forest floor with plentiful canopy cover becomes available, the red mangroves are able to outcompete the white and emerge as the dominant species. The result of this competition between the mangrove species is that the red mangrove in a mature forest occupies the intertidal zone—the most favourable part of the ecosystem. The white and black mangroves are confined to the higher and less frequently flooded parts of the coast (Bell 1980).

The second coastal sector covers a stretch of c. 300 km to the north of the mangrove forests and swamps, and includes Charlotte Harbor. This area is generally too cold for mangrove to become a dominant species, although mangrove forests survive in sheltered areas such as Tampa Bay. This sector is dominated instead by sandy barrier islands and tidal inlets, with the

Fig. 7.3. The mature mangrove forest of the Ten Thousand Islands, with red mangroves dominant on the seaward end of the forest.

northernmost island being Anclote Keys. These barrier islands are largely constructed from reworked older deposits through wave action, with very little material being transported either longshore or downstream from rivers. They have been studied in considerable detail. Research has established that the underlying geology, in the form of Miocene limestone, defines the presence of barrier islands. The barriers developed on eroded—and from reworked—Pleistocene sediments. On top of these sediments, vegetation developed on the landward side of the coast during the period 3000–1000 cal BC; this vegetation, recorded at a depth of between –6.0 and 2.3 m (Davis et al. 2003), would have helped to keep the sediments in place. The current barriers were formed by low-energy wave action and deposition of sediments, a process that commenced only after c. 1000 cal BC and continues to this day (Stapor, Mathews, and Lindfors-Kearns 1991). There is no evidence for the movement of the barrier islands either seawards or landwards since the time of their formation, although the older barriers are geomorphologically more complex than the younger ones (Davis et al. 2003). In the twentieth century, many of the barrier islands were developed for housing and tourism, and further development continues apace.

The third coastal sector, extending from Anclote Key to the delta of the Apalachicola River, is characterized by a vegetated, marshy coast that developed on top of a predominantly bare limestone geology. This very low-energy coast, with minimal wave activity, provided ideal conditions for the formation of extensive oyster beds on top of limestone bedrock ridges (Fig. 7.4). These oyster

Fig. 7.4. An oyster reef in Crystal Bay. The mixing of fresh water from the Crystal River and seawater provides an environment that is particularly well-suited to oysters.

beds, which frequently developed parallel to the coast, have migrated landwards with the slowly rising sea level over the last 3,000 years. Oyster beds, or reefs, develop in near-shore and offshore parts of the coasts where the depth of water is less than 10 m. Oysters are especially productive in brackish environments, and the oyster beds are at their most extensive in estuaries and where rivers discharge their fresh water into the sea, such as at Crystal River. Because oysters and oyster larvae require hard surfaces for their development and growth, successive generations of oysters settle on top of the shells of previous generations. This encourages the formation of complex oyster beds, which in turn provide feeding grounds and shelter from predators for a range of fish species (Coen et al. 2007).

The fourth and final coastal sector concerns the Panhandle Coast. This is dominated by the Apalachicola River, which brings significant amounts of fine sediment to the Gulf Coast. These sediments form barrier islands with tidal inlets and a well-developed beach and dune system (Davis 1997: 160–8). The sediment- and nutrient-rich outflow of the Apalachicola River has contributed to the very high biomass production of this part of the Florida Gulf Coast. The existence of abundant plankton in this coastal ecosystem has resulted in an extremely biodiverse resource (e.g. Marcus 1991).

The tides on the coastline of the Gulf of Mexico are predominantly diurnal, with a daily single high and low tide, determined by their geometric relationship to the moon and sun. Determined by the size of the Gulf of Mexico, and the narrowness of the straits connecting it to the Atlantic Ocean, the tidal range on Florida's Gulf Coast is relatively small. The coast is affected, however, by the semi-diurnal tides which are dominant in the Atlantic Ocean. As a result there is a mixed tidal regime, with two high and two low tides each day. The influence of the Atlantic tides on the tides in the Gulf of Mexico is small but measurable. For example, the average tidal range at Cape Sable in the very south of Florida is 1.5 m; in Tampa Bay it is 0.8 m; and in the Apalachicola Bay in northern Florida the tidal range is around 0.5 m. Within the context of the Wetlands Change Model, this low tidal range limits the ability of coastal wetlands to adapt to rapidly rising sea levels (see chapter 4).

The environmental history of the Everglades, as the largest non-coastal wetland in Florida, deserves some additional attention. Increased precipitation after c. 4000 cal BC, and the ensuing higher groundwater table, resulted in the creation of Lake Okeechobee around 2000 cal BC. When the water spilled over from this lake, the surrounding land became waterlogged. The natural run-off of rainwater from the c. 1.5 million ha of the Everglades had always been limited by the very low gradient of the lie of the land, along with the presence of coastal barriers such as the Atlantic Ridge (Griffin 2002: 12); but the combination of increased precipitation, overflows from Lake Okeechobee, and a higher sea level turned the former prairie into an enigmatic wetland: the onset of accumulation of calcitic mud and peat deposits date to the period after 2500 cal BC (Gleason et al. 1984: 321). The presence of calcitic mud deposits indicates extended periods of drought—or absence of waterlogging—and appears to date predominantly to the period between 2500 and 500 cal BC; after this point, peat accumulation is the main soil-forming activity, reflecting a wetter environment without droughts. Increased precipitation causing more frequent overflows of Lake Okeechobee, together with a higher sea level with stable coastal barriers (see above) causing further impediments to freshwater run off, is the likely cause (Griffin 2002: 28–49). The vegetation was one of extended areas of sawgrass and spike rush, with Cyprus domes and stands of buttonwood, bay, and willow (Fig. 7.5).

Shellfish are an important part of the coastal ecology of Florida, and the horse conch (*Triplofusus giganteus*) is one of Florida State's official symbols. The role of oyster reefs in coastal evolution processes has already been

Fig. 7.5. A view of the inside of a Cyprus dome on the Big Cypress Seminole Indian Reserve in the Everglades.

mentioned. The eastern oyster (*Crassostrea virginica*) is the most significant species, but other species play a role in reef construction off Florida's Gulf Coast, including conchs (*Melongena* spp. and *Strombas* spp.), clams (e.g. *Mercenaria mercenaria*), scallops, and mussels. The mangrove swamps and oyster beds provide protective habitats and food-rich ecosystems for a very wide range of fish species.

Ongoing discussions about the nature of the sea level during the Holocene in Florida continue to focus on two alternative hypotheses. It is important to deal with this debate in some detail, because it continues to have a significant influence on current sea-level debates, and on the discussions focusing on human adaptation to sea-level change.

The first hypothesis—which retains significant support in archaeological circles in Florida (see Thompson and Worth 2011: 53–4, and the many references cited here)—is one where the sea level is seen to oscillate during the middle and late Holocene. According to this hypothesis, there were times in the middle and late Holocene when the sea level on Florida's Gulf Coast was higher by 1 to 2 m than at present. The key evidence supporting this hypothesis is the existence of ridges which have been interpreted as relict beach ridges. It has been argued that low beach-ridge sets were formed during sea-level low-stands, and high beach-ridge sets were formed during sea-level high-stands (Tanner 1992). On the evidence provided by these relict beach-ridge sets, sea-level 'high-stands' on the Gulf Coast would have occurred four times in the last 6,000 years according to Rhodes Fairbridge (1992), and twice in the last 2000 years according to Stapor and Tanner (1977). This concept continues to have a legacy in archaeological research on the Gulf Coast, and many scholars have linked palaeoenvironmental and geomorphological observations with sea-level oscillations and middle and late Holocene sea-level high-stands. For example, William Marquardt and Karen Walker (2012) argue that Florida experienced two periods of sea-level high-stands—100 BC to AD 500, and AD 850–1200, coinciding with the Roman Warm Period and the Medieval Warm Period—and two periods with a sea-level low-stand—AD 500–850, and AD 1200–1850, corresponding to the Vandal Low and Little Ice Age. This approach to reconstructing sea-level changes in the Holocene is based, in essence, on building regional models of sea-level change using a range of local observations, including land-based and marine-based sediments and geomorphological features, which are used as sea-level proxies. Proponents of this approach point out that the ESL curve is too generalized to be of value when studying the behaviour of coastal communities.

The second hypothesis is of an initially very rapid rise of the ESL in the Early Holocene, which decelerated in the Middle Holocene, producing a smooth sea-level graph without significant oscillations; this view is accepted by the IPCC and supported by most geographers and climate change scientists working in the Gulf of Mexico (e.g. Törnqvist et al. 2004; Wright et al. 2005;

Milliken, Anderson, and Rodriguez 2008). In essence this approach is based on the reconstruction of the ESL using a single type of proxy from similar locations—for example the coral reefs in tropical regions. Regional and local variations in coastal development that produced phases of marine transgressions and regressions are explained in the context of the ESL curve. These local and regional variations can be caused by an array of factors, including:

- glacio-isostatic adjustment
- the sediment loading of continental shelves
- the impact of barrier islands
- changes in the marine, estuarine, and riverine sedimentation processes
- significant changes in prevailing weather patterns such as the El Niño-Southern Oscillation effect
- a range of anthropological processes including large-scale sediment compaction and the construction of artificial barriers (e.g. Kennedy et al. 2007).

Regional climatic 'optima', such as the Roman Warm Period, were not periods of global warming, but represent a change in the redistribution of the sun's radiation in the Earth's atmosphere and hydrosphere (see chapter 4). As such, regional climatic warm or cold periods would have no impact on the total amount of water held as ice and snow, or on the global sea temperatures. Proponents of this approach point out that the uncertainties that come with the proxies derived from these local and regional variations are too great to allow the reconstruction of past sea-level change.

The geologist Ervin Otvos (1995; 2001; 2004; 2011) and others (e.g. Donnelly and Giosan 2008) have argued that the interpretation of the elongated ridges as being formed by wave action is simply mistaken. The analysis by Otvos of all available borehole data and the increased application of Ground Penetrating Radar coupled with Optical Stimulated Luminescence dating shows convincingly that these ridges were carved by repeated wind and storm events, and that in terms of geomorphology and lithostratigraphy they are not the same as the genuine relict beach ridges found in areas of glacio-isostatic uplift. Subsequent research has identified periods of increased hurricane activity in the middle and late Holocene which could have caused the synchronous creation of these elongated ridges over relatively large areas in the Gulf of Mexico (e.g. Liu 2004; Donnelly and Woodruff 2007; Donnelly and Giosan 2008). In addition, no evidence for middle and late Holocene sea-level high-stands has been encountered in any of the biological and geomorphological studies of the coastal evolution of mangrove forests and barrier islands presented earlier in this section. Thus, the very basis of the first hypothesis has been shown to be erroneous and no sedimentary or other evidence for the proposed middle or late Holocene sea-level high-stands exists. Some archaeologists, such as John Griffin (2002: 36–7), have accepted this, noting further that although cultural material such as ceramics provide a *terminus ante quem* for the date of the creation of these beach ridges, they are not necessarily contemporaneous as has been argued in the past.

On the basis of the latest findings, it is pertinent to understand the sea-level change during the Holocene in the Gulf of Mexico—and therefore the coastal evolution of Florida's Gulf Coast—as one that closely followed and follows the ESL (see chapter 4), without any middle or late Holocene sea-level high-stands. In terms of glacio-isostatic adjustment, the most recent meta-analyses of all sea-level index points for the Atlantic Coast of North America identified the location of the forebulge during the last glacial stage in the Carolinas. The glacio-isostatic adjustment diminishes rapidly further south and there is no evidence for such adjustment of the land surface for Georgia and Florida (Engelhart, Peltier, and Horton 2012). The ESL and RSL are therefore expected to be very similar for Florida, although the Panhandle Coast nearest to the Mississippi delta may have isostatically subsided. However, it should be remembered that what matters to coastal communities is how the ESL change affects short-term fluctuations at local and regional scales—and how this translates in terms of local marine transgression and regression—and this deserves detailed attention (see below).

What about future sea-level change? Tide-gauge observations from Florida suggest that sea-level rise is possibly accelerating (e.g. FOCC 2009: 21), although the relative short period of operation of many of the tide gauges—usually less than 100 years—leaves much room for debate on which mathematical method is best suited to determine and predict long-term changes (e.g. Houston and Dean 2011; Baart, Van Koningsveld, and Stive 2012). Nevertheless, an acceleration of sea-level rise, as anticipated by each of the SRES scenarios, will have far-reaching consequences for Florida's coastal wetlands (Fig. 7.6).

The effects of sea-level rise on Florida's Gulf Coast have already been noted. For example, shoreline retreat is already happening along much of the coastline; barrier islands are developing new inlets and there has been a noted increase in the number of dissection events of some barrier islands. There are more than 50 artificial beach nourishment projects along the Atlantic Coast and Gulf Coast of Florida, without which barrier islands would lose their tourist attraction in the short term, and their existence in the long term. It has also been noted that the mangrove swamps are expanding, both landward in south Florida and northwards on the coast, with pioneer mangrove swamps currently extending as far north as Crystal River on the vegetated marsh coast. Autogenic upward and landward movements of some coastal wetlands have been noted, but others cannot migrate landwards where development and hard defences prevent this relocation of the wetlands. There are other changes taking place, such as saltwater intrusion of the freshwater aquifers, which are essential not only for domestic, industrial, and agricultural use, but also

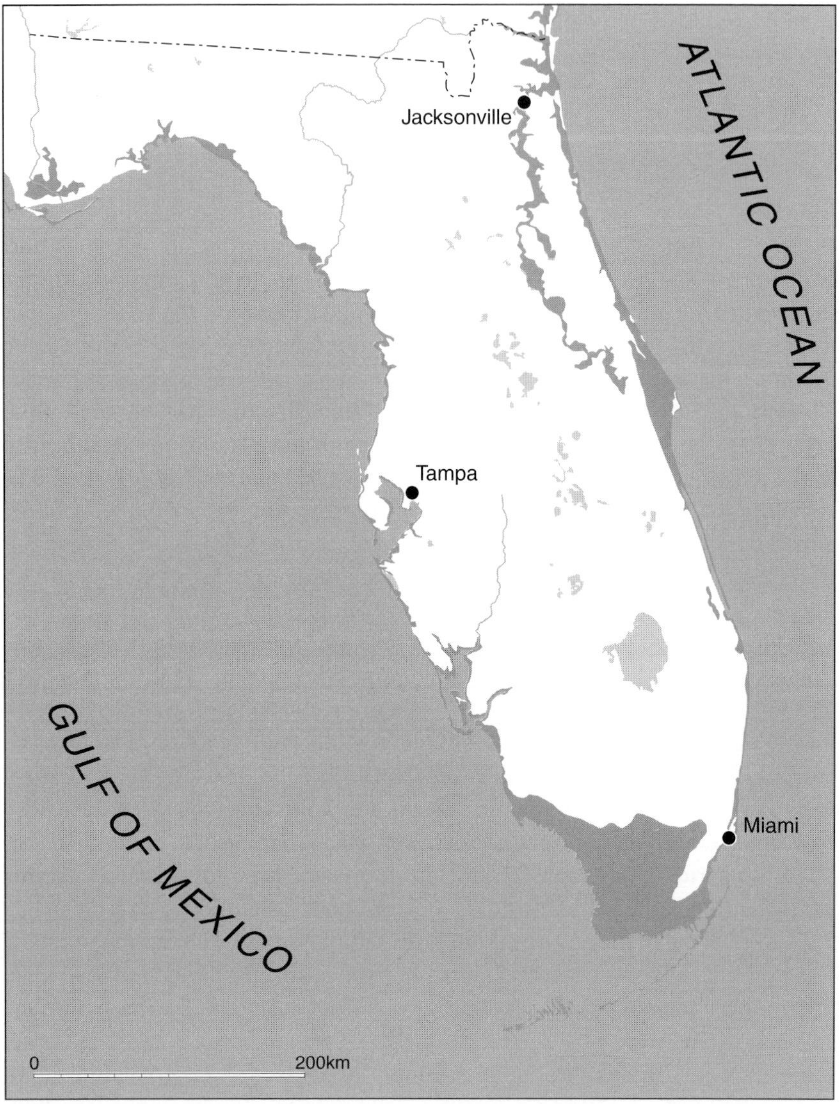

Fig. 7.6. Projected marine transgression of the Gulf of Mexico and Atlantic Ocean on Florida's coasts (dark-grey tint) based on a rise of the RSL of 1 m and excluding the impact of constructed coastal defences (based on the model presented in Weiss, Overpeck, and Strauss 2011).

for maintaining freshwater ecosystems such as the Everglades (FOCC 2009). Different coastal wetlands have different degrees of resilience to sea-level change, and different tipping points that are specific to each wetland ecosystem.

In the context of the Wetland Change Model (McFadden, Spencer, and Nicholls 2007), the resilience of coastal wetlands in dealing with sea-level rise is dependent on three aspects: tidal range, sediment budget, and accommodation space (see chapter 4). The coastal wetlands of Florida's Gulf Coast will all be severely limited by the low tidal range in their ability to adapt autogenically to sea-level change; in effect, there is little tidal and wave energy available for the transportation of sediment which would allow the upward and landward retreat of these wetlands in a time of accelerating sea-level change. The limited availability of sediments will adversely affect the mangrove swamps, barrier islands, and tidal inlets, and especially the vegetated marshy coastal sectors; but the sediment supply on the Panhandle Coast may in fact increase due to the accelerated frequency of extreme weather events, which can add sediments to the Apalachicola River. Accommodation space is, generally speaking, less of a problem for the mangrove coast, the vegetated marshy coast, and the Panhandle Coast, but following the extensive development on the barrier islands and along the margins of the tidal inlets, practically no accommodation space is available in this sector.

In summary, the factors underpinning the development of the coastal wetlands on Florida's coasts in the last 3,000 years have been very stable, and it is this very stability that reduces the resilience of the coastal wetlands to climate change-driven sea-level change in the twenty-first century. The tidal and wave energy and the sediment budgets are low, and this will limit the ability of wetlands to adjust autogenically when the sea level rises. The most critical factor is the absence of accommodation space, especially on the barrier islands and tidal inlets, where the impact of sea-level change will be the most adverse. Sea-level rise will affect the communities that live on these barrier islands, or around the tidal inlets, the most severely.

PAST ADAPTIVE PATHWAYS TO CLIMATE, ENVIRONMENTAL, AND SEA-LEVEL CHANGE

The earliest inhabitants of Florida, the Palaeoindians (before c. 9000 cal BC), found a cool and arid landscape, with a scarcity of fresh water. In the so-called 'oasis model' of such a landscape, animals' watering holes are considered to have been of key importance to the subsistence strategy of hunter-gatherers (Dunbar 1983). Archaeological evidence for coastal exploitation on the Florida Plateau in the early Holocene is scarce, as a result of the rapid sea-level rise during the Early Holocene. Any coastal settlement would have been of a very transitory nature, and submerged evidence of such activity is very difficult to identify (Faught 2004). In addition, the role of coasts, and marine and estuarine resources, in the Palaeoindian and Early Archaic period (9000–5500 cal BC), remains a matter of debate, although shell-midden sites dating to the Palaeoindian period have been identified during research elsewhere in the Gulf of Mexico (cf. Thompson and Worth 2011). On the Florida Gulf Coast, the remains of widespread submerged oyster beds of this period have been extensively dredged—with the shell used in roadworks—for much of the twentieth century. Occasionally, such as in Tampa Bay, ex situ artefacts of both Palaeoindian and Early Archaic periods have been found from contexts associated with these early oyster beds (e.g. Goodyear and Warren 1972; Goodyear et al. 1983). This implies that, at the very least, the earliest inhabitants explored sporadically the marine resources that existed here, but in situ archaeological evidence will be very difficult to find.

Our understanding of the importance of estuarine and marine resources to coastal communities in Florida improves for the period after c. 6000 cal BC. The deceleration of the sea-level rise led to the development of the extensive coastal wetlands (described in the previous section). Around this time the climate ameliorated further, becoming warmer and wetter, resulting in the development of extensive freshwater wetlands on what is now the land mass of Florida. Both types of wetland landscapes became extensively utilized by local communities, as shown by the significant increase of sites dated to the Middle Archaic (5500–3700 cal BC) in all parts of Florida, implying that something of a population boom had taken place (Weisman 2003: 216). The utilization of coastal wetlands on both the Atlantic and the Gulf Coasts concerned a range of shellfish, with oyster (*Crassostrea virginica*) shells frequently dominating the food refuse found at coastal and estuarine sites (Fig. 7.7). There is also evidence for the consumption of fish from estuaries and shallow marine environments, caught with nets, and of terrestrial animals (Russo 1998). This activity is evident principally from middens constructed on Pleistocene dunes, or on other higher grounds, that have remained above current sea levels; examples include the Horr's Island and Useppa sites, Tick Island, and Tomoko sites.

Analysis of very large faunal samples from the extensive shell middens at Horr's Island in south-west Florida has produced convincing evidence that this Middle and Late Archaic (3700–500 cal BC) site represents year-round collection of shellfish and catching of estuarine fish. In the absence of similar large-scale analysis from other midden sites, it is not clear whether all Archaic midden sites represent year-round habitation and utilization of estuarine and marine resources. The site itself was located on a Pleistocene sand dune overlooking an estuary, the present-day Barfield Bay. The remains of the settlement at Horr's Island extend over an area of 1,000 by 300 m, with up to 4 m of accumulated living floors which include post moulds representing small circular structures, organically stained soils with ash, and in situ hearths, fish remains, and tools made of shell and fragmented shell. The settlement was

Fig. 7.7. A midden of shellfish remains on Weedon Island, St Petersburg, Florida.

surrounded by shell ridges, rising between 7 and 9 m above the mangrove swamp. Mound B on Horr's Island has produced what is still the oldest radiocarbon date for an estuarine midden in Florida with a midpoint of c. 5800 cal BC (Russo 1991; 1994: 97–9).

The prehistory of Useppa Island, on the Gulf side of Pine Island Sound, goes back to 5600 cal BC. The rising sea level had turned the environment of Useppa into a newly created estuary and early inhabitants utilized the natural resources, especially during the spring and summer. Between 5600 and 3700 cal BC, a range of shellfish—including oysters, whelks, conchs, and clams—and fish—notably catfish, pigfish, pigfish, rays, and sharks—were eaten. Some of the oldest shell tools from Florida were found here, including shouldered celts, hammers, and cutting tools made of *columellae* of lightning whelks, and anvils made from quahog clam shells. Useppa Island remained inhabited throughout much of Florida's prehistory into the twelfth century AD (see below; Marquardt 1999: 241–53).

Within the context of the prevailing linear cultural evolution model in Floridian archaeology, it has been widely accepted that early shell mounds were simply refuse mounds—the result of the discard of the remains of shellfish and fish after consumption—which may also have been used for habitation. However, some of the oldest of these shell mounds appear to have been deliberate constructions serving ritual or ceremonial purposes. For example, the Middle Archaic period mound at Tick Island, in the valley of the St John's River in north-east Florida, comprises a shell midden of pre-ceramic Archaic age (i.e. predating the production of pottery c. 2000 cal BC) containing 175 flexed human internments, on top of which a pre-ceramic Archaic burial mound of sand was constructed. The nearby Tomoko mound complex, on a marine terrace that separates the freshwater Tomoko River from the brackish Halifax River, incorporates pre-ceramic Archaic conical mounds of shell alternating with sand deposits, which have been attributed ceremonial functions (Piatek 1994). The extensive shell middens at Horr's Island not only produced evidence for year-round coastal presence during the Archaic period (see above), but also for four mounds which were constructed deliberately during the Late Archaic stage. These mounds comprise thick layers of predominantly oyster shells alternating with sand deposits, in some cases overlying earlier habitation surfaces which had been cleansed with fire. It has been argued that these mounds, too, were constructed for ritual or ceremonial purposes (Russo 1994).

The period from c. 2500 cal BC has produced many of the shell middens that characterize so much of the archaeology of the Florida Atlantic and Gulf Coasts, including the large number of shell midden rings and arcs on the contemporaneous barrier islands which are now submerged (Thompson and Worth 2011: 55). The presence of pottery in many of these middens emphasizes the point that they were occupied, with house floors and post holes/moulds providing direct evidence of the existence of houses on top of the flattened middens. The absence of any pottery in Florida in the period before c. 2500 cal BC could mean that a number of middens whose formation commenced earlier have not been recognized as such.

It has been argued by some that, commencing around 2200 cal BC, many middens on the barrier islands on Florida's Atlantic Coast ceased to be developed further, and that by c. 1200 cal BC most had been abandoned (Thompson and Worth 2011: 56). This phenomenon appears to be not exclusive to Florida; a similar pattern has been observed for most of the South Atlantic Coast (Thompson and Turck 2009). This change is attributed to 'major fluctuations in sea level' between 1400 and 400 cal BC (Thompson and Worth 2011: 56). However, as already argued above, this idea of an oscillating sea level in the middle and late Holocene is not supported by modern research. More importantly, whilst the enlargement and construction of shell middens may have been discontinued, coastal habitation and the utilization of marine and estuarine resources carried on throughout the Late and Terminal Archaic, and indeed throughout the period 500 BC–AD 500 as evidenced by Glades-1 and Caloosahatchee-1 pottery (e.g. Marquardt 1999: 249). The lack of midden construction could, therefore, be explained with equal logic as indicating a period with only very 'minor' fluctuations in sea level.

From c. AD 700 onwards, the famous large shell-mound complexes along the Gulf Coast of Florida were constructed. The largest of these extend over 10 ha in area, with some mounds 9 m high and in excess of 35 m across. The mounds consist of thick layers of oyster shells over natural oyster bars. According to the accounts of the sixteenth-century Spanish explorers, the largest of these shell-mound complexes were the seats of local or regional elites. The shell-mound complexes were villages with a water court, plaza areas, and interconnecting canals. Burials were normally located at some distance from these settlements, in sandy burial mounds (Weisman 2003: 217). One of the most remarkable shell-mound complexes on Florida's Gulf Coast is that of Key Marco, where Frank Hamilton Cushing excavated in 1895 a small part of the water court, retrieving a large collection of carved and painted wooden figurines, tablets, and masks, alongside cups, bowls, toy boats, and a large number of other wooden and shell artefacts (Cushing 1897). Historically, this site was associated with the Calusa. Radiocarbon dates, combined with the ceramics retrieved, reveal this high-status settlement to have been inhabited between c. AD 650 and the first contact with the Spanish in the 1510s (see below). Cushing referred to the site as the 'Court of the pile dwellers', although this reflects the remarkable impact on archaeologists across the world of Ferdinand Keller's work on the Swiss lake settlements, and its translation into English in 1866, rather than a sanguine archaeological interpretation (Gilliland 1975; 1989; Purdy 1991: 23–54).

The study of the Calusa Indian settlement on Pine Island has left little doubt that the adaptive pathways chosen by coastal communities were linked directly to local marine transgressions and regressions. These environmental changes were acutely felt in the shallow Pine Island Sound, which provided by far the largest proportion of food resources for the Pine Island dwellers. The leaders of the long-standing research on Pine Island and Pine Island Sound, William Marquardt and Karen Walker (2001; 2012), have interpreted these local marine regressions and transgressions in the context of global climate change. However, as already argued in some detail (see above), this is based on a misunderstanding of the impact of regional optima and cold phases—such as the Medieval Warm Period and the Little Ice Age—on global temperatures and global sea levels. Nevertheless, their research is important and informative as it provides some excellent examples of how coastal communities in Florida adapted to locally experienced environmental changes. The causes of the local marine regressions and transgressions in Pine Island Sound are most likely connected to the shoreline evolution of the nearby barrier islands, and even minor changes would have had disproportionate impacts on the shallow Sound. Because of the shallowness of the Sound, regressions and transgressions affected the salinity of the water, and therefore the suitability of the Pine Island Sound's inshore ecosystem for marine fish, the most important food source for the coastal communities.

Around AD 850, Pine Island Sound experienced a marine regression and marine fish populations deserted the Sound in favour of deeper water in the Gulf to the west. The observed adaptive pathways included a resettling of (some of) the Pine Island residents further to the west, possibly after a short phase of intensified gathering of shellfish before this mode of subsistence became unsustainable. After 900, a local marine transgression resulted in the Sound becoming once more a suitable habitat for marine fish. For the next 300 years, Pine Island's community benefited from this abundant resource; the material culture reflects the fishing technology—both net and hook-and-line fishing—alongside the continued gathering and consumption of shellfish, which is likewise well represented archaeologically. Later in the tenth century, the extensive shoreline settlement on Pine Island was replaced by a nucleated settlement, with the enlarged and heightened midden mounds becoming the foci for settlement. Marquardt and Walker (2012: 40) have explained this activity by connecting it to 'an abruptly rising . . . sea level'. Noting that the midden mounds were up to 7 m high—well above even the wildest estimate of the late Holocene sea-level high-stands—such an explanation has to be rejected as too overtly environmentally deterministic. It is more likely that the reasons for nucleation and increased construction activity on top of the midden mounds were political and socio-economical in origin, reflecting the evolving complexity of the Calusa, rather than being the direct result of environmental change. A short period of marine regression around AD 1100 is correlated to a period during which the midden mounds were not enlarged, but the situation reverted after c. 1150. For the period from 1200 to 1850, a low sea level associated with the Little Ice Age has been purported (Marquardt and Walker 2012: 42–4), but this time the lower level had only minimal effects on the subsistence basis of the Pine Island inhabitants.

The study of the Calusa and their predecessors on Pine Island demonstrates that at various times these coastal communities responded to local environmental changes that affected their key sources of food, and that their adaptive pathways included both resettling and the utilization of alternative food sources. However, their actions and impacts on the environment are not limited to adaptations that followed environmental change; the increasing political and socioeconomic complexity of the Calusa Indians cannot be explained by reference to environmental change alone. This is also demonstrated with reference to the construction of large canals. Around AD 1000, a series of canals were dug in south-western Florida, including the Pine Island Canal, the Ortona Canals and the Naples Canal. These canals, up to 9 m wide and several kilometres long, connected the Gulf Coast of Pine Island with the Thousand Islands and, through the River Caloosahatchee, with Lake Okeechobee and the interior of Florida. Research on the Pine Island Canal demonstrated the presence of large shell middens at either end of the canal, with a mound at the easternmost end containing some 250 burials, near the shell works of Indian Fields. In this context, it is noted that mangrove swamps provide poor landing places for boats, and shell mounds would have served as natural ports. The canals would have been used by logboats and reflect an increase in the interregional exchange and trade by the Calusa—an indication of the increasing complexity of their political organization—around the beginning of the second millennium. The material culture, in the form of Belle Glade Plain pottery, marine shell artefacts, and food remains, supports this (Luer 1989). The canals illustrate the ability of coastal communities to shape their landscape, just as the landscape they inhabited shaped them.

The historical landscape of Crystal Bay, on the Gulf Coast in central Florida, provides another excellent—and landscape-wide—example of the time-transgressive chronology of adaptive pathways adopted in times of environmental change. Furthest west into the Gulf, one can find a large number of shell rings that have been submerged by the rising sea. Although many of these, when investigated in the early twentieth century, produced pottery of prehistoric date, their formation could well predate the introduction of pottery in Florida around 2500 cal BC (Fig. 7.8). Moving eastward towards the current coast, some shell rings are partially submerged, such as Mullet Key: this island is now covered by red mangrove, which in recent years has extended its habitat northwards as temperatures have increased (Richard Estabrook and Garry Ellis, personal communication). Non-diagnostic late prehistoric pottery fragments can be found embedded within the shell. Further eastward still are very large shell midden rings with mounds constructed behind the shell ring. At Shell Island, in the mouth of Crystal River, this constructed mound is some

Fig. 7.8. A shell ring made of the refuse of the shellfish overlying an oyster reef in Crystal Bay. The presence of the mangroves this far north on Florida's Gulf Coast is a recent phenomenon.

7 m high. A small freshwater wetland, receiving water from an aquifer, occupies the centre of the island. Diagnostic Swift Creek-type (c. AD 100–800) complicated stamped pottery fragments and a decorated rim piece of Weeden Island-type (c. AD 500–1200) pottery were identified on the island. Nearing the coast, more elaborate complexes of midden material present themselves. On Roberts Island—formed by two channels of the Crystal River—a 20 by 13 m flattened mound made of midden material was located, overlooking a shell midden containing pottery of Weeden Island and Safety Harbour (AD 900–1600) types, and a shell ridge which includes pottery types representative of the first and second millennia. The island may have included a small plaza or ballcourt, and landing places for canoes, all formed from midden material. The site served both as a village and as a ceremonial centre (Bullen 1953; Weisman 1995; Thompson and Pluckham). Finally, on the other side of the Crystal River and on the mainland is the famous Crystal River site excavated by Clarence B. Moore around the turn of the last century (e.g. Moore 1907). The site comprises two burial mounds, and ceremonial mounds of which the largest overlooks the Crystal River, along with shell middens and platforms, and two 'stelae', loosely arranged around a large plaza. The site has two construction phases extending approximately over a 1,600-year period, from c. 200 BC to AD 1400.

To what extent do the shell mounds represent a deliberate adaptation to sea-level rise? At the end of the nineteenth century, Frank Hamilton Cushing (1897) was already arguing that the 'vast shell ridges of the monumental shell-mound complexes of the south-western coast . . . [functioned] as sea walls protecting against storm surges. He thought other shell-mound villages to have been abandoned half-finished in the wake of some spectacular, unnamed, hurricane' (cf. Weisman 2003: 216). Whilst it is debatable whether the shell ridges acted as sea walls, they certainly provided safety for coastal communities from storms and surges, and the enormous investment represented by the shell-mound complexes represent a sustainable adaptive pathway by the Florida Indians. The debate in Florida whether the construction of shell mounds and shell-mound complexes represents a deliberate adaptive pathway to sea-level change is ongoing, but there is no doubt that these mounds and complexes sustainably supported local coastal communities for centuries and, sometimes, millennia.

The Spanish landfalls in Florida include those led by Ponce de Léon (1513), Panfilo de Narvaez (1528), and Hernando de Soto (1539). The Spanish brought with them not only diseases to which the Calusa Indians had no resistance, but also new forms of political alliances leading to social structures with greater differentiations, as evidenced by the many gold and silver pendants and tablets from burial mounds in south and central Florida (Griffin and Smith 1948; cf. Weisman 2003: 214). The first permanent settlement, St Augustine, was founded in 1565 by Pedro Menendez de Aviles.

Archaeological excavations show a European-style gridded layout, but a material culture dominated by aboriginal artefacts. Such artefacts have been interpreted as indicating the presence of local women in Spanish households, possibly through marriage (Deagan 1981; 1983; 1985; cf. Weisman 2003: 214). A 'Missionary' phase followed the Contact period, and only during this time did the long-standing dependency on marine, estuarine, and riverine resources for subsistence diminish. Corn agriculture was completely absent from Florida till the eighth century AD, but it was the subsistence mainstay in the Panhandle away from the coast after 1000. It was only as a result of changing settlement structures and economy, under the influence of the Spanish, that corn—alongside maize—became a staple food, but agriculture did not extend southwards of modern Tampa before the eighteenth century. During the following century the territory of Florida became a front line for competing British and Spanish interests, with Sir Francis Drake taking St Augustine in 1586.

The nature of the coastal communities changed dramatically after the beginning of the eighteenth century. The complex societies of the Calusa Indians in central and southern Florida had, by that time, fragmented due to the impact of European diseases on the population. Yamassee and Uchise Indians of the Creek nation—under British administration and with British firearms—raided Florida during the first half of the eighteenth century, progressively extending their raids further south. Armed with firearms, the Creeks defeated and enslaved the remaining Calusa and many other Indians, who were sold to plantations in Carolina. Only small vestiges of the former proud Calusa nation remained, on the islands and in the Florida Keys. The Creeks settled themselves in many coastal and inland locations, and the name 'Seminole' was adopted towards the end of the eighteenth century for all Native American inhabitants in Florida.

At the end of the War of Independence in 1784, British rule in Florida was transferred to Spain. The US claimed Florida in 1803, establishing a united government of Florida in 1822. In 1845 Florida became the 27th American State. Three Seminole Wars followed, resulting in the partial decimation of the Seminole Indians and their forced transport to Oklahoma, after which only small communities of Seminoles remained in Florida. Some escaped into the very southern part of Florida, and found safety in the wilderness of the Everglades, where they adapted to the opportunities provided by these extensive freshwater wetlands.

The drainage history of the Everglades deserves some additional attention, as this reflects the ways in which wetland landscapes in Florida have been perceived and treated over the last 150 years. The drainage of the Everglades commenced in earnest early in the twentieth century, led by the entrepreneurial governor Napoleon Bonaparte Broward who intended to turn this wasteland into productive agricultural land. His and subsequent attempts to drain the Everglades relied on drainage engineers who based their advice on 'state-of-the-art scientific knowledge of his day' (McCally 2005: 143). Such scientific knowledge was nearly always found wanting and the early drainage attempts were unsuccessful. Hurricanes in 1926 and 1928 caused Lake Okeechobee to overflow, with devastating consequences for the farmers who had settled in the Everglades: over 2,600 people died and large tracts of land were inundated. The subsequent construction of the Herbert Hoover Dike (1932–8), which enclosed Lake Okeechobee to the west and south, along with the enlargement of earlier canals to a size that could carry off even hurricane floodwater, provided the missing parts of the puzzle to drain the Everglades effectively. The post-war drainage of the Everglades, masterminded by the US Army Corps of Engineers, as well as their subsequent environmental degradation and the attempts made to restore parts of the natural ecosystem, is well documented elsewhere (e.g. Davis and Ogden 1994), and does not need repeating here. The environmental history of the Everglades, and of the adaptive pathways followed in dealing with environmental change, reflects the limited capability of scientists and engineers—often coupled with an unbridled optimism in the belief of human ingenuity—to predict with accuracy the outcome of large-scale changes to the landscape. We would do well to recognize that science and engineering are socially embedded activities, closely linked to, and biased by, the socio-economic and political *Zeitgeist* (McCally 2005: 155).

Florida, and in particular the southern Gulf Coast, remained underdeveloped and underutilized during the eighteenth and most of the nineteenth centuries. Cuban fishermen set up fishing ranchos, sometimes employing Seminoles or small numbers of Europeans to process the fish for consumption in the towns such as Havana. According to the Historical Census Browser at the University of Virginia (mapserver.lib.virginia.edu) and the United States Census (2010.census.gov), Florida's population grew steadily from the 1830s, when the total number of inhabitants was recorded as c. 35,000. The population expanded more rapidly between 1900 and 1930, at which date the number of Floridians exceeded 1 million, but the real acceleration occurred only after the Second World War (Fig. 7.9). Currently, some 19 million people live in Florida. The population is concentrated along the coast, with coastal urban conurbations such as Jacksonville and Miami on the Atlantic Coast, and Tampa-St Petersburg-Clearwater on the Gulf Coast. The attraction of the beaches and of living on the barrier islands goes back to the late nineteenth century, with early developments such as Pass-a-Grille in modern St Petersburg; but the near-continuous construction of houses, hotels, and condominiums that dominate the barrier islands and tidal inlets north of the Ten Thousand Islands dates to the post-war period (Fig. 7.9).

To accommodate the rapidly growing population of Florida, a significant part of the wetlands was drained and converted into urban and suburban housing estates, industrial parks, and agricultural land to feed the Floridians.

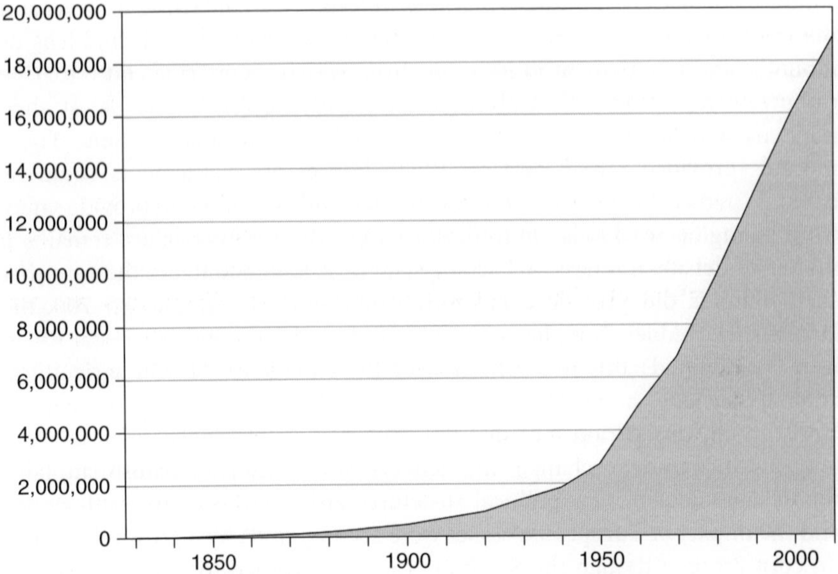

Fig. 7.9. Florida's population boom, 1830–2010. The population of Florida has increased tenfold since 1950, and the main concentrations are on its coasts.

Because of the absence of reliable maps detailing the original extent of the wetlands, an accurate assessment of the total loss is not possible. An innovative way of dealing with coastal wetlands was the 'dredge-and-fill' method, developed initially in the 1920s. This method used the drift-geology spoil dredged from rivers, lakes, estuaries, and the shallow marine zone to fill low-lying and frequently inundated parts to create habitable land. The technique was refined in the 1960s to produce new canals and 'small peninsulas that reached into the water like grasping fingers' (Pittman and Waite 2009: 11). This form of land reclamation, which provides for premium-value domestic properties with direct access to the water, can be found everywhere along the Florida Gulf Coast, especially in the sections dominated by mangrove swamps and barrier islands with tidal inlets. The environmental damage caused by these developments includes the degradation of water quality, which affects corals, oysters, and barnacles in particular, along with the removal of nursery and feeding grounds for many fish and shellfish species, and the reduction of flood storage capacity and increased saline incursion (Johnston 1981).

In summary, Florida's coastal wetlands have attracted people since the Palaeoindian phase through to the present day. The near-limitless marine and estuarine resources, especially fish and shellfish, were the main attractions throughout the Holocene. There is no evidence for unsustainable utilization of these resources before the twentieth century. This overview has identified a number of adaptive pathways followed in response to the rising sea level and the frequent hurricanes: these include the construction of shell middens or mounds, and resettlement in locations from which the marine and estuarine resources were most effectively or conveniently accessed. The oldest shell middens date to the end of the Early Archaic or Middle Archaic. These mounds provided raised land on which houses and settlements could be based. Whether the mounds were constructed with the intent to provide safety from changing sea levels, or from sea surges during hurricanes, remains a matter of debate amongst archaeologists, but it is undoubtedly the case that shell mounds did provide a protective environment. After c. AD 700, the largest shell midden complexes, such as on Pine Island and at Crystal River, were 're-sculpted'; this is a reflection of the emergence of non-agricultural complex societies along Florida's Gulf Coast. Other changes to the natural landscape in this period included the construction of canals which enabled long-distance trade, exchange, and delivery of tribute. The Spanish landfalls, the development of new political structures around missionary settlements, and the impact of European diseases were accompanied by a growing reliance on corn crops, although the Seminole Indians returned to a dependence on using natural resources—both those of the coast and the estuaries, and those of the freshwater wetlands of the Everglades. In the last 150 years or so the land has become the main focus for food production. But the coast and coastal wetlands continue to attract very large numbers of settlers, drawn by the warm weather and beaches.

STRENGTHENING THE RESILIENCE OF COASTAL COMMUNITIES

The political concern with Florida's coastal and freshwater wetlands has only relatively recently included an environmental dimension. Five years after Florida became the 27th of the United States, the US Congress passed the Swamp and Overflowed Land Act, handing the federal wetlands over to the states for reclamation. By 1854, the young State of Florida was in control of c. 8 million hectares of wetlands (Pittman and Waite 2009: 9). These wetlands were sold to developers and passed into private ownership, with no control or effective regulation in place until the Clean Water Act was passed in 1972. The most important element of the Act in regulating the development of Florida's Gulf Coast wetlands was Section 404, which determined that permits were required for the deposition of dredged or fill material when used to turn wetlands into dry lands. A permit could be denied or restricted if the effects of the proposed dredge-and-fill proposal 'on municipal water supplies, shellfish beds and fishery areas (including spawning and breeding areas), wildlife, or recreational area' were considered adverse (Clean Water Act, Section 404, C).

The operation of Section 404 has not, however, led to the effective protection of Florida's wetlands. In part, this is due to the reluctance on behalf of the United States Army Corps of Engineers, whose responsibility it is to issue the permits, to prevent development on private lands. The Corps is also under political pressure at various levels; during periods of economic downturns and demands for free-market economies, permits appear to be more readily issued (Pittman and Waite 2009: 77). Key Marco or Marco Island, where Frank Cushing excavated the 'Court of the pile dwellers', is a case in point. The Deltona Corporation bought and commenced development of the near-pristine beach-fronted and mangrove forest-protected Marco Island in the 1960s, before the Clean Water Act had been passed. It sold yet-to-be-built house allotments to fund the work, but by the time it needed further dredge-and-fill work, the Clean Water Act had been passed. The Corporation applied for permits under Section 404; the compromise attained provided the permit for the development of one of the three proposed subdivisions, despite this meaning the destruction of a part of the mangrove forest, but the development of two further subdivisions was rejected in favour of the protection of the coastal wetlands (Pittman and Waite 2009: 46–60). Section 404 has, undoubtedly, prevented the destruction of some wetlands in Florida, but has certainly not halted the piecemeal destruction of Florida's wetland habitats.

In 1984, Florida passed the Warren S. Henderson Wetland Protection Act. This law introduced the concept of 'mitigation' into the protection of wetlands, permitting development to take place but offsetting the loss in acreage through the artificial creation of wetlands twice the size. This concept became federal policy. In 1988, President George H. W. Bush declared that 'all existing wetlands, no matter how small, should be preserved' (Pittman and Waite 2009: 91). This statement evolved into the more nuanced policy of 'no net loss', which allowed for wetlands to be filled and developed as long as new wetlands of equal size were created in their place. This policy was adopted by subsequent incumbents of the White House and by many other countries around the world. However, the policy of mitigating the destruction of wetlands through the creation of new ones is not without controversy. Undoubtedly, some mitigation projects have been successful in that a 'like-for-like' replacement of wetland habitat has been achieved, and the biodiversity has been protected. However, the consensus amongst wetland ecologists is that these achievements pertain to a small minority of the new wetlands created under the no-net-loss policy. In the majority of cases, opinions range from the notion that the gains obtained from the creation of new wetlands fail to compensate fully for the loss of the destroyed wetland, to the more outspoken conclusion that 'no net loss' is an unmitigated disaster and a waste of taxpayers' money (Burgin 2010). This is the case for Florida, as well as elsewhere (Pittman and Waite 2009).

During my visit to Florida in April–May 2012, it was evident that very few freshwater wetlands had been created or restored in the last three decades. However, those in existence—created under the no-net-loss policy and Florida's Wetland Protection Act—were evidently thriving, at least in terms of their plant communities and especially the mangrove forests. This comes as no surprise, because coastal wetlands develop rapidly under the right conditions. The sea and humanity continue to be co-creators of the coastal landscape (cf. Van de Noort 2011b: 228–31).

The urgency with which the impact of climate change and sea-level change on Florida is addressed is closely linked to the political inclination of the state governor and the Florida Senate towards modern climate change. Under Governor Charlie Crist (2007–11), the Florida Senate adopted the Florida Climate Protection Act. As a result, the Florida Energy and Climate Action Team, and later Commission, acted as the formal focus for considering the impact of climate change on Florida. Between 2008 and 2011, it developed an ICZM under the auspices of the Florida Oceans and Coastal Council (FOCC). The management of Florida's coastal zone is the responsibility of several departments of the State of Florida, including the Florida Department of Environmental Protection, the Florida Fish and Wildlife Conservation Agency, and the Florida Department of Agriculture and Consumer Services. These departments work together within the framework provided by the FOCC. A wide range of academics and other specialists are consulted, but only a handful of local stakeholders to represent coastal communities.

The FOCC (2009) report, *The Effects of Climate Change on Florida's Ocean & Coastal Resources*, was based on the findings presented in the IPCC's *Fourth Assessment Report*, adopting similar language and terminology in its statements and displaying the same confidence. The report concluded that:

- there is no clear tide-gauge evidence that the sea-level rise around Florida has accelerated in the twentieth century, but the sea has certainly warmed over the last 100 years;
- shorelines are retreating and there is an observed increase in the dissection of barrier islands;
- Florida's reef-building corals are 1–1.5 °C closer to their upper thermal limit than they were a century ago; the occurrence of coral diseases has significantly increased, and the limits of some coral species, including staghorn and elkhorn corals, are moving further northwards along Florida's Gulf Coast as a direct consequence of higher sea-surface temperatures;
- fish and shellfish species distributions are changing, either as a direct effect of warmer sea-surface temperatures or because of changes in habitat caused by warmer ambient and water temperatures, such as the northward extension of the limits of the mangrove forests;

- whilst some coastal wetlands have responded autogenically to the rising sea level, other coastal wetlands have been drowned because of the absence of accommodation space; some have undergone changes in type, for example where estuarine and coastal forests with high biodiversity are replaced by salt marshes with lower biodiversity;
- saltwater incursion has affected a number of freshwater aquifers, but the exact relationship between rising sea levels and increased water abstraction—for domestic, agricultural, and industrial use—has not been established.

Following the election in 2010 of Governor Rick Scott, who has on several occasions publicly denied the impact of the burning of fossil fuels on climate change, the Florida Energy and Climate Commission was abolished on 1 July 2011 and its functions were redistributed to a number of departments, leaving no organization in the State of Florida to consider the impact of climate change. References to climate change on the State Department's websites have been confined to the 'back pages'. For example, the beach nourishment programme of the Department of Environmental Protection is explained without any reference to climate change, sea-level rise, or sustainability (http://www.dep.state.fl.us/beaches).

This absence of action on climate change at state level has to be viewed in the context of the extraordinary impacts that sea-level change will have on the low-lying coastal regions and population of Florida. So, what does the study of the adaptive pathways from the past have to offer the current communities on Florida's Gulf Coast?

The first and most obvious contribution that the study of past adaptive pathways to sea-level change can make to modern communities concerns the long-term evolution of coastal wetlands. This has shown that all current coastal wetlands developed only once the Holocene sea-level rise had decelerated, and that their existence is threatened by a future acceleration of the sea-level rise and by the warming of the oceans and seas. Some coastal wetlands are more resilient to sea-level rise and rising sea-surface temperatures than others. For example, the mangrove forests' resilience to a higher sea level is considered to be great, as long as the sediment budget remains positive and sufficient accommodation space exists. The warmer sea-surface and ambient temperatures benefit this ecosystem, and it has already been noted that the mangrove forests' habitat appears to be extending increasingly northwards along Florida's coast. However, the barrier islands' resilience to sea-level rise is much smaller. The relatively low levels of tidal and wave energy and the limited sediment budgets in the Gulf make this type of coastal wetland highly vulnerable to an accelerating sea-level rise. The development of the barrier islands has, effectively, removed any accommodation space for the islands to migrate landwards with a rising sea level, even though there is very little evidence of their natural ability to do this (Fig. 7.10). Already, artificial

Fig. 7.10. View across Boca Ciega Bay to the St Petersburg Beach barrier. The beach barrier is completely overbuilt, and there is no accommodation space for it to respond to sea-level rise. Beach nourishment provides a short-term solution, but this damages the coastal ecosystem.

beach nourishment is required to ensure the survival of the majority of barrier islands, on both the Atlantic and Gulf Coasts (www.dep.state.fl.us/beaches/). The long-term sustainability of this practice is being questioned, however, both from a geological perspective—whether Florida's Continental Shelf can provide suitable material in years to come (e.g. Duane and Meisburger 1969)—and from a biological perspective—whether sand extraction has an adverse impact on marine ecosystems (e.g. Courtenay, Hartig, and Loisel 1980). An improved understanding of how coastal wetlands have developed over the last millennia is an essential element in strengthening the resilience of modern coastal communities.

The second useful contribution from the study of past adaptive pathways comes in the form of concepts of long-term and sustainable coastal living. The prehistoric way of living by the sea, with settlements and modes of subsistence surviving for centuries, offers a mirror up to the short-term practices of today. An exploration of this adaptive pathway can be divided into three components: the sociopolitical nature of societies; knowledge and understanding; and public and private ownership.

In Florida's prehistoric past, in complex societies of the Calusa—and similar societies further north during the Woodland and Mississippi periods—political power rested with chiefs who were likely to have inherited their position, and who were expected to pass their role on to their heirs. It provided a basis for long-term and sustainable developments, including ensuring that marine and estuarine resources were not exhausted. In modern US politics, the length of a term between two elections determines how far-sighted politicians can afford to be, and some of the examples presented above illustrate this point. This is not, of course, to suggest that the US should turn its back on the democratic process. Rather, it serves to emphasize the need for cross-party, bipartisan, and stakeholder support for the protection and long-term, sustainable development of coastal wetlands. Without sustained and effectively executed policies, short-lived attempts to protect or to construct new wetlands are not going to be worthwhile or, worse, are a waste of taxpayers' money.

A closely connected issue relates to the knowledge base of the Floridians past and present. Prehistoric Floridians may not have fully understood the interconnectedness between the mangrove forests and oyster reefs and their function as fish nurseries; neither would they have appreciated the details of the sedimentary dynamics that created the barrier island. However, when prehistoric Floridians changed the landscape, they did so within the context of their knowledge of it. That landscape carried the impact of earlier generations, and a deep-founded respect for the landscape of the ancestors would have been a limiting factor in any attempt to change the coastal wetlands. In twenty-first-century Florida, however, many of the residents have limited knowledge of the landscape and its history, or of the value of the coastal wetlands in offering natural protection against the impacts of climate change. In other words, the sense of place that would have been integral in prehistoric societies is effectively missing from the modern coastal communities in Florida. Educating the modern coastal communities on the importance of interconnected wetlands and ecosystems is clearly a priority.

Detailed knowledge about the value of coastal wetlands now rests with specialists, notably coastal managers, ecologists, geologists, and archaeologists, and these are frequently distrusted by the wider public. To make matters worse, the specialists often fail to agree on how to interpret their findings, as illustrated by the discussion on whether there were middle and late Holocene sea-level high-stands (see above). If the specialists cannot agree on what climate change means, how it will impact on sea-level change, and how modern communities should adapt to this, then the public and politicians are unlikely to adopt a consistent or long-term policy approach to wetland protection.

This connects directly to the third component, one that sets this case study apart from the other case studies in this book: the private ownership of the coastal wetlands. The sale of coastal wetlands to private investors and developers in the second half of the nineteenth and first half of the twentieth centuries has created a situation in which catchment-wide coastal management is difficult to achieve. Individual developers seek to maximize the economic value of their investment in, principally, domestic residences for the rapidly growing population of Florida. These developments continue to use the 'dredge-and-fill' approach established early in the twentieth century, whereby canals are dredged to ensure that the maximum number of residences have direct access to navigable watercourses, in order to optimize their sale value, whilst the lower-lying grounds and wetlands are filled in. The protection of wetlands, or their 'mitigation', nearly always depresses the profit margins; the overwhelming majority of developers seek to minimize the costs associated with this. The Key Marco development by the Deltona Corporation provides an example of this practice, but the phenomenon can be observed everywhere along Florida's Atlantic and Gulf Coasts. Developers have little interest in the long-term sustainability of the landscape after the sale of the plots has been completed.

These residential developments and the construction of associated infrastructures frequently interrupt the interconnections that exist between the different ecosystems, wet and dry. This has adverse effects on the sustainability of coastal wetlands. For example, many dredge-and-fill developments affect the natural hydrogeology and water quality and this, in turn, can have adverse effects on the fish and shellfish populations and threaten the sustainability of fishing tourism. More importantly, all such developments remove accommodation space, and with every new development the natural ability of coastal wetlands to adjust to a rising sea level is reduced. The situation has already prompted government agencies to take action to avoid coastal erosion and the destruction of these (recent) developments. But although most beaches are conserved through beach nourishment programmes, the long-term sustainability of this practice is doubtful (see above). Increasingly, 'hard' defences are being constructed as the natural ability of the coastal wetlands to protect the human infrastructure is diminished and their resilience decreased; in turn, the construction of embankments exacerbates the problem of lack of accommodation space for coastal wetlands.

In short, private ownership of coastal wetlands, along with the unbridled development of the wetlands to accommodate the growth of Florida's population, precludes the emergence of some sort of catchment-wide masterplan for Florida's Gulf Coast. The situation prevents both state and federal government agencies from acting proactively on the impacts of climate change-driven sea-level change. The role of these agencies has become one of protecting unsustainable developments from inevitable environmental change, whereas what they should be doing is taking a lead in sustainable development. The long-term and sustainable ways of living on Florida's coast in the past could form a starting point for reconsidering the unsustainable developments of the twenty-first century.

The third adaptive pathway from the past that can help in building the resilience of coastal communities is the sustainable utilization of marine resources. There is no archaeological or palaeoenvironmental evidence to suggest that coastal wetland ecosystems were ever damaged or overutilized to the point that the fish and shellfish species selected for human consumption could not autogenically recover. Many of the settlements on the Gulf Coast of Florida were in use for centuries, and sometimes millennia, as shown in the accumulation of the shell mounds and shell-mound complexes; such a situation could only be achieved if the main food sources, fish and shellfish, were sustainably utilized. This contrasts with modern practices towards coastal wetlands. A few examples from the very extensive literature on this topic serve here merely as an illustration: the fishing of marine sponges for use as bath sponges in the early twentieth century resulted in the complete disappearance of the *Hippospongia* and *Spongia* species from the Gulf (McClenachan 2008); increased fishing practices have caused, directly or indirectly, ecological near-extinctions of turtlegrass and *Thalassia* seagrass in Florida Bay, with detrimental consequences for the turtles and fish species that depended on these (Jackson et al. 2001); and a majority of the snapper-grouper species are over-fished around the Florida Keys, as evidenced by their progressively smaller average length (Ault, Smith, and Bohnsack 2005). The modern mindset that the sea's resources are inexhaustible is, evidently, not justified.

The fourth and final adaptive pathway from the past comes from the shell middens and shell-mound complexes. The deliberate deposition of the shells in mounds created elevated areas for settlement, above the wetlands and sea levels. This practice did not require that the wetlands were filled, and the palaeoenvironmental and archaeological evidence available suggests that the interconnected ecosystems remained unaffected by these structures. This contrasts with the modern dredge-and-fill approach, where coastal wetlands that have grown over centuries are altered, often with the sole purpose of maximizing the number of residences that have water access. Maybe a return to older and more sustainable ways in which more people can be accommodated on the coast, building in a manner that does not require the destruction of the coastal wetlands in the process, is required.

CONCLUSION

Sea-level rise in the twenty-first century will affect the full length of Florida's 2,000 km of coastline. In some parts of the state, a 1 m rise will flood extensive parts of the low-lying land, including the southern half of the Everglades and the Big Cypress, extensive tracts of urban developments around Charlotte Harbor and Tampa Bay, and along many other estuaries and deltas, as well as deep into the St John's River catchment. Higher sea levels will impede the run-off of fresh water from the mainland, causing upstream flooding after periods of heavy precipitation more often than is the case now. In addition, higher sea levels will cause saline intrusion into the freshwater aquifers that provide modern Floridians with drinking water. All coastal wetlands will need to adjust. Where sufficient accommodation space and positive sediment budget exist, it seems likely that the mangrove forests will cope with the expected sea-level rise in the twenty-first century. However, where insufficient accommodation space remains, coastal wetlands will be flooded. The barrier islands on the Gulf Coasts are particularly vulnerable to higher sea levels because of the low-energy environment, near-absence of accommodation space, and limited sediment budgets. By the end of the twenty-first century, the coastal wetlands—including the beaches—will be much smaller; fish and shellfish populations will be under severe stress, and coastal communities will have less protection from, and resilience against, the impact of hurricanes and other extreme weather events.

Archaeological evidence for sustainable living on the Gulf Coast goes back some 6,000 years. Coastal communities have been able to utilize the rich resources of the rivers, estuaries, and seas for centuries without adversely affecting the ecosystems. From as early as 4000 cal BC, early Floridians deliberately constructed mounds from the shellfish remains, creating elevated areas for settlement and ceremonies. Complex sedentary societies developed, based largely on the utilization of marine resources; this way of coastal living continued till the earliest Europeans arrived in Florida in the early sixteenth century. With the Europeans came disease and a disruption of the socio-political and economic order. Nevertheless, the wetlands continued to provide for those who sought to live away from the emerging towns, and many of the traditional methods survive amongst the Seminoles in the Everglades today. In Florida's prehistoric past, the coast provided not only the main source of food; shells were used too, both as tools and as the material for the construction of the shell middens and shell-mound complexes.

The privatization of the wetlands of Florida in the second half of the nineteenth and the beginning of the twentieth century has turned out to be a critical point in their sustainability. Privatization encouraged the development of wetlands, initially for agricultural production but increasingly for residential developments, providing houses for the influx of people who wish to live in Florida's sunshine. Privatization has led to the dredge-and-fill approach being implemented in many coastal wetlands, thus replacing the natural ecosystems with artificial landscapes designed to accommodate the maximum number of residences with direct access to the water or the beach. With an acceleration of sea-level rise in the twenty-first century, private landowners will demand that the federal and state governments 'hold the line' and protect coastal communities from the inevitable damage to their property. Mitigation, or the creation of new wetlands to replace the old ones, has been largely unsuccessful in terms of maintaining biodiversity or ecosystem services. In both the short and long terms, this is not a sustainable adaptive pathway: the recent history of the Everglades, where taxpayers' money was used first to drain the wetland and latterly to restore the ecosystem, still provides the best illustration of why long-term thinking and action are essential. This case study has suggested that the palaeoenvironmental and archaeological study of the past can strengthen the resilience of coastal communities in a time of rising sea levels. The role of such research can be summarized thematically as follows:

First, as has been observed in other case studies, knowledge of the interactions between climate change, sea-level change, and coastal wetland development—on the required millennia scale—is poorly developed. This knowledge is also not shared between the institutions and communities that are stakeholders in the future coastal zone management of Florida's coast. Academic researchers sometimes rely on outdated concepts of sea-level change in the past and so misinform the public. Meanwhile the opinions of politicians play too great a role for long-term sustainable solutions to be developed and adopted, and scientific facts are ignored. Both the study of how the coastal wetlands developed in the past and the sharing of this knowledge will be fundamental prerequisites for a sustainable future.

Second, the effective management of Florida's coast requires cooperation and collaboration on a range of levels between local communities and the institutions, including the federal and state governments, insurance companies, and NGOs that seek to protect the coast and coastal wildlife. This poses a significant challenge. Private landowners demand from the institutions that their property is protected from the impacts of sea-level rise and extreme weather events, even when the subdivisions are located in places where sustainable living was never an option. Coastal protection projects that aim to 'hold the line' are destined to become more expensive under any scenario of sea-level rise, requiring more from the taxpayer. Such practices will also have increasingly negative impacts on their environment. Hard defences will make it impossible for coastal wetlands to respond autogenically to sea-level rise and will eventually drown, whilst the cost and adverse environmental impact of the many beach nourishment projects can only increase over time. This is likely to lead to a polarization of the positions taken by the coastal communities and the institutions that seek either the best use of taxpayers' money or to protect the coastal wetlands. Political opportunism may attempt to exploit this polarization, weakening the basis for building social capital amongst the parties. In short, the future of collaboration needs to be based on an acknowledgement of the legitimacies of the wide range of stakeholders, but there are few signs that this is actually happening. A look at the past reveals something of the community-wide collaboration that made living by the sea a venture from which all involved could benefit.

Third, bearing in mind that the coast is the key attraction for many US citizens who seek to move to Florida or visit for their spring break, research into the past indicates that for Florida to remain a sought-after paradise, modern coastal communities—alongside physical adaptation to the impacts of sea-level rise—will have to find ways in which to reformulate their relationship with the sea and coast. Examples from the past of how communities defined and refined their sense of place, and learnt to live with the dynamic interaction of sea and land, may offer some guidance in this respect.

Fourth and finally, the coast of Florida, from c. 4000 cal BC, has been a highly valued economic landscape. Fish and shellfish are still important for Florida's economy. The value of fish is significantly higher now than it ever has been, with the annual income from recreational fishing adding some $8 billion, and commercial fishing over $5 billion, to Florida's economy, the equivalent of some 250,000 jobs (http://www.st.nmfs.noaa.gov). Healthy fish and shellfish populations require that their habitats, and especially the ecosystems where fish spawn, are protected. Tourism is Florida's largest source of employment and many tourists are attracted by the coasts. The sustainability of the coastal wetlands, the mangrove forests, oyster reefs, and barrier islands is essential for Florida's well-being, as it has been for thousands of years.

8

The Iraqi Marshlands

INTRODUCTION

The Iraqi Marshlands, or Al-Ahwar, developed during the sixth millennium BC in the delta of the rivers Tigris, Euphrates, and Karun (Fig. 8.1). The Marshlands formed at the deposition centre of these sediment-rich rivers, at the point where they met the Persian Gulf. The Gulf, as a direct result of Holocene sea-level rise, had already transgressed onto the lower Mesopotamian Plain, but during the late Holocene the rate of accretion of sediments outpaced sea-level rise and the Gulf Coast prograded south-eastwards, towards modern Basra. Some of the world's earliest cities settled on this coast after its transgression halted, and the Mesopotamian towns of Eridu, Uruk, Ur, Lagash, and Girsu have all provided ample evidence for the usefulness of the coastal wetlands in providing food and building material for their inhabitants.

In more recent millennia, the Al-Ahwar was home to the Ma'dan, or Marsh Arabs, whose traditional culture was based on the sustainable utilization of the Marshlands, characterized by the use of the milk of water buffalo and fishing, and with an architectural tradition based around reed constructions. The ecosystem was maintained by the seasonal pulse-floods delivered by the rivers Tigris, Euphrates, and Karun, which submerged the landscape for several months each year. This water provision was deliberately interrupted during the final years of the reign of Saddam Hussein, with devastating consequences for the ecosystem and the Ma'dan. Since the early twenty-first century, attempts to restore the Marshlands and the traditional lives of the Ma'dan have had some success, but the long-term and sustainable future of the Al-Ahwar remains in serious doubt.

Research into the Iraqi Marshlands provides a case study that is distinct from those presented in the preceding three chapters. Whilst the origin of the Al-Ahwar is closely linked to Holocene sea-level change, the threat to the marshes from sea-level rise in the twenty-first century is limited. In fact, the predictions suggest that the rise in sea level during this century will affect only the south-eastern margins of the Al-Ahwar, along with the narrow floodplain of the Shatt al-Arab and the coastal regions of the Gulf. Because this wetland is

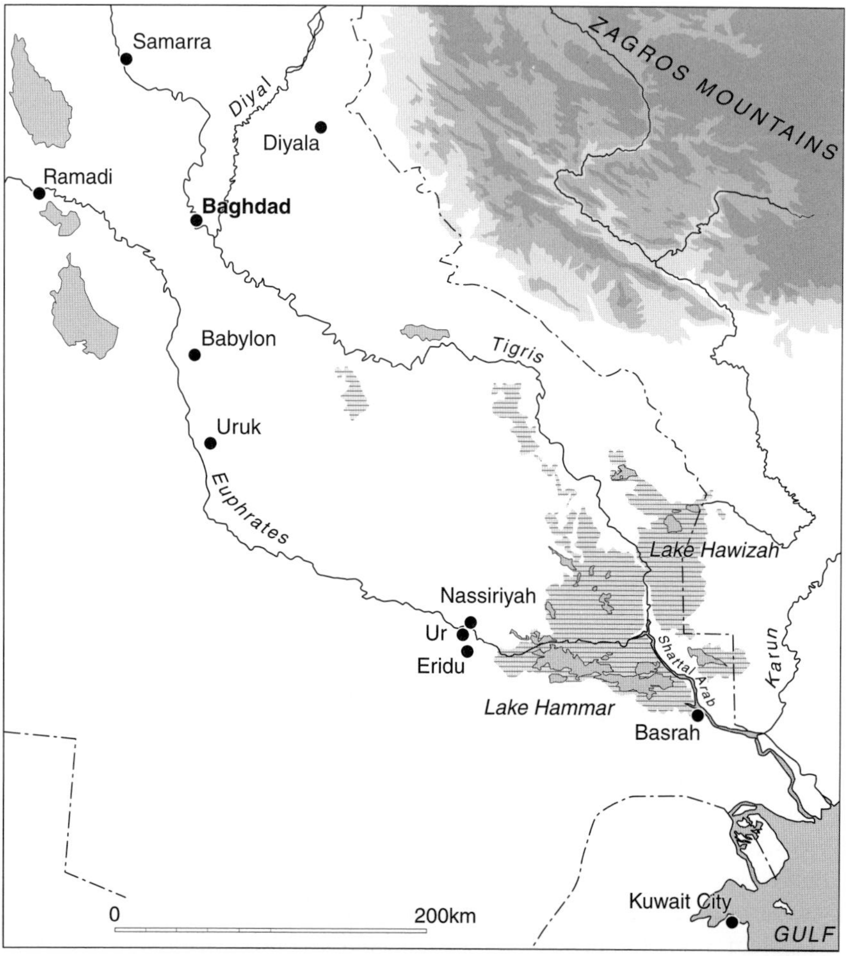

Fig. 8.1. Map of the Iraqi Marshlands and key sites mentioned in the text.

already suffering from water shortage, the predicted sea-level rise will have predominantly positive impacts on the ecosystem and the communities that live in the Marshlands. Nevertheless, the inclusion of this region into this study is justified by the knowledge that can be gained from the recent work to restore the marshes following the deliberate attempts to degrade this ecosystem in the later part of the twentieth century. In this way, it may serve as a proxy for coastal landscapes that are flooded by a rising sea, or severely damaged by an extreme weather event, in the near future.

This chapter follows the same structure as the previous case studies. It starts with a description of the interaction and connectedness of climate change, sea-level change, and the development of the Iraqi Marshlands. This is followed by a consideration of the adaptive pathways developed by communities in the context of these environmental changes. It concludes with a consideration of how this information can be used to build the resilience of modern coastal communities. However, because the specific interest in this particular study lies in the wetlands' restoration, an emphasis is placed throughout on relatively recent developments. This contrasts with the other case studies. The attempts to re-wet the Marshlands provide us with some early lessons—ecological, but more importantly social and political—which are pertinent in this comparative study.

HOW CLIMATE, ENVIRONMENTAL, AND SEA-LEVEL CHANGE SHAPED THE IRAQI MARSHLANDS

The Al-Ahwar developed relatively late in the Holocene on the lower Mesopotamian Plain, defined here as the area in Iraq and Iran bounded by the Al-Jezira Plain to the north, the foothills of the Zagros Mountains to the east, the Widian Limestone Plateau to the west, and the Persian Gulf or Gulf of Arabia to the south. The lower Mesopotamian Plain measures approximately 600 km from Samara on the Tigris and Ramadi on the Euphrates to the current coast, and 200 km at its widest between the Zagros Mountains and the Widian Plateau. The Plain receives its waters from three main rivers: the Tigris, the Euphrates, and the Karun. The nature of these three rivers is quite distinct. Whilst both the Euphrates and the Tigris have their sources in the mountains of Anatolia, the River Euphrates receives no tributaries after it is joined by the River Khabur in Syria, and consequently delivers relatively modest amounts of water to the lower Mesopotamian Plain. The Tigris, however, is joined by the numerous rivers that come off the Zagros mountain range in Iran, including the rivers Little Zab, Adhaim, and Diyala. These greatly swell the Tigris especially during the spring months. The Karun rises on the southern slopes of the Zagros mountain range, and has a shorter journey than the other two rivers to the lower Mesopotamian Plain.

Geologically, the Mesopotamian Plain and the Persian Gulf form the edge of the Arabian Plate, which subducts beneath the Iranian Plate to the east, lifting the Zagros mountain range in the process. During the time that has elapsed since the Last Glacial Maximum, some 21,000 years ago, the tectonic influence on the region is considered to have been very minor, and earthquakes are rare (e.g. Teller et al. 2000). The extensive low-lying area of the Mesopotamian Plain has received sediments from the surrounding higher grounds from at least the Pliocene onwards. A very large conglomeratic alluvial fan was formed during the Pleistocene, to the south and south-west of modern Basra by the Wadi Batin. During the Holocene another extensive fan was created to the north-west of Basra, formed by the deposits brought by the rivers Karun and Kharka, which originate in the Zagros mountain range. These two huge fans on either side of Basra constrict the connection between the Mesopotamian Plain and the Gulf. The regions are linked by the Shatt al-Arab (Sanlaville 2002: 133–4).

The development of lower Mesopotamia during the Late Glacial and Holocene is not as well researched as that of the coastal wetlands in the previous chapters, and this situation has not been helped by the inaccessibility of the region in recent decades due to the various wars, although the position is improving. In particular, it has not yet been possible to apply modern fieldwork techniques and radiometric dating to this region to the same extent as noted in the other case studies. Nevertheless, an outline of the evolution is presented here, with the areas of ongoing debate acknowledged.

During the Last Glacial Maximum, with the Eustatic Sea Level (ESL) some 120 m below the current level, the Gulf was effectively a dryland extension of the Mesopotamian Plain. Borehole evidence from the lower Mesopotamian Plain indicates that the beds of the rivers Euphrates and Tigris incised deeply into the sedimentary strata, to a depth of between 25 and 30 m. These rivers would have flowed into the Ur-Schatt or ancient Schatt—recognizable on bathymetric maps as a submerged canyon—that followed the deepest part of what is now the Gulf, towards its confluence with the Indian Ocean in the Gulf of Oman (Teller et al. 2000; Kennett and Kennett 2007: 233). This relatively steep decline of the rivers of the lower Mesopotamian Plain towards the sea facilitated the transport of large amounts of sediment into the Gulf of Oman. As was the case elsewhere, the Late Glacial was a period of very limited precipitation, and Mesopotamia would have been a cold and arid landscape with limited vegetation (Sarntheim 1972; Al-Ameri and Jassim 2011: 445).

The continuing rise of ESL in the early Holocene caused the transgression of the sea into the present Gulf and the corresponding gradual filling of the deeply incised beds of the rivers Euphrates and Tigris. The date for this marine transgression remains a matter of some debate, but around 6500 cal BC the coastline could have reached modern Basra (Aqrawi 2001: 275; Kennett and Kennett 2007: 238; Hritz et al. 2012: 74). The subsequent marine transgression onto the lower Mesopotamian Plain would have been very rapid due to the flat nature of the landscape. Deposits from Lake Hammer, some 90 km north of Basra, containing marine gastropods and marine bivalve molluscs including oysters, have been dated to c. 5000 cal BC (Purser et al. 1982). Deposits of a similar age with proxies of estuarine ecosystems have been retrieved from the vicinity of the ancient city of Ur (Kennett and Kennett 2007: 240). Furthermore, recent research in the immediate vicinity of Eridu has established that marshlands developed here towards the end of the sixth millennium cal BC (Hritz et al. 2012: 70–3). Taking these findings together, a broad sixth millennium BC date for the marine transgression stage will be used here. In the absence of extensive coring programmes, the position of the new shoreline is not known, but one is postulated between 150 and 200 km north of the present confluence of the rivers Euphrates and Tigris, near the modern towns of Amarah and Al Nasiriyah. The ancient cities of Ur, Ubaid, and Eridu are believed to have been located immediately landward of this hypothesized ancient coast, whilst the cities of Lagash and possibly Girsu would have been situated on islands or dry areas within the extended Gulf (Sanlaville 2002: 142, 148; see below).

Sea levels continued to rise throughout the Holocene. At some time before 4000 cal BC the discharge of the main rivers would have been impeded, leading to extensive overbank floods during the spring months and sediment deposition on the Mesopotamian Plain north of the new coastline. The region was also more humid in the first half of the Holocene, as a result of changes in the monsoon circulation (Kennett and Kennett 2007: 236–7; Preusser 2009). These conditions produced sediments of silty sand rich in organic matter (Aqrawi 2001: 272).

The subsequent development of the lower Mesopotamian Plain was dominated by the process of progradation. At the confluence of the rivers Euphrates, Tigris, and Karun with the newly established coastline, a wide delta formed which became the deposition centre of the sediments transported by these rivers. As these rivers received their water predominantly from melting snow from distant mountain ranges, and from spring rain, their peak flows occurred in the spring months, especially April and May. During these months, the rivers transported large amounts of sediments, the vast majority of which were deposited in the delta. The result of this was coastline progradation and a regression of the marine environment. Just as marine transgression onto the lower Mesopotamian Plain in the period between 6500 and 5000 cal BC had been rapid, the process of progradation in the period after 5000 cal BC is likely to have occurred with similar speed, as a result of the decelerating rate of the ESL rise, along with the flatness of the landscape and the large sediment load of the rivers. The River Tigris, for example, had an annual sediment load estimated at 40 million m^3 prior to the construction of the first dams (Cressey 1958). The process of progradation has been poorly dated and there are contrasting views on when the current coastline was established, with dates ranging from the fifth century BC to the early centuries AD (cf. Hansman 1978).

The southward extension of the coastline, combined with the deposition of riverine sediments in the prograding delta, created an extensive flatland with low riverbed gradients. During spring peak-flow times, the Euphrates, Tigris, and Karun rivers left their courses and flooded extensive areas, creating the Iraqi Marshlands (Fig. 8.2). The onset of the creation of the Marshlands was a time-transgressive process. The oldest marshes developed near the 5000 cal BC coastline, younger ones nearer modern Basra. Silty deposits containing

Fig. 8.2. A view of the Iraqi Marshlands, a landscape dominated by reed and reed-mace. Photo by Wilfred Thesiger © Pitt Rivers Museum, University of Oxford.

freshwater molluscs from the area between Larsa and Oueili, north-east of Nassiriyah, have been radiocarbon-dated to the middle of the second millennium BC (Geyer and Sanlaville 1996), and lacustrine and freshwater marsh deposits dating to around 1000 cal BC were found throughout the region of the modern Iraqi Marshlands prior to their drainage in the twentieth century (Aqrawi 2001: 276–7).

The climate warmed up quickly during the early Holocene. In the wetter conditions before c. 6000 cal BC palm trees were a common sight, alongside the extensive grasslands (Al-Ameri and Jassim 2011: 445). The climate in the period after 6000 cal BC was one of increased aridity, as implied by a range of palaeoenvironmental proxies, including pollen analysis (Kennett and Kennett 2007: 240). This new arid climate led to the increased movement of dunes that had become active once the vegetation ceased its hold on the sand deposits, and resulted in a greater reliance of human communities on the rivers and wetlands (Aqrawi 2001; see below).

The evolving landscape can be described physiographically as follows (Sanlaville 2002: 135–9). On the Mesopotamian Plain north of Baghdad, the rivers created wide levees as the result of deposition of the coarsest material alongside the rivers during periods of flood. These levees stand up to 3 m in height and can be as much as 3 km wide. Further to the south-east is an area described as an interior fluviatile delta, where the Tigris and Euphrates become braided, multi-channel rivers, with smaller levees and marshes developing in basins and along former channels. Much water is lost here (see below), resulting in the creation of deltaic fans. Further again to the south-east is the area of the marshes and lakes proper. Here, the land is without significant slope, and the water flow is very slow. By the time they arrive in the marshes the rivers have retained little sediment, and sedimentation and siltation have therefore only a limited impact on the landscape. Even further to the south-east is the outflow of the rivers into the Shatt al-Arab that connects the Al-Ahwar with the Gulf, and which is tidal as far upstream as Qurna, north of Basra.

As already noted, neotectonic activity during the Holocene was rather limited. Paul Sanlaville (2002: 144) has argued that the main subsidence in lower Mesopotamia occurs as the result of sediment loading, noting that the lowest areas are now within the interior fluviatile delta rather than in the marshes proper. This observation is supported by fieldwork by Adnan Aqrawi (2001) who has noted that sediment accumulation in the Marshlands attributable to the last 3,000 years is typically less than 1 m in thickness, whereas further north on the Mesopotamian Plain more than 2.5 m of sediments have accumulated over the last three millennia.

Before the construction of the dams and reservoirs on the upstream reaches of the rivers Euphrates and Tigris in the twentieth century, the ecosystem of the marshes was dominated by reed (*Phragmites australis*) and reedmace (*Typha angustifolia*), with the former found in the permanently flooded marshes, and the latter in the seasonally flooded parts. The reed prospered on the nutrient-rich river waters and grew to a height of 8 m. The very extensive reedbeds removed many of the nutrients, resulting in the creation of several large and clear lakes, such as Lake Hammar and Lake Hawizah. The marshes were renowned for their abundance of fish and waterfowl, with water buffalo also present. The latter were either introduced by humans around 3500 BC or existed in the wild in the marshes before that date (Evans 2002).

Water management has a very long history on the Mesopotamian Plain. The earliest irrigation practices date to the sixth millennium BC, with larger-scale irrigation projects dating to the third millennium BC (e.g. Sherratt 1980; see below). Throughout the centuries, water from the rivers Euphrates and Tigris has been diverted to optimize agricultural production, and this practice continues to the present day. However, in the twentieth century, larger-scale hydro-engineering projects sought to control the natural seasonal flow of these rivers and to harness their power for the generation of hydroelectricity. A large number of dams and barrages were constructed during the second half of the twentieth century, on the Tigris and Euphrates in Turkey and Iraq, and on the Euphrates in Syria, for the purposes of flood control, agricultural irrigation, domestic and industrial water storage, and the generation of hydroelectric power (Naff and Hanna 2002: 180). Many of these projects were envisaged as far back as the 1950s, but it was the oil revenues—which increased significantly after 1973—that allowed them to be executed.

The projects had two immediate impacts. The first of these was a reduction in the volume of river water reaching the Iraqi Marshlands, and the regulation of its flow, which effectively evened out the natural seasonality. The lowering of the groundwater table and the absence of seasonal floods have made it possible for the lower reaches of the Euphrates and Tigris to be embanked and large tracts of the margins of the Iraqi Marshlands to be taken into permanent agricultural cultivation. The second impact concerns the water quality, with polluted waters being returned to the Euphrates and Tigris from agricultural and artificially fertilized lands. Engineering solutions for the pollution problem include the construction of open drains—such as the Main Drain or Third River—designed to remove the polluted run-off from the natural river system and divert it to the lowest reaches of the Euphrates and Tigris.

However, in the 1990s the drainage of the marshes was politically motivated. Following the first Gulf War in 1990–1, the Marshlands became a safe haven for many Iraqis, especially Shi'ite Muslims, who opposed the regime of Saddam Hussein. The Iraqi Army was unsuccessful in its pursuit of those who sought refuge in the wetlands. With devastating impact on the Al-Ahwar, Saddam Hussein ordered the implementation of existing plans to drain the marshes through the construction of several diverting canals and drainage channels, including the ominously named Crown of Battles River, Loyalty to the Leader Canal, and the Mother of Battles River (Kubba and Jamali 2011: 15; see below).

The expected rise in the ESL in the twenty-first century according to the various SRES scenarios is unlikely to have a significant impact on the Iraqi Marshlands. The prediction for lower Mesopotamia is that some coastal areas will be affected, and the tidal range in the Shatt al-Arab will extend further inland which may affect the salinity of the water in the southern Marshlands. The impact of the rise on the wetland ecosystem will be minor, especially when compared to the effects of changes in the flood regimes of the rivers Euphrates, Tigris, and Karun precipitated by anthropogenic activity such as dam construction.

PAST ADAPTIVE PATHWAYS TO CLIMATE, ENVIRONMENTAL, AND SEA-LEVEL CHANGE

Because of the extensive masking of old land surfaces by younger marine, fluvial, and aeolian sediments—the latter an ongoing process, as land dunes in the deserts of the Mesopotamian Plain continue to shift—archaeological evidence dated to the late Palaeolithic and early Holocene is rather limited (Wright, in Adams 1981: 323). A very recent synopsis for these periods on the Arabian Peninsula shows a complete absence of late Palaeolithic and early Holocene sites in lower Mesopotamia, but a significant number of early Holocene sites on the current Gulf coast (Groucutt and Petraglia 2012).

The Mesopotamian Plain is part of the area known as the 'Fertile Crescent', which has played a key role in some of the most important cultural, political, and social changes in the human past, notably the introduction of sedentary communities and the transition to farming. Debates on these important changes have been linked, at least since the 1930s, to changes in climate and environment in the postglacial and Early Holocene periods.

V. Gordon Childe's (1936) 'oasis hypothesis' is one of the earliest attempts to explain sedentism and farming within the context of climate change. Childe argued that the earliest domestication of animals in South-west Asia, following warming during the early Holocene, was the inevitable consequence of wild ungulates and people interacting near the increasingly rare freshwater locales. But subsequent research showed that the early Holocene was in fact a time of increased global precipitation, thus ruling out the oasis hypothesis as a possible explanation for the onset of sedentism. In the 1960s and 1970s, the interplay between climate change and population growth was thought to provide the backdrop against which human communities in South-west Asia became sedentary and famers (e.g. Binford 1968; Cohen 1977; cf. Matthews 2003: 72–4). Kent Flannery (1969) observed that the earliest sedentary settlements, such as those in the foothills of the Zagros mountain range, preceded the move to farming. The transition to more intensive hunting, fishing, and gathering—what is now known as the 'broad-spectrum revolution'—enabled the population to grow in niche environments and to become sedentary, with farming a consequence of sedentism. A variation on this approach is the more recent hypothesis that animal domestication and the gathering of wild cereals originated in the south Levant in the postglacial period, and spread during the cold Younger Dryas stadial around 10,000 cal BC to other regions in the Near East, including Mesopotamia. The communities in these regions had experienced population growth in the postglacial period, but now had to cope with environments where productivity was reduced. The more intensive use of crops was the answer (Bar-Yosef and Meadow 1995). Even more recently, Peter Richerson, Robert Boyd, and

Robert Bettinger (2001) published their understanding of the origin of farming in the evocatively titled paper 'Was agriculture impossible during the Pleistocene but mandatory during the Holocene? A climate change hypothesis'. In this paper, it is argued that the very cold and highly variable climate of the Pleistocene, with low levels of precipitation and atmospheric CO_2, made reliance on a few domesticated plant resources simply impossible. In the Holocene, the climate became more stable; it was warmer, wetter, and provided higher concentrations of atmospheric CO_2, which stimulated plant growth. Because human populations tend to grow rapidly to the carrying capacity of the environment, it is argued, a more intensive use of the environment allowed farming communities to grow and be more successful than non-farming communities. In the last three decades, alternative explanations for the shift to sedentism include arguments that this is principally a sociopolitical shift, driven not by the need to produce more food or feed more mouths, but by a sociopolitical or religious aspiration to live together in larger groups (e.g. Hodder 1990; Hayden 1995; Cauvin 2000; cf. Matthews 2003: 89–92).

These early developments passed lower Mesopotamia largely by, and before c. 7000 cal BC, the region remained one inhabited by hunter-gatherers. It seems that the earliest settlements known from the region date to the seventh and early part of the sixth millennium BC (Kennett and Kennett 2007: 249–53). This was a time of relatively favourable climatic and environmental conditions, due to the predictable monsoons. The dispersed settlements of this period are located in river valleys, in the wadi systems, in the Ur-Schatt River, and along the coast of the Gulf. These settlements are thought to have been seasonally occupied, part of a mixed foraging-farming subsistence with the herding of sheep and goats practised alongside the hunting of wild animals, foraging for wild grasses, and fishing (Beech et al. 2005).

The final stages of the Mesopotamian Neolithic and the Chalcolithic are the Ubaid period, from 6500 to 4200 BC, and the Uruk period, from 4200 to 3200 BC. The Ubaid period in lower Mesopotamia is characterized by the appearance in the archaeological record of elevated and permanent settlements, many of them located on the edge of estuaries and near wetlands. The importance of such locations in terms of food production has long since been recognized. The lower parts of the extensive river levees, and the margins of freshwater marshes, were ideally suited to the cultivation of wheat and barley. The alluvium was ideal for the planting of gardens and orchards, and both the cereal crops and gardens were irrigated at this time. Pastureland included semi-arid land, marginal wetlands, and fallow fields, and reeds from the Marshlands provided fodder for livestock (Adams and Nissen 1972: 86). Finally, fish were caught from canoes using nets in the rivers and marshes (Lloyd and Safar 1948: 118; Bottéro 2001). The close association of Ubaid I pottery with the ceramics of Samarran settlements in the foothills of the Zagros Mountains implies close contact between the Samarran and lower

Mesopotamian people (Oates 1960). These Samarran communities kept domesticated cattle and operated small-scale irrigation agriculture at this time.

Amongst the most important of the Ubaid-period towns is Eridu—Tell Abu Shahrain—one of the oldest proto-urban settlements in the world and recognized as the oldest town in Sumerian literature. Excavations at Eridu in the nineteenth and early and mid twentieth century produced evidence of sedentary habitation throughout the Ubaid period and beyond, including the town wall, the temple with 18 successive phases, and extensive funerary remains (Taylor 1855; Hall 1919; Campbell Thompson 1920; Safar, Mustafa, and Lloyd 1981). The landscape beyond the tell presented a well-developed system of irrigation canals of substantial proportions, and a range of smaller settlements, identified through aerial photography and field surveys. Irrigation agriculture produced wheat, six-row barley, and dates. Cattle were domesticated. Marine fish found in Eridu's temple complex provide evidence of fishing in the Gulf (Wright, in Adams 1981: 324).

The story of a great Flood, brought about by a deity to deal either with the problem of overpopulation or with the sinfulness of humanity, is mentioned in a number of Mesopotamian myths, including the *Epic of Gilgamesh*. Eridu is one of the cities named in Sumerian literature as predating this Flood (e.g. Dalley 1989: 4–8). Many commentators have suggested that the annual flooding of the land between the rivers Euphrates and Tigris may have been the physical manifestation that produced this myth. However, the story of the Flood may have been inspired by the marine transgression of the Gulf onto the lower Mesopotamian Plain, dated to the sixth millennium BC, after the earliest beginnings of Eridu, which resulted in a shoreline just south of this proto-urban settlement. The story of the Flood continues with the re-emergence of the land, and the rapid progradation of the coastline or marine regression (see above) could, conceivably, have been the event described in the Mesopotamian myths.

During the Uruk period an urban society emerged, with towns in lower Mesopotamia, such as Ur, Lagash, and Uruk, growing from their more modest size in the Ubaid period. The hinterlands of these towns were well-developed agricultural lands with extensive and organized irrigation systems. Many of the smaller settlements in the landscape beyond the cities seem to have been abandoned, with their former inhabitants swelling the population of the cities (Wright, in Adams 1981: 325–8). The best-known city of this period is Uruk—Warka—located on the left bank of the River Euphrates. Excavated in the middle of the nineteenth and throughout the twentieth century, this town was in use from around 4000 cal BC through to the seventh century AD. The walls held a town with monumental ceremonial architecture. The oldest archaeological evidence for writing comes from Uruk, from remains dated to the second half of the fourth millennium BC (Boehmer 1991). The 'temples' are thought to have played a role in the storage and distribution of grain, and the location of great cities such as Eridu, Uruk, Lagash, Larsa, and Ur near the rivers and coast made the efficient transport of agricultural goods possible through the use of boats. The archaeological evidence from these cities also points to the longer-distance exchange of exotic goods, presumably using boats, as shown by the presence of lapis lazuli and carnelian with a likely provenance in modern Iran or Afghanistan (Matthews 2003: 104).

The sheer size of these early cities, with large populations to feed, has encouraged the view that the lower Mesopotamian Plain was of an exceptional fertility, producing surpluses that supported the political elite. The irrigation of agricultural land through a network of canals—revealed in aerial photographs and field surveys (Adams 1981)—appears to have made urbanization possible.

The Bronze Age in Mesopotamia sees a further flourishing of large urban centres with administrative hierarchies. Ur—Tell el-Maquvyar—located some 24 km north-east of Eridu, is the key site for the first half of the third millennium BC. Investigations, including those directed by Leonard Woolley, took place here in the mid-nineteenth century and first half of the twentieth century. Ur's origin dates back to the later part of the Ubaid period, and the city continued to the end of the fourth century BC. Over this long time period, Ur was dominated by its ziggurat (Woolley 1938). It may have been the cultic centre in the second millennium BC, especially during its Third Dynasty, which ruled after the short-lived Akkadian Dynasty (Matthews 2003: 115). The use of mud bricks in the ziggurat provides a direct link with the Marshlands. Apart from the use of reed as a temper in mud bricks, woven reed mats were used to stabilize the ziggurat, with every twelfth or thirteenth mud brick course alternating with a course of reed matting (Pournelle, Hritz, and Smith 2010: 12). The volume of reeds required for this construction could only have been provided by the vast expanses of *Phragmites* stands in the Marshlands.

The relative decline of the towns on the lower Mesopotamian Plain in the second half of the third millennium BC has been attributed to the same reason identified for its success: irrigation. Sodium and magnesium from the rocks in the Anatolian and Zagros Mountains are, in relatively low concentrations, transported by the Tigris and Karun, and in much smaller concentrations by the Euphrates, to the lower Mesopotamian Plain. Here the waters are diverted through the many canals to irrigate the agricultural land. Evaporation and transpiration of the waters causes the soluble sodium and magnesium to precipitate as carbonates, and this causes both groundwater and soil salinity. Increased groundwater salinity inhibits the uptake of water by plants and impedes germination. Increased soil salinity in fine alluvial soils—such as clay—produces structureless soils that are effectively impermeable. The earliest historical records of reduced fertility as a direct consequence of increased groundwater and soil salinity date to the period 2400–1700 BC. The records refer to the southernmost part of the region, where a shift from salt-intolerant wheat to salt-tolerant barley crops was undertaken. The decline of the cities of Eridu, Uruk, Lagash, Larsa, and Ur, and the shift of power northwards, would have been an inevitable consequence of this salinity crisis (Jacobsen and Adams 1958).

This understanding of the impact of salt in Mesopotamia is not without its critics, and some have argued that leaving land fallow and leaching reduces the salt concentrations in the soil (e.g. Powell 1985). Nevertheless, the salinization theory retains broad support. It is recognized, however, that good land stewardship was also essential for long-term success (e.g. Artzy and Hillel 1988). For example, in certain areas such as the Diyala Basin, which joins the Tigris north-east of Baghdad, a sustainable irrigation practice was in operation over a period of 5,000 years, even though the canals had to be recut on many occasions to deal with the problem of siltation (Jacobsen and Adams 1958). Recent modelling suggests that leaving land fallow for sufficiently long periods holds the key to avoiding the build-up of salt in soils, and that agricultural crises may have been caused by population pressure which prevented land from being allowed to lie fallow for adequate periods of time (Altaweel and Watanabe 2012). It is also worth noting that the only sustainable solution to the problem of increased groundwater salinity is to route the water directly to the sea, something not managed in Mesopotamia before the end of the twentieth century through the Third River Project (see above).

There is little doubt that the relative status and importance of the cities on the lower Mesopotamian Plain declined after c. 2400 BC. Politically, the regional 'empires' in the second and first millennia BC had their core further north in Mesopotamia—in Babylon, Assyria, and Mitanni—or outside Mesopotamia altogether. Examples are the Kassites from beyond the Zagros Mountains, the Achaemenids in southern Iran, the Macedonian Empire in northern Greece, and the Parthians in north-eastern Iran (Matthews 2003: 143–52; see above). Whilst the advantage of being located on the edge of the Marshlands, in terms of food production, may have diminished somewhat during the second half of the third millennium BC because of salinization and siltation problems, the reasons for decline are more likely to be found in the sociopolitical sphere: whilst the major cities declined, the number of smaller settlements in lower Mesopotamia actually increased (Wright in Adams 1981: 336). The great cities located on the northern and western edges of the Marshlands, such as Lagash, all continued to be occupied. Some experienced phases of renaissance in subsequent periods, most notably Ur during the Third Dynasty towards the end of the third millennium BC, and others during the reign of the Babylonian Nebuchadnezzar I in the twelfth century BC. Surveys of lower Mesopotamia clearly show that canal systems for irrigation continued to be maintained throughout the first millennium BC and the early centuries of the first millennium AD, by which time the region was under Sassasian control. A number of new towns emerged, but rural settlement was rare at this time.

Historical texts and archaeological surveys confirm the widespread construction and maintenance of irrigation canals, but also indicate problems relating to the salinization of agricultural land throughout this period (e.g. Jacobsen and Adams 1958; Adams 1956; Wright in Adams 1981: 328–36).

The role of irrigation agriculture in lower Mesopotamia towards the end of the Sassasian rule is a matter of some debate. Some commentators, including those who have undertaken archaeological surveys, see a break in the widespread practice of irrigation agriculture, but historians have argued that this is not the case, and that a significant agricultural intensification took place in the sixth century AD (Morony 2004). According to historical data, the pattern of land use in lower Mesopotamia following the establishment of the Arab Abbasid Caliphate—AD 750 to 1258—continued to embrace irrigation agriculture, a practice which has continued through Islamic time.

In general, the success or failure of irrigation agriculture in Mesopotamia has been closely linked to the changing political context. Periods of strong central control, it has been argued, coincided with sustainable and highly productive irrigation agriculture, which included not only the maintenance of canals and sluices but also the practice of leaving fields fallow for sufficiently long periods, in order to prevent salinization. In times when strong central control was absent, irrigation canals were not adequately maintained, which limited the options for leaving land fallow. This led to salinization, reduced productivity, and a shift towards pastoral agriculture (e.g. Adams 1969; cf. Pournelle 2003: 134). Whilst such a cause-and-effect argument is oversimplifying a complex and changing situation, with diverse sociopolitical, economical, ritual, and environmental factors involved, it is most likely that political stability and effective irrigation agriculture were closely interrelated.

Jennifer Pournelle (2003) has recently suggested an alternative economic model for lower Mesopotamia, which provides a central role for the Iraqi Marshlands. In her PhD thesis, Pournelle (pp. 213–14) shows a close spatial correlation between the progradating delta and the establishment of new settlements during the late sixth, fifth, fourth, and third millennia BC. The reason for this, she argues, is the myriad opportunities that the Marshlands offered, which included grain cultivation and horticulture on the levees, fishing and fowling, extensive grazing opportunities and fodder for water buffalo, cattle, and sheep, and the cash-crop value of reed and fish. The settlement development in the Iraqi Marshlands from the sixth to the third millennium BC is described by her as 'beginning with Ubaid opportunistic dependence on littoral [or coastal] biomass which, by the Early Dynastic, had proceeded to intensive cultivation of wetland forelands as "agriculturalized" marsh zone' (p. 213).

Much of our understanding of the Mesopotamian people who lived within the Iraqi Marshlands is informed by the ethnographic studies of the Ma'dan, or Marsh Dwellers. The Ma'dan were studied during the late nineteenth and twentieth centuries by a number of travellers and ethnographers, including John Henry Haynes (cf. Ochsenschlager 2004: 251–69), Fulanain (1927), Gavin Maxwell (1957), S. M. Salim (1962), and Wilfred Thesiger (1964). The accounts and photographs left by these intrepid explorers are invaluable for any study of the Iraqi Marshlands.

However, there is very little evidence for the presence of the Ma'dan in the Iraqi Marhslands in earlier millennia. Some commentators explicitly or implicitly assume that the Ma'dan have lived in the Iraqi Marshlands ever since their formation in the sixth millennium BC (e.g. Kubba and Jamali 2011: 25). Others take it for granted that because of the similarities in environment and unchanging animal behaviour, the way of life of the Ma'dan must have been very similar to that of the Mesopotamians who lived here during the fifth and later millennia BC (e.g. Pournelle 2003). Some archaeologists, noting the evidence for the apparent continuity of aspects of material culture—for example, the architecture using arched reed bundles and the boat design—consider the possibility that the Ma'dan assimilated certain customs and practices from an indigenous people whom they encountered here in the first millennium AD (e.g. Ochsenschlager 2003). Whichever of these assumptions is correct, any extrapolation drawn from the study of the Ma'dan onto the Mesopotamians who lived in the marshes has to come with the caveat that the relationship between these peoples remains ambiguous.

Based on these ethnographic studies and more recent research, the following picture emerges of the Iraqi Marshlands in the century prior to the 1960s, when the Ba'ath regime in Iraq set out to undermine the socio-political structures of the Ma'dan and their traditional way of life. The Ma'dan were one of three self-defined groups who utilized the environment of the Iraqi Marshlands for their needs. The Ma'dan lived year-round in the Marshlands, the Beni Hasan lived on the Marshlands' edge, and the Bedouin visited the Marshlands between August and December. For the Ma'dan, the water buffalo was the most important livestock animal, providing milk and dung for fuel, despite the fact that it was never fully domesticated. Fish were caught using five-pronged fish spears from boats. Rice was cultivated on seasonal marshland. Reeds were harvested for animal fodder, for reed mats and baskets, and for the construction of guest houses, familial houses, and huts. The reed mats and baskets, along with dairy products, were sold to outsiders. The Beni Hasan cultivated a number of crops on the Marshlands' margins: wheat, barley, rice, and millet, supplemented with the produce from gardens, small herds of sheep and cattle, and flocks of chickens. Fish were caught using nets, and sold to nearby towns. The Bedouin came to the Marshlands when the floods receded, by late August, and used the grasslands as pasture for their herds of sheep and goats, alongside the camels and horses used for transport and travel. All three groups were led by hereditary sheiks, who upheld traditional norms, values, and laws. In the late 1960s, however, the authority of the sheiks was undermined by the ruling Ba'ath regime, and change accelerated with the onset of the Iraq–Iran war in 1980 (Ochsenschlager 2004: 13–33).

The traditional values and costumes of the Ma'dan were illustrated by many of the ethnographers and researchers who studied them, and compared with the archaeological evidence (e.g. Postgate 1980; Ochsenschlager 2004). The architecture, in particular, has received much attention. Each sheik had a guest house, or *mudhif*, providing room for tens of men. The *raba* combined the function of guest house and family house, and was found in smaller villages, and the single-room *bayt* was the home of ordinary families. Reed was also the main building material for the *sarifa* huts (Fig. 8.3). The construction of the *mudhif* involved columns made of the longest reeds, up to 8 m in height, bundled together and with their bases lowered in holes dug in the ground. Pairs of reed bundles were tied together to form the arches of the *mudhif* (Fig. 8.4). A series of ribs, also made of reed bundles, connected the arches along the length of the building, and these were covered with woven reed mats. A second set of ribs was then placed over the mats, securing them in place, and a latticework of reed closed off the ends, leaving a door at one end. Sometimes, the entrance was given further reed-bundle columns, which were not functional but added monumentality and a tribal identifier to the structure. The great antiquity of the reed *mudhif* appears confirmed by its detailed depiction on a limestone drinking trough from Uruk, dated to c. 3300–3000 BC and currently on display in the British Museum (Fig. 8.5) (BA registration number

Fig. 8.3. A *sarifa* at Kubaish. Photo by Wilfred Thesiger © Pitt Rivers Museum, University of Oxford.

1928, 0714.1; Hall 1929; Ochsenschalger 2004: 145–69). The trough, which also depicts sheep and lambs, is described by the British Museum as being impractical for everyday use, because the image can only be seen if the trough is elevated to a level too high for sheep to drink from. Based on the symbols

Fig. 8.4. The interior of a *mudhif.* Photo by Wilfred Thesiger © Pitt Rivers Museum, University of Oxford.

Fig. 8.5. The fourth-millennium BC limestone drinking trough from Uruk, showing a large building made of reed (see detail), and a procession of sheep and lambs. © Trustees of the British Museum.

used in the carving, the trough is believed to be a cult object from the temple of the goddess Inanna. Arched reed buildings are also known from cylinder seals and seal impressions found in excavations in Ur, Lagash, and Uruk (Kubba 2011: 31).

Boats, too, have been given ample attention by ethnographers and archaeologists. In terms of their outline shape, the boats of the Ma'dan resemble Mesopotamian model boats and depictions of boats on cylinder seals. Amongst the Ma'dan specialist boat-builder communities existed, who could be found along the Euphrates and Shatt al-Arab. The generic type of boat built by the Ma'dan is called the *mashuf*, and its canoe shape is well suited to travel through the extensive reedbeds. Different types are used for travelling through the Marshlands, for fishing, for waterfowling, and for the transport of larger cargos such as woven reed mats. The sheiks' *mashuf*, or the *tarada*, is the longest *mashuf*-type craft, and is typically over 10 m in length with greatly extended bows and sterns, and decorated with bronze round-headed nails and wooden carvings (Fig. 8.6). The one-man *mashuf* is called a *mataur*, and is particularly suited for duck hunting. The wider of the *mashuf* is the *burkasha*, which is used to transport reeds and reed mats, and the *giood* for long-distance journeys. The boats are propelled using a reed pole (*mardi*) or paddle. Sails are used only by professional sailors on the *giood*. The Ma'dan made their boats from hard wood of the acacia or mulberry tree found outside the Marshlands, since local trees such as the date palm and willow are unsuited for this. The

Fig. 8.6. The *tarada* used by Wilfred Thesiger during his travels through the Iraqi Marshlands. Photo by Wilfred Thesiger © Pitt Rivers Museum, University of Oxford.

construction of the *mashuf* deployed the frame-first technique, with the keelplank fitted with frames and the shell of smaller planks nailed to the frames. The Ma'dan covered their boats in bitumen, which was obtained in towns outside the Marshlands and was removed annually for repairs and fully recycled (Ochsenschalger 2004: 176–85; Haykal 2011: 68–70). Archaeological finds from Mesopotamia and surrounding areas include a number of model boats, usually of pottery, depictions of boats on cylinder seals and seal impressions, and lumps of bitumen, all dating back as far as the Ubaid period (e.g. Hall and Woolley 1927; Safar, Mustafa, and Lloyd 1981). The outline shape of these model boats and the boats on seals are similar to the *mashuf*, with their canoe shape and extended bow and stern. However, the impressions and details leave no doubt that reed bundles, not wooden planks, were used for the hulls. Lumps of bitumen from archaeological excavations show impressions of reed bundles and some have barnacles attached to them, attesting to the use of reed boats in marine contexts. The seaworthiness of these craft is evidenced by the distribution of Ubaid pottery of the later sixth century BC on coastal sites in the Gulf as far as the Hormuz Strait (Carter 2006).

The subsistence of the Ma'dan is principally based on fish, mainly from the carp family, and on dairy products from the milk of the water buffalo. The most important function of the water buffalo is to provide dung for fuel. Its meat is eaten only in times of acute food shortage. The Ma'dan sell fish, dairy products, and reed mats at markets outside the Marshlands, and buy grain, rice, wool, and any metal they need for tools. Despite the invocation of the biblical 'Garden of Eden' (e.g. Ochsenschlager 2004), the Iraqi Marshlands did not provide the Ma'dan with a generous diet or an easy existence, and ethnographers such as Wilfred Thesiger and also more recent commentators have recognized the harshness of life in this environment (Thesiger 1964; Kubba and Jamali 2011; Kubba 2011). The Marshlands did provide a landscape ideal for those who sought isolation or refuge from authority. The customs and values of the Ma'dan, including their traditional Shi'ite Muslim faith, remained protected from the authorities by the relative isolation of the Marshlands. This function has a long history: the marshes are mentioned as a place of hiding for the Chaldean kings from the Assyrians in the eighth century BC, as a refuge for the Zanj in the ninth century AD, and as a safe haven in the seventeenth century, when Arab tribes resisted Ottoman rule (Kubba and Jamali 2011: 10–12).

In summary, the Iraqi Marshlands have been inhabited by people ever since the Gulf transgressed onto the lower Mesopotamian Plain in the sixth millennium BC. The diversity and richness of the resources offered by the marine, estuarine, and freshwater wetlands were comparable to those found in the other case studies, but in this instance it provided the economic basis for the development of the first urban settlements. The management of water through irrigation made expansion of land for cereal cultivation possible. Nevertheless, climate change induced sea-level rise in the sixth and fifth millennia BC and created the preconditions for the proto-urban development of places such as Eridu, Uruk, Lagash, Ur, and Girsu. The coastal wetlands provided the necessary resilience for the new settlements in the form of fish and fodder for livestock—Pournelle (2003) refers to this as 'deltaic resilience'—in years when the harvest failed or periods when agricultural productivity diminished. Much has been made of the problem of salinization, but this problem was successfully dealt with during the millennia that the lower Mesopotamian cities were in existence, in some cases over 6,000 years. There is, therefore, no evidence of a collapse of the Mesopotamian civilization caused by unsustainable land use, or 'ecocide' as Jarred Diamond noted in his *Collapse: How Societies Choose to Fail or Succeed* (2005: 48, 174, and 424); rather the so-called 'rise-and-fall' of the Mesopotamian cities is one principally concerning the political and religious elites (Yoffee 2010).

The study of the Ma'dan shows us a tradition that could have survived, sustainably, for many millennia. The Ma'dan way of fishing, using the five-pronged spears rather than nets, prevented over-fishing; there are no reports of declining fish populations before the 1990s (Kubba and Salim 2011). Water buffalo provided milk for dairy products and dung for fuel. The reedbeds provided animal fodder, building materials, and the material for reed mats, which could be sold or bartered for objects and food not obtainable in the Marshlands. Life in the Al-Ahwar was never easy, but the Ma'dan utilized their

Fig. 8.7. Kubaish village life in the Iraqi Marshlands. Photo by Wilfred Thesiger © Pitt Rivers Museum, University of Oxford.

landscape in a sustainable manner (Fig. 8.7). It was deliberate political action during the 1900s that disrupted the ecosystem and this traditional way of life.

STRENGTHENING THE RESILIENCE OF COASTAL COMMUNITIES

The Iraqi Marshlands are not expected to be threatened much by sea-level rise in the twenty-first century. Modelling of a 1 m sea-level rise suggests that floods will be restricted to the coast and islands in the northern Gulf region, and along much of the Shatt al-Arab, and will not extend deep into the Marshlands (Weiss, Overpeck, and Strauss 2011). The tidal reach of the Gulf is expected to lengthen into the Marshlands, and a north-eastwards extension of saltwater and brackish water is anticipated. This may have adverse impacts on agricultural activities in the south-west parts of the Al-Ahwar, but the influx of more seawater could improve fish populations and dilute the polluted waters that come into the Marshlands.

However, it is not the impact of twenty-first-century sea-level rise on the Marshlands that is the reason for selecting this region as a case study. As noted above, this chapter presents an opportunity to study the impacts of a human-induced environmental disaster, and the subsequent attempts to restore the landscape, on the communities that live here. It is a proxy of what may happen when sea-level rise or extreme weather events disrupt the lives of coastal communities elsewhere in the world.

The traditional lives of the Ma'dan started to change in the 1960s—well before the Marshlands were drained—when the Ba'ath regime set out to bring the tribal system under its direct control by removing the political power of the sheiks (Ochsenschlager 2004: 14). The Ma'dan were not the only people targeted, and the other communities who utilized the Marshlands—the Beni Hasan and the Bedouin—experienced similar pressures. Initially, such changes followed the introduction of the 'benign' benefits of living in a modern world, such as the greater availability of goods and foodstuff, improved means of transport, schooling, and education, and better health care. Improved transport meant greater access to the urban markets; some Marshland crafts were no longer in demand by the 1970s, and skills disappeared. The pace of change quickened during the 1980s Iraq–Iran war, when many young Ma'dan were conscripted. The main changes happened after the first Gulf War, following the short-lived Shi'ite uprising against Saddam Hussein in 1991. After the uprising had been crushed, the Ma'dan retreated into the Marshlands. Even though the army could not follow them, helicopter gunships could, and the ensuing loss of life was great. Following the instigation of the UN 'no-fly zone' in 1992, the decision to drain the Marshlands was made by the Ba'ath regime, leading to the demise of all but the core of the Marshlands (Kubba 2011; see above). The impact on the Ma'dan can be illustrated by the demographic changes. The number of Ma'dan living in the Marshlands in 1988 is estimated at c. 500,000; this number fell to 192,000 in 1997, but of these only 85,000 were living in, or on the edge of, the Iraqi Marshlands (Coast 2002; Kubba 2011: 18).

In the mid-1990s, the United Nations Environmental Programme (UNEP), the AMAR International Charitable Foundation, and other NGOs alerted the world to the deliberate and politically motivated drainage of the Iraqi Marshlands by Saddam Hussein's regime. Even though the impact of the drainage could not initially be quantified, the consequences for the Marshland ecosystem were nevertheless considered disastrous (e.g. Maltby 1994; Nicholson and Clark 2002). Quantification of the exact impact of these hydrological changes on the wetland ecosystem became possible with the release of satellite imagery. This showed the adverse impact on the southern part of the marshes, from 1977. The combined extent of permanent wetland in the Hammar March, the Qurna or Central Marsh, and Hawizah March in 1977 was estimated at just over 10,000 km^2. By 2000, less than 1,500 km^2 survived as permanent marsh or open water, and in 2003 only 600 km^2 remained. It was noted that the main changes in the land cover had occurred between 1985 and 2000 (Brassington 2002: 161; UNEP 2003: 161). The Hammar Marsh, which received its water from the Euphrates, was effectively isolated from the river following the construction of embankments and with the river water being channelled into the Main Drain. The Qurna Marsh, which depended on floodwater from both the Euphrates and the Tigris, was adversely affected by the embankment of the Euphrates and the diversion of the Tigris into the Mother of Battles River. The Hawizah Marsh was affected by the diversion of floodwaters from the Tigris, but appeared to receive an increased freshwater supply from its Iranian catchment (UNEP 2003: 166).

Following the second Gulf War in 2003, UNEP and other NGOs promoted the restoration of the Iraqi Marshlands. UNEP's 'Support for Environmental Management of the Iraqi Marshlands Project' was announced in 2004. The aim of this project was 'the sustainable management and restoration of the Iraqi Marshlands by facilitating strategy formulation, monitoring marsh conditions, raising the capacity of Iraqi decision makers, and providing water, sanitations and wetland management options on a pilot basis' (UNEP 2009: 7). The work was undertaken by a large number of local communities and organizations—some set up with the specific purpose of restoring the marshes, such as the Centre for Restoration of the Iraqi Marshlands—and overseen by the Iraqi Ministries of Environment, Water Resources, and Municipalities and Public Works. The funding for UNEP's role came from donations by the governments of Japan and Italy, and with the USA and Canada providing financial support for other elements of the restoration of the Iraqi Marshlands.

The restoration of the marshes was hampered by the polluted water found in many of the recently constructed drains and outflows. The Main Drain, for example, which drained wastewater from Baghdad and surrounding areas, was unsuitable for use in wetland restoration. Improvement in the quality of water coming into the marshes from the rivers and the constructed drains became, therefore, a prerequisite to any attempts to reflood the land. Reeds and other wetland vegetation were utilized at a number of pilot sites, and the capacity of these plants to achieve significant reductions in the concentration of suspended solids, nitrates, and phosphates was demonstrated (UNEP 2009: 36). The UNEP project included the creation of clean drinking water and sanitation provision, which were essential in convincing the local residents and former inhabitants that life in the Iraqi Marshlands was a genuine option.

The initial results from the restoration project were promising. The rapid expansion of the area covered with wetland vegetation was noted, along with the return of a range of fish species and wetland macrophytes such as algae, mosses, and ferns. The reasons for this early success, however, had little to do with the restoration projects. Very large amounts of water held behind dams and in reservoirs on the Tigris and Euphrates in Iraq were released during and in the immediate aftermath of the second Gulf War, when field stations were abandoned. This release of vast amounts of water flooded the Marshlands. The floods were intensified by the destruction of many dams and sluices by local communities seeking to undo the damage to the Marshlands. In 2005, when the reservoirs on the major rivers were in the process of being refilled, the region experienced a relative 'wet' year, following heavy snowfall in the Anatolian mountains (Warner, Douabul, and Abaychi 2011: 218). Significant improvements in the water quality of the rivers and canals draining into the Marshlands were observed in these years, as the pollution was diluted. The extent of marsh vegetation identified on satellite images shows that the Marshlands expanded from c. 600 km^2 in January 2003 to nearly 4,000 km^2 by July 2006 (UNEP 2009: 55).

However, the subsequent years have seen a reversal in the Marshlands' fortunes. The ongoing construction of dams and reservoirs on the rivers Euphrates, Tigris, and Karun in Turkey, Syria, Iraq, and Iran has significantly reduced the flow of fresh water into the marshes. This has effectively removed the seasonal flood-pulses caused by the natural spring peak flows. Droughts in 2007 and 2008 exacerbated this situation. Consequently, by the end of 2009, the Iraqi Marshlands had shrunk to less than 3,000 km^2, and have since continued to deteriorate and shrink at a significant pace (Warner, Douabul, and Abaychi 2011: 218).

In the context of the increasing control of water in the upper and middle reaches of the rivers Euphrates, Tigris, and Karun—including the ambitious

South-eastern Anatolia Project—the Iraqi Marshlands have no sustainable future. An international agreement on the release of significant amounts of water from the dams on the Tigris and Euphrates, which is the single most important prerequisite for the restoration of the Al-Ahwar, is highly unlikely in the current political climate. Climate change in the twenty-first century will make these freshwater resources even scarcer as communities along the length of the river courses continue to extract large amounts of water. With a growing human population, it seems increasingly unlikely that the pleas from the remaining Ma'dan to protect their Marshlands will be heeded.

Nevertheless, what sustains the Ma'dan, the Iraqi government, and international NGOs in their attempts to restore, or rehabilitate, the wetlands is their acute sense of place as regards the Iraqi Marshlands. The motivations are different for each of these groups. The sense of place of the Ma'dan has its roots in the memories of the older generation, and in a sense of tradition; despite the fact that younger generations do not always share this sense of place, the Marshlands are seen as a distinctive political and religious community, with a history that goes back some 1,500 years or more. The importance attributed to the Al-Ahwar by the Iraqi government is additionally influenced by concepts of national heritage. In particular the Marshlands are seen as the landscape context of the famous Mesopotamian cities. The NGOs seeking to help restore the Marshlands share a sense of place that views the Marshlands as the biblical Garden of Eden: some NGOs use the concept as a metaphor to stress the great age, biodiversity, and ecological richness of the Marshlands, whilst others believe the Marshlands to be the actual location of the Garden (e.g. Richardson and Hussain 2006; Al-Asadi 2011; Al-Asadi, Knutsson, and Ali 2012).

These stakeholders have joined forces in their ongoing attempts to restore the Iraqi Marshlands. Wholesale restoration is no longer a realistic option, but a core or cores of the marshes could be restored, providing biodiverse ecosystems and landscapes for the remaining Ma'dan who want to live their traditional lives. The sustainability of this community could be strengthened by developing 'green tourism', aimed at visitors interested in the culture of the Ma'dan and in the wetland flora and fauna. Plans for a museum have already been drawn up, and a combination of a restored ecosystem and cultural heritage could be an attractive offering. Tourists could be charged premium prices for traditional crafts and food, and for overnight stays in traditional reed buildings (Al-Asadi 2011: 264–8).

How can this case study indicate how the social resilience of the Ma'dan and other stakeholders might be strengthened? First, although the importance of understanding the long-term geological, geographical, and environmental history of coastal wetlands development remains undiminished, this study has shown the value of including the impact of recent changes brought about by human action. The long-term natural history explains the processes that produced the Iraqi Marshlands as a unique ecosystem through the interplay of sea-level rise, riverine progradation, and seasonal floods. However, the construction of dams on the Tigris and Euphrates rivers has changed the region's hydrology fundamentally. If the Marshlands were to be restored to their former glory, then the spring flood-pulse would need to be reinstated. It has been suggested that as much as 75 per cent of the Marshlands could be restored if the natural hydrological regime was re-established (Al-Asadi, Knutsson, and Ali 2012). The original wetland flora would naturally emerge because the seedbanks have proven to be highly resilient, as shown in the early restoration success in the period 2003–6 (Richardson and Hussain 2006). The full restoration of the Al-Ahwar is, however, an improbable scenario. This is partly because the dams are constructed with the aim of providing an even discharge throughout the year, so that the freshwater resources can be optimized for consumption, irrigation, and the generation of hydroelectricity. Furthermore, the dams limit the downstream transport of sediments, curbing the process of progradation. Thus, the long-term perspective and an understanding of current issues show that the large-scale and permanent restoration of the Al-Ahwar is not possible, at least not without affecting the livelihoods of communities in the upper and middle reaches of the Tigris and Euphrates rivers.

Second, the adaptive pathways from the distant and not so distant past provide us with a number of examples of successes and failures of sustainable living that could provide guidance for the future. For the area outside the Iraqi Marshland proper, the principal concern is the necessity of allowing sufficient time for land to lie fallow, in order to avoid the problem of salinization. Water from the rivers Tigris and Euphrates has supported irrigation agriculture for over 7,000 years, but careful management is required to keep productivity high. In the context of growing populations and climate change, it seems highly likely that fresh water will become an increasingly contested resource. The managed and careful use of irrigation land in ancient Mesopotamia provides an antecedent for managed and careful use of fresh water through international cooperation in the twenty-first century. How the Iraqi Marshlands proper were utilized in the past provides an important example of living sustainably in a wetland environment. There is no evidence for a degradation of the flora or fauna in this ecosystem before the 1990s.

Third, the power of a sense of place is strong both amongst the Ma'dan and within the Iraqi government, and equally among those external organizations that seek to support them in their desire to return to the Al-Ahwar. Although each group has a different point of view on what makes the Marshlands important from a historical or heritage perspective, without this sense of place there would have been no attempts to restore the wetlands, and the Marshlands would have been lost.

CONCLUSION

The origin of the Iraqi Marshlands lies in the sixth millennium BC, when sea-level rise caused the transgression of the Gulf onto the lower Mesopotamian Plain, creating a shallow sea. Around 5000 cal BC, the coastline regressed as a result of the process of progradation, caused by the deposition of sediments by the rivers Tigris, Euphrates, and Karun. A deceleration of the ESL rise in the late Holocene, combined with the ongoing progradation, meant that during the following millennia the freshwater wetlands moved increasingly towards modern Basra. Further movement of the Marshlands was prevented by the existence of two very large alluvial fans near Basra.

Archaeological research has shown the interconnections between the natural development from the sixth millennium BC of the resource-rich landscape in the lower Mesopotamian Plain, and the emergence of some of the world's oldest cities. Whilst irrigation agriculture was a critical aspect in the emergence of the proto-urban settlements such as Eridu, Uruk, Lagash, Ur, and Girsu, the delta and Marshlands provided important additional foodstuff in the form of fish and reed as animal fodder, and essential resilience in times when the grain harvest was poor. They also provided the reed mats used in the monumental architecture of these cities. Additionally, archaeological research has shown the close spatial correlation between the location of the delta and the establishment of new settlements in lower Mesopotamia. The Ma'dan, who were living in the Marshlands probably as early as the middle of the first millennium AD, developed a sustainable lifestyle based on fishing and water buffalo, and the sale of reed and reed products to the cities outside the Al-Ahwar.

The study of the lower Mesopotamian Plain has shown not only how climate change and its impact on the environment provided the basis for the development of the first cities, but also how communities could live, sustainably, within the Marshlands themselves. These adaptive pathways from the past have already been used in building the resilience of communities in the present, and despite the significant challenges that will have to be overcome if the Marshlands, or parts thereof, are to be restored, without this historical antecedent and a strong sense of place, the Marshlands would have been lost forever. On this basis, the desire in Iraq for a World Heritage Site designation for the Iraqi Marshlands (John Curtis, personal communication) is more than justified.

9

Conclusions

To me, as a townsman, the tide country's jungle was an emptiness, a place where time stood still. I saw now that this was an illusion, that exactly the opposite was true. What was happening here, I realized, was that the wheel of time was spinning too fast to be seen. In other places, it took decades, even centuries, for a river to change course; it took an epoch for an island to appear. But here, in the tide country, transformation is the rule of life: rivers stray from week to week, and islands are made and unmade in days. In other places, forests take centuries, even millennia, to regenerate; but mangroves can recolonize a denuded island in ten to fifteen years. Could it be that the very rhythms of the earth were quickened here so that they unfolded at an accelerated pace? *(Amitav Ghosh 2004: 224, on the Sundarbans)*

INTRODUCTION

Climate change archaeology views the past as a repository of adaptive pathways, with the aim of gathering ideas and concepts that can help build the socioecological resilience of communities in a time of rapid climate change. Research into the climate of the Earth's past provides the scientific basis of our understanding of how climate will change in the twenty-first century. However, when it comes to understanding the impact of climate change on the environment, and how this will affect human communities, the past is usually ignored by the climate change science community. In addressing this paradox, and through this study, I have sought to reposition archaeological research within current climate change debates, and to give archaeologists a voice that has relevance for the present world. The decision to focus the study on coastal wetlands was based, primarily, on the realization that these landscapes, and the communities that live here, are amongst the first that will be affected significantly by climate change-induced sea-level rise.

This study has focused on the significance of strengthening the socioecological resilience of coastal communities through the study of their archaeological and historical pasts. Socioecological resilience is defined as the ability of communities to cope with the impacts of external environmental change, without losing the characteristics that define the distinctiveness and values of that community (see above, chapters 1 and 2). A socioecological resilience that is connected to the past will be stronger, as well as being more relevant, than one without such a connection. It will be stronger because communities' customs, values, beliefs, traditions, technologies, and histories define the way in which they deal with external changes. It is more relevant because a socioecological resilience that is disconnected from the past effectively disregards what makes communities distinctive and meaningful.

This study has sought to describe the long-term geological, geographical, and environmental processes that formed the coastal wetlands in the past and will determine their sustainability in the near future. How coastal communities dealt with these changes has been termed 'adaptive pathways'. The concept of adaptive pathways recognizes that past communities had to deal with climate change and its environmental impacts, including sea-level change, but that they also sought to change the landscape for economic, sociopolitical, or religious reasons, all of which impacted on the environment. Thus, although the concept of adaptive pathways does not presuppose that changes in coastal communities were caused or determined by environmental change alone, it accepts that climate change and environmental change had—and continue to have—an undeniable impact on communities, past and present. The concept also recognizes that many changes emerge from within the communities themselves, or through interaction with other groups, for reasons completely unconnected with environmental change.

This book has employed a comparative approach, involving four case studies: the North Sea, the Sundarbans, the wetlands of Florida's Gulf Coast, and the Iraqi Marshlands. Comparative studies in archaeology are, however, uncommon outside neo-evolutionary approaches. The reason for this is that most modern archaeologists do not consider the comparative approach a fruitful or even valid exercise, because it denies the historical-specific context in which past societies developed. I concur. The purpose of this comparative study was not to seek high-level generalizations pertaining to social structure or complexity, nor to seek a single (environmental) explanation for why coastal communities change. Instead, the comparative approach was adopted in order to explore the *diversity* of environmental histories, adaptive pathways, and modern responses to the impacts of climate and sea-level change. Whilst climate change is a global phenomenon that affected and will affect everyone living by the sea, its impact on coastal wetlands and coastal communities varies greatly by region, by wetland type, and by community.

A COMPARATIVE OVERVIEW

The findings from this research can be summarized by answering four questions.

Question 1: Is a long-term understanding of sea-level–coast interaction applied in coastal management?

The importance of understanding in detail the long-term perspective of past development and future sustainability in the coastal wetlands under study cannot be exaggerated. Sustainability can only be achieved once the natural processes have been taken fully into account, but in their planning for a future with an accelerating Eustatic Sea Level (ESL) rise many stakeholders seem unaware of these. Wetlands such as mangrove swamps and salt marshes can cope under natural circumstances with the predicted rise in sea level. However, few coastal wetlands in any of the four case studies exist in their natural context: sedimentation budgets have been reduced and much of the accommodation space removed, which will result in many coastal wetlands changing or drowning.

In and around the North Sea, something of a paradigm shift is taking place. Realizing the interconnectedness of the long-term geological and environmental coastal processes, the practice of constructing hard defences of ever greater size and height has been challenged, and a range of softer or hybrid solutions, such as managed realignment of the defences or tidal barrages, is increasingly identified as the preferred alternative. This development has emerged from the political context of the European Union and has been enforced by legal frameworks such as the European Union Habitats directive of 1992. At international and national levels, the Integrated Coastal Zone Management (ICZM) approach has become the norm, shown for example in the transnational cooperation on the Waddensea. The ICZM approach brings together the various stakeholders, and whilst this in itself is not a remedy for overcoming all differences, it does encourage the discussion of sustainable solutions that reflect the detailed understanding of long-term sea-level–coast interaction.

Such an integrated approach to coastal zone management does not exist in the Sundarbans. The ongoing dispute between Bangladesh and the Indian state of West Bengal over the Farakka Barrage not only thwarts international cooperation but also precludes any attempt to apply the emerging long-term understanding of sea-level–coastal interaction to coastal management. This is illustrated in WWF-India's (2011) plan for the Sundarbans, which ignores altogether the long-term geological and environmental perspectives and the interconnectedness of the Sundarbans as an ecosystem across the Indian–Bangladeshi border. Meanwhile coastal communities continue to be at risk from flooding, and the long-term perspective detailed in this study indicates that the risk of erosion and flooding will increase significantly in the twenty-first century.

On Florida's Gulf Coast, the application of the understanding of the long-term sea-level–coast interaction, and its use within an ICZM approach, has been frustrated by the primacy of private land ownership and the right to develop one's property with only limited regard for environmental impacts. The situation is aggravated by the political non-acceptance of the findings from climate change science, along with the ongoing disagreements amongst coastal managers, researchers, and politicians on the impact of climate change on sea-level rise. As a consequence, coastal defence schemes are often not sustainable and do not address the concerns of all stakeholders. This is most appropriately illustrated in the increasing requirement for beach nourishment projects. These are instigated to ensure that beaches and beach barriers are not degraded to the point that they fail to function as coastal defences. The main source of sand, however, from the submerged part of the Florida Plateau, is rapidly diminishing; additionally, the grain size is different from that of aeolian blown sand, leading to a degeneration of the coast's ecological value.

In the case of the Iraqi Marshlands, the understanding of long-term geological and environmental processes is fully integrated in post-Ba'ath Iraq. The restoration of the Marshlands involves all stakeholder groups, brought together when the wetlands were on the brink of destruction. However, despite the fact that the seedbank proved to be resilient to the impact of drainage in the 1990s, and that the natural *Phragmites* stands are demonstrably effective in their ability to clean polluted water, the recent changes to the hydrogeology of the wider region makes a successful outcome of the restoration project doubtful.

Question 2: Is an understanding of past successes and failures applied in coastal management?

Since the Last Glacial Maximum about 21,000 years ago, human communities have developed a number of adaptive pathways in dealing with the impact of sea-level rise. The most common response has been to migrate to higher and drier lands. This was the only option available to communities living on the North Sea Plain, the Florida Plateau, and the lower Mesopotamian Plain, where the rising sea level submerged large tracts of land. However, because marine resources such as fish and shellfish continued to draw communities to the coasts, adaptive pathways had to be revised to make this possible. The most successful adaptive pathways in the middle and later parts of the Holocene were those involving the development of coastal settlements that were safe from flooding, but which did not seek to disrupt the dynamic interaction between sea-level rise and coastal development. Examples include the *terps* of the North Sea, the mud-walled towns of the Sundarbans and the shell rings, shell mounds, and shell complexes of Florida. On the lower Mesopotamian Plain, the early towns rose above the land in the process of tell formation, making them safe from the annual floods, whilst the Ma'dan moved their houses—which could be easily dismantled and reassembled—in times of flood.

Adaptive pathways that failed have been observed in all regions and tend to be of more recent age. Unsurprisingly, they reflect solutions that are the opposite of the successful adaptive pathways. In the North Sea, for example, failures include the degeneration of the marine and intertidal ecosystem through the combined impacts of over-fishing and the effective separation of the sea from the land, following the construction of continuous dikes. In the Sundarbans, the reclamation of the coastal wetlands over the last three centuries included the construction of bunds around the islands, and a clearing of the mangrove swamps to make way for agriculture and settlements. As a direct consequence of this activity the natural ability of the islands to respond and adjust to rising sea levels has been weakened, and floods are now an annual occurrence involving significant loss of life. In Florida, the ecological degeneration of the Everglades provides a clear warning that coastal wetlands cannot simply be engineered; this is also the message from the Iraqi Marshlands.

Coastal management in the North Sea has only just begun to seek to apply an understanding of past successes and failures, most notably through the shift from hard defences to softer or hybrid solutions (see above). Continuous dikes have been constructed here for over 1,000 years and it is only recently that their adverse impacts on the marine and intertidal ecosystems have become apparent. Furthermore, in a time when sea level rises faster, the use of soft defences and coastal wetlands that can adjust autogenically are not only a more cost-effective solution but can also provide habitats for migrating birds, as well as helping to restore the natural ecosystems.

In the Sundarbans, an understanding of past successes and failures is not applied in coastal management. The rapidly growing population in India and Bangladesh, and the political movement giving underprivileged groups the right to land and a homestead discourage the concentration of the rural population in centres away from the islands' edges; the restoration of the mangrove forests is therefore prevented. There are early warnings of over-fishing, and of the adverse impact of aquaculture on the environment.

In Florida, little or no evidence exists for the application of past successes and failures in coastal management. Despite the recent history of environmental disaster in the Everglades, and the clearly unsustainable nature of beach nourishment (see above), politicians, developers, and the public continue to put faith in the ability of the US Army Corps of Engineers and the State of Florida to keep the sea from flooding the land or eroding the beach barriers. However, a notable exception should be made for the communities of

Seminole Indians in the Everglades, who retain many aspects of living sustainably in a wetland landscape.

In the Iraqi Marshlands, the restoration of the Marshlands has become a cause célèbre for a range of stakeholders. The devastating human and environmental impacts of the deliberate drainage of 1990s made many people realize that a future without the Marshlands was not a desirable one. The traditional way of life of the Ma'dan, held up as a paragon of how to live sustainably in this wetland landscape, has become symbolic of a highly successful adaptive pathway to environmental change. Much has been learnt here from past failures.

Question 3: Is attention given to communities' sense of place in coastal management?

The sense of place among coastal communities is an essential component of their socioecological resilience. After all, socioecological resilience has been defined as the ability of communities to adapt to environmental changes without losing the characteristics that define the distinctiveness and values of each community. Recognition of local communities' sense of place is also important for combating climate change itself. The scientific findings of the IPCC, and the protocols of the United Nations Framework Convention on Climate Change (UNFCCC), are frequently contested or resisted not least because of their very high level of abstraction, both geographically and temporally. The local experience of climate change can, in fact, be the opposite of what has been projected in the various SRES scenarios. For communities to 'think global, act local', such abstractions need to be expressed in terms of the challenges that communities will face, including an estimation of how climate change will affect their sense of place.

Around the North Sea—where the self-reflected social identity is frequently defined by the closeness to, dependence on, or fight against the sea—communities' sense of place is well defined, and continues to play an important part in developing adaptive pathways. Monuments that keep the social memory of great engineering feats and great flood disasters alive can be found in many places. Despite the risks associated with living below the sea level, and the confidence placed in strong dikes, communities value their physical connection to the sea. The relationship is important for general well-being and the (tourism) economy, whilst increased connectivity—through softer coastal management solutions and tidal barrages—supports healthier marine and estuarine ecosystems. The sense of place of coastal communities, and an acknowledgement of its significance have helped to build the social capital that has made the inclusive ICZM approach possible.

In the Sundarbans, the continued veneration of Bonbibi and the persisting values of the forests in the *punthi* literature show that some elements of the traditional sense of place survive. However, the hunger for land has put this traditional belief system under severe pressure, and there is little evidence that everyday coastal management takes much account of the sense of place of local communities. Institutions based outside the Sundarbans have sought, and continue to seek, the exclusion of local people from a large number of islands for the purpose of nature conservation. This exclusion fails to respect the traditional way of life or sense of place of local communities, who cherished the forests as a vital resource for woodcutters, honey collectors, salt makers, and forest fishers, especially in times of hardship.

On Florida's Gulf Coast, there is no evidence that coastal communities' traditional sense of place plays a role in coastal management, again with the notable exception of the Seminole reserves. One could suggest that a new sense of place is emerging amongst the modern Floridians and the 'snowbirds' or retirees who overwinter on Florida's Gulf Coast. This is a sense of place that values the sun and the magnificence of the coast, but one that has few roots in the past and is definitely not sustainable in a time when the ESL rise is accelerating.

In the Iraqi Marshlands the sense of place of the Ma'dan is now highly regarded. Having experienced the devastating impacts of the drainage of the Marshlands, the traditions of the Ma'dan have become a byword for living sustainably in a fragile wetland landscape. National and international institutions work with the Ma'dan to provide basic sanitary provisions such as clean drinking water, but this is never intended to undermine the traditions and sense of place.

Question 4: Are logistical solutions from the past adopted in coastal management?

In this study, not many logistical solutions from the past that could be adopted in coastal management in the present have been identified. This is, principally, a direct consequence of the population growth in recent times, which places such a high premium on land by the sea that the more expansive solutions from the past are no longer options. Nevertheless, one particular concept deserves further attention.

Intertidal zones are resource-rich areas, and this was very well understood in the past. In the North Sea, salt marshes were used as grazing ground from at least 1500 cal BC, and between 500 cal BC and AD 1000 specialist communities lived on the *terps*, optimizing the salt marshes as grazing grounds. In the Sundarbans, the intertidal sandbanks that emerged in the delta were used successfully for rice cultivation, with salt-tolerant species such as *boro* rice being selected. All intertidal waters are rich in fish and shellfish because of the very high levels of nutrients that supply the food web, released by the action of the tides; this has attracted many communities to settle by the sea, most extensively on Florida's Gulf Coast. Intertidal zones also provide materials for building shelters, and there is no better example of this than the reed-built houses in the Iraqi Marshlands.

Around the world, coastal wetlands are being recognized for their value in nature conservation and coastal protection, but this frequently seems to involve the exclusion of people. For example, the majority of managed realignment projects in the North Sea and in the newly created forest reserves on tidal islands in the Sundarbans have no place for local communities, despite the fact that the symbiotic relationship of communities with these landscapes stretches back millennia. In the case of the salt marshes of the North Sea, it has been shown that their managed utilization as pasture increased their biodiversity. In the Sundarbans, where people entered the forest for many centuries, there is no evidence for resource depletion of the forests or of the decline in numbers of the Royal Bengal tiger. In the Iraqi Marshlands, the harvesting of reed and reedmace helps to create a diverse mosaic of wetlands within the marshes, which increases biodiversity. Thus, one practical solution from the past is to allow coastal communities to benefit economically from coastal wetlands. Such a solution would not only help to reconnect a broader diversity of stakeholders with sustainable coastal wetland management but would also deliver greater biodiversity.

The results of this comparative overview have been summarized in Fig. 9.1. Three cross-cutting findings are revealed.

First, this comparative study has shown that Nicholas Stern's warning—that the poorest coastal communities are likely to have the least resilience to sea-level change, as adaptation is often a costly matter (Stern 2006; see chapter 1)—is wholly justified. The coastal communities in the Sundarbans

	North Sea	Sundarbans	Florida	Iraqi Marshlands
Is long-term understanding of sea-level–coast interaction applied in coastal management?	yes, or frequently	no, or rarely	no, or rarely	yes, or frequently
Is an understanding of past successes and failures applied in coastal management?	sometimes	sometimes	no, or rarely	yes, or frequently
Is attention given to the sense of place in coastal management?	yes, or frequently	sometimes	no, or rarely	yes, or frequently
Are logistical solutions adopted in coastal management?	sometimes	no, or rarely	no, or rarely	yes, or frequently

(Legend: dark shading = yes, or frequently; light shading = sometimes; blank = no, or rarely)

Fig. 9.1. A summary of the main findings of the comparative study involving the North Sea, Sundarbans, Florida's Gulf Coast, and the Iraqi Marshlands.

are already experiencing extensive floods, with loss of life, displacement, and loss of livelihoods an annual occurrence. There is little prospect of a plan that will safeguard their future. In contrast, the relative wealth of institutions in and around the North Sea and on Florida's Gulf Coast has ensured that costly engineering works—ranging from the construction of dikes and barriers to managed realignment and beach nourishment—have staved off the short-term impacts of sea-level rise.

Second, the value of collaboration and the adoption of an ICZM approach to coastal change has been highlighted. In and around the North Sea, this has become a standard way of working. Here, collaboration extends across national borders and includes a great diversity of stakeholders who will have to recognize each others' rights and responsibilities. This collaborative process strengthens the adaptive pathway of coastal communities and builds their socioeconomic resilience. In contrast, the coastal management of Florida is of a piecemeal and fragmented nature. The quality of the relationship between private ownership and public responsibilities has been recognized as a key determinant in successful and sustainable management of coastal wetlands (e.g. Adger and Luttrell 2000), but on Florida's Gulf Coast a good working relationship is largely absent.

Third, the case of the Iraqi Marshlands was included in this study to provide an example of a post-environmental disaster. Whilst the Marshlands were deliberately drained for political reasons, their loss, and the loss of the traditional way of life of the Ma'dan, was recognized as environmentally and humanely insupportable. The cost of restoration of the Al-Ahwar to date has been high, and its longer-term prospects are not good when the regional changes in hydrogeology are taken into consideration. The lesson to learn from this is to acknowledge that the proactive protection of coastal wetlands such as salt marshes, mangrove swamps, and reedbeds while they still exist should always be the preferred option; once they are destroyed the cost of restoration will be much greater, assuming it is possible at all.

SOME FINAL THOUGHTS

Through developing the concept of climate change archaeology, and through this study, I hope to have given archaeologists an example of how we can have a stronger voice within current climate change debates. I hope also to have made archaeology a little bit more relevant for the challenges we face in the twenty-first century. I do not expect that archaeological research will significantly alter the findings from climate change science, nor necessarily provide big solutions for the problems we will encounter. But there is definitely a role for archaeology in climate change debates. This role could be to communicate the implications of climate change at the local and regional levels, and explain what such change has meant and will mean for flora, fauna, familiar landscapes, and communities. It could be to show the long-term perspectives on the dynamic nature of sea–coastal interaction, and make the point more forcefully that sustainability can only be achieved by understanding the past as well as looking to the future. Or it could be by telling the stories of the courage and inventiveness of the generations that went before us, who had to find adaptive pathways that allowed them to deal with climate change and its impacts on the environment. In these ways, we can strengthen the socioecological resilience of communities in a time of rapid climate change.

Ultimately, there is no doubt that climate change in the twenty-first century will cause the ESL to rise faster than has been the case for the last 6,000 years, and that this will have far-reaching consequences for the world's coastal wetlands. It is not at all clear how humanity will respond to climate change and to its impact on coastal communities. Nevertheless, it is undoubtedly the case that—to paraphrase Amitav Ghosh's quote at the beginning of this chapter—the very rhythms of the world's coastal wetlands are quickening and they will unfold at an accelerated pace.

Bibliography

Adams, P. N., Opdyke, N. D., and Jaeger, J. M. (2010). 'Isostatic uplift driven by karstification and sea-level oscillation: modeling landscape evolution in north Florida'. *Geology* 38: 531–4.

Adams, R. McC. (1965). *Land behind Baghdad: A History of Settlement on the Diyala Plains.* Chicago, IL: University of Chicago Press.

—— (1969). *The Evolution of Urban Society.* Chicago, IL: Aldine.

—— (1981). *Heartland of Cities: Surveys of Ancient Settlement and Land Use on the Central Floodplain of the Euphrates.* Chicago, IL: Chicago University Press.

—— and Nissen, H. J. (1972). *The Uruk Countryside: The Natural Setting of Urban Societies.* Chicago, IL: University of Chicago Press.

Adger, W. N., Dessai, S., Goulden, M., Hulme, M., Lorenzoni, I., Nelson, D. R., Naess, L. O., Wolf, J., and Wreford, A. (2009). 'Are there social limits to adaptation to climate change?'. *Climate Change* 93: 335–54.

——, Hughes, T. P., Folke, C., Carpenter, S. R., and Rockström, J. (2005). 'Social-ecological resilience to coastal disasters'. *Science* 309: 1,036–9.

—— and Luttrell, C. (2000). 'Property rights and the utilisation of wetlands'. *Ecological Economics* 35: 75–89.

Adhémar, J. A. (1842). *Révolutions de la Mer.* Paris: Carilian-Goeury et Dalmont.

Agassiz, L. (1840). *Études sur les Glaciers.* Neuchâtel: Jent et Gassmann.

Al-Ameri, T. K. and Jassim, S. Y. (2011). 'Environmental changes in the wetlands of southern Iraq based in palynological studies'. *Arabian Journal of Geoscience* 4: 443–61.

Al-Asadi, J. (2011). 'Future of the Marshlands: a return to Eden?', in S. A. A. Kubba (ed.) (2011), *The Iraqi Marshlands and the Marsh Arabs: The Ma'dan, their Culture and the Environment.* Reading: Ithaca Press, 247–69.

——, Knutsson, S., and Ali, A. A. (2012). 'Restoring the Garden of Eden, Iraq'. *Journal of Earth Sciences and Geotechnical Engineering* 2: 53–88.

Algaze, G. (2001). 'Initial social complexity in southwestern Asia: the Mesopotamian advantage'. *Current Anthropology* 42: 199–233.

Allen, J. R. L. (1990). 'Salt-marsh growth and stratification: a numerical model with special reference to the Severn Estuary, southwest Britain'. *Marine Geology* 95: 77–9.

Allison, M. A., Khan, S. R., Goodbred Jr, S. L., and Kuehl, S. A. (2003). 'Stratigraphic evolution of the late Holocene Ganges–Brahmaputra lower delta plain'. *Sedimentary Geology* 115: 317–42.

Alongi, D. M. (2008). 'Mangrove forests: resilience, protection from tsunamis, and responses to global climate change'. *Estuarine, Coastal, and Shelf Science* 76: 1–13.

Alt, D. and Brooks, H. K. (1965). 'Age of the Florida marine terraces'. *Journal of Geology* 73: 406–11.

Altaweel, M. and Watanabe, C. E. (2012). 'Assessing the resilience of irrigation agriculture: applying a social-ecological model for understanding the mitigation of salinization'. *Journal of Archaeological Science* 39: 1,160–71.

Anthony, A., Atwood, J., August, P., Byron, C., Cobb, S., Foster, C. Fry, C. Gold, A., Hagos, K., Heffner, L., Kellogg, D., Q. Lellis-Dibble, K., Opaluch, J. J. Oviatt, C., Pfeiffer-Herbert, A., Rohr, N., Smith, L., Smythe, T., Swift, J., and Vinhateiro, N. (2009). 'Coastal lagoons and climate change: ecological and social ramifications in U.S. Atlantic and Gulf coast ecosystems'. *Ecology and Society* 14: 8.

Aqrawi, A. A. M. (2001). 'Stratigraphic signatures of climatic change during the Holocene evolution of the Tigris-Euphrates delta, lower Mesopotamia'. *Global and Planetary Change* 28: 267–83.

Archaeological Survey of India (1956–7). 'Early historical sites in West Bengal'. *Indian Archaeology: A Review 1956–7*: 80–1.

—— (2000–1). 'Excavations at Chandraketugarh, North 24 Parganas'. *Indian Archaeology: A Review 2000–1*: 156–60.

Arrhenius, S. (1896). 'Über den Einfluss des atmosphärischen Kohlensäurengehalts auf die Temperatur der Erdoberfläche'. *Proceedings of the Royal Swedish Academy of Science* 22: 1–101.

Artzy, M. and Hillel, D. (1988). 'A defence of the theory of progressive soil salinization in ancient southern Mesopotamia'. *Geoarchaeology* 3: 235–8.

Atwater, B. F. (1987). 'Evidence for great Holocene earthquakes along the outer coast of Washington State'. *Science* 236: 942–4.

Ault, J. S., Smith, S. G., and Bohnsack, J. A. (2005). 'Evaluation of average length as an estimator of exploitation status for the Florida coral-reef fish community'. *ICES Journal of Marine Science* 62: 417–23.

Baart, F., Van Koningsveld, M., and Stive, M. J. F. (2012). 'Trends in sea-level trend analysis'. *Journal of Coastal Research* 28: 311–15.

Baker, S. (2006). *Sustainable Development*. London: Routledge.

Bakker, J. P., De Leeuw, J., Dijkema, K. S., Leendertse, P. C., Prins, H. H. T., and Rozema, J. (1993). 'Salt marshes along the coast of the Netherlands'. *Hydrobiologia* 265: 73–95.

Bar-Yosef, O. and Meadow, R. H. (1995). 'The origins of agriculture in the Near East', in T. D. Price and A. B. Gebauer (eds.), *Last Hunters, First Farmers: New Perspectives on the Prehistoric Transition to Agriculture*. Santa Fe, NM: School of American Research Press, 39–94.

Barange, M. and Perry, R. I. (2009). 'Physical and ecological impacts of climate change relevant to marine and inland capture fisheries and aquaculture', in K. Cochrane, C. De Young, D. Soto, and T. Bahri (eds.) (2009), *Climate Change Implications for Fisheries and Aquaculture: Overview of Current Scientific Knowledge*. Rome: FAO Fisheries and Aquaculture Technical Paper 530, 7–106.

Barriopedro, D., Fischer, E. M., Luterbacher, J., Trigo, R. M., and Garcia-Herrera, R. (2011). 'The hot summer of 2010: redrawing the temperature record map of Europe'. *Science* 332: 220–4.

Barui, N. C. (2011). 'Floral diversity of mangrove plants with reference to the palaeoenvironment during the Holocene in Bengal Basin, India'. *Quaternary International* 229: 89–93.

—— and Chanda, S. (1992). 'Late Quaternary pollen analysis in relation to dating, vegetational history and biostratigraphy of Calcutta peat'. *Proceedings of the Indian National Science Academy*, B 58: 191–200.

Bazelmans, J., Gerrets, D., and Vos, P. (1998). 'Zoden aan de dijk; kleinschalige dijkbouw in Friesland van de Romeinse tijd'. *Noorderbreedte* 6: 1.

Beech, M., Cuttler, R., Moscrop, D., Kallweit, H., and Martin, J. (2005). 'New evidence for the Neolithic settlement of Marawah Island, Abu Dhabi, United Arab Emirates'. *Proceedings of the Seminar for Arabian Studies* 35: 37–56.

Beekman, F. (2007). *De Kop van Schouwen onder het Zand.* Utrecht: Matrijs.

Beets, D. J. and Van der Spek, A. F. J. (2000). 'The Holocene evolution of the barrier and the back-barrier basins of Belgium and the Netherlands as a function of late Weichselian morphology, relative sea-level rise and sediment supply'. *Geologie en Mijnbouw/Netherlands Journal of Geosciences* 79: 3–16.

Behre, K.-E. (2003). 'Coastal development, sea-level change and settlement history during the later Holocene in the Clay District of Lower Saxony (Niedersachsen), northern Germany'. *Quaternary International* 112: 37–53.

Bell, M. C. (1980). 'Patterns of secondary succession in a mangrove forest of southern Florida'. *Oecologia* 44: 226–35.

Bender, B. (ed.) (1993). *Landscape: Politics and Perspectives.* Oxford: Berg.

Berkes, F. and Folke, C. (eds.) (1998). *Linking Social and Ecological Systems.* Cambridge: Cambridge University Press.

Bernhardi, A. R. (1832). 'Wie kamen die aus dem Norden stammenden Felsbruchstücke und Geschiebe, welche man in Norddeutschland und den benachbarten Ländern findet, an ihre gegenwärtigen Fundorte?'. *Jahrbuch für Mineralogie, Geognosie, Geologie und Petrefaktenkunde.* Heidelberg: Georg Reichard, 257–67.

Bernstein, L., Bosch, P., Canziani, O., Chen, Z., Christ, R., Davidson, O., Hare, W., Huq, S., Karoly, D., Kattsov, V., Kundzewicz, Z., Liu, J., Lohmann, U., Manning, M., Matsuno, T., Menne, B., Metz, B., Mirza, M., Nicholls, N., Nurse, L., Pachauri, P., Palutikof, J., Parry, M., Qin, D., Ravindranath, N., Reisinger, A., Ren, J., Riahi, K., Rosenzweig, C., Rusticucci, M., Schneider, S., Sokona, Y., Solomon, A., Stott, P., Stouffer, R., Sugiyama, T., Swart, R., Tirpak, D., Vogel, C., and Yohe, G. (2007). *Climate Change 2007: Synthesis Report.* Cambridge: Cambridge University Press.

Beveridge, C. M. and Little, D. C. (2002). 'The history of aquaculture in traditional societies', in B. A. Costa-Pierce (ed.), *Aquaculture: The Evolution of the Blue Revolution.* Oxford: Blackwell, 1–29.

Bidgood, S. (2002). *A Reconstruction of the Walton Heath Hurdle Trackway.* MA thesis, University of Exeter.

Bigano, A., Bosello, F., Roson, R., and Tol, R. S. J. (2008). 'Economy-wide impacts of climate change: a joint analysis for sea level rise and tourism'. *Mitigation and Adaptation Strategies for Global Change* 13: 765–91.

Bijlsma, L., Ehler, C. N., Klein, R. J. T., Kulshrestha, S. M., McLean, R. F., Mimura, N., Nicholls, R. J., Nurse, L. A., Pérez Nieto, H., Stakhiv, E. Z., Turner, R. K., and Warrick, R. A. (1995). 'Coastal zones and small islands', in R. T. Watson, M. C. Zinyowera, and R. H. Moss (eds.), *Climate Change 1995: Impacts, Adaptations and Mitigation of Climate Change: Scientific-Technical Analyses. Contribution of Working Group II to the Second Assessment Report of the Intergovernmental Panel on Climate Change.* Cambridge: Cambridge University Press, 289–324.

Bindoff, N. L., Willebrand, J., Artale, V., Cazenave, A., Gregory, J., Gulev, S., Hanawa, K., Le Quéré, C., Levitus, S., Nojiri, Y., Shum, C. K., Talley, L. D., and Unnikrishnan,

A. (2007). 'Observations: oceanic climate change and sea level', in S. Solomon, D. Qin, M. Manning, Z. Chen, M. Marquis, K. B. Averyt, M. Tignor, and H. T. Miller (eds.) (2007), *Climate Change 2007: The Physical Science Base. Contribution of Working Group 1 to the Fourth Assessment Report of the Intergovernmental Panel on Climate Change*. Cambridge: Cambridge University Press, 395–432.

Binford, L. R. (1968). 'Post-Pleistocene adaptations', in S. R. Binford and L. R. Binford (eds.), *New Perspectives in Archaeology*. Chicago, IL: Aldine: 313–41.

—— (1972). *An Archaeological Perspective*. New York: Seminar Press.

Binford, M. W., Kolata, A. L., Brenner, M., Janusek, J. W., Seddon, M. T., Abbott, M., and Curtis, J. H. (1997). 'Climate variation and the rise and fall of an Andean civilization'. *Quaternary Research* 47: 235–48.

Boehmer, R. M. (1991). 'Uruk 1980–1990: a progress report'. *Antiquity* 65: 465–78.

Bottéro, J. (2001). *Everyday Life in Ancient Mesopotamia*. Baltimore, MD: Johns Hopkins University Press.

Bradley, R. S. (1999). *Palaeoclimatology: Reconstructing Climates of the Quaternary*. San Diego, CA: Academic Press.

Brander, K. M. (2007). 'Global fish production and climate change'. *Proceedings of the National Academy of Sciences* 104: 19,709–14.

Brandt, R. W. and Van der Leeuw, S. E. (1987). 'Conclusions', in R. W. Brandt, W. Groenman van Waateringe, and S. E. Van der Leeuw (eds.), *Assendelver Polder Papers 1*. Amsterdam: University of Amsterdam, 339–52.

Brassington, J. (2002). 'Monitoring marshland degradation using multispectral remote sensed imagery', in E. Nicholson and P. Clark (eds.) (2002), *The Iraqi Marshlands: A Human and Environmental Study*. London: Politico Publishing, 151–68.

Braudel, F. (1949/1972). *La Méditerranée et le Monde Méditerranéen à l'Epoque de Philippe II/The Mediterranean and the Mediterranean World in the Age of Philip II*. Paris: Armand Colin; London: Collins.

Brennand, M. and Taylor, J. (2003). 'The survey and excavation of a Bronze Age timber circle at Holme-next-the-Sea, Norfolk, 1998–9'. *Proceedings of the Prehistoric Society* 69: 1–84.

Bronk Ramsey, C. (2006). *OxCal Program v4.0* (http://c14.arch.ox.ac.uk/).

Brook, E. J. (2005). 'Tiny bubbles tell all'. *Science* 310: 1,285–7.

Brooks, D. R. (1985). 'Historical Ecology: a new approach to studying the evolution of ecological associations'. *Annals of the Missouri Botanical Garden* 72: 660–80.

Brown, K., Tompkin, E. L., and Adger, W. N. (2002). *Making Waves: Integrating Coastal Conservation and Development*. London: Earthscan.

Bruun, P. (1962). 'Sea-level rise as a cause of shore erosion'. *Journal of the Waterways and Harbors Division, American Society of Engineers* 88: 117–30.

Buckland, P. C. and Sadler, J. (1985). 'The nature of late Flandrian alluviation in the Humberhead Levels'. *East Midlands Geographer* 8: 239–51.

Bullen, R. P. (1953). 'The famous Crystal River site'. *Florida Anthropologist* 6: 9–37.

Burgin, S. (2010). '"Mitigation banks" for wetland conservation: a major success or an unmitigated disaster?' *Wetland Ecology and Management* 18: 49–55.

Burroughs, W. J. (2005). *Climate Change in Prehistory: The End of the Reign of Chaos*. Cambridge: Cambridge University Press.

Butzer, K. (1972). *Environment and Archaeology*. London: Methuen.

Cameron, T. D. J., Crosby, A., Balson, P. S., Jeffrey, D. H., Lott, G. K., Bulat, J., and Harrison, D. J. (1992). *The Geology of the Southern North Sea: British Geological Survey United Kingdom Offshore Regional Report*. London: HMSO.

Campbell Thompson, R. (1920). 'The British Museum excavations at Abu Shahrain in Mesopotamia in 1918'. *Archaeologia* 70: 101–44.

Carmiggelt, A., Guiran, A. J., and Van Trierum, M. C. (eds.) (1997). *BOORbalans 3; Archeologisch onderzoek in het tracé van de Willemsspoortunnel te Rotterdam*. Rotterdam: Bureau Oudheidkundig Bodemonderzoek Rotterdam.

Carpenter, S., Walker, B., Anderies, J. M., and Abel, N. (2001). 'From metaphor to measurement: resilience of what to what?'. *Ecosystems* 4: 765–81.

Carter, R. A. (2006). 'Boat remains and maritime trade in the Persian Gulf during sixth and fifth millennia BC'. *Antiquity*: 52–63.

Cassar, M. (2005). *Climate Change and the Historic Environment*. London: English Heritage.

Cauvin, J. (2000). *The Birth of the Gods and the Origins of Agriculture*. Cambridge: Cambridge University Press.

Chakrabarti, D. K. (1992). *Ancient Bangladesh: A Study of the Archaeological Sources*. Delhi: Oxford University Press.

—— (2001). *Archaeological Geography of the Ganga Plain: The Lower and Middle Ganga*. Delhi: Permanent Black.

—— (2010). *India, an Archaeological History: Palaeolithic Beginnings to Early Historic Foundations*, 4th edn. Delhi: Oxford University Press.

Chaudhuri, A. B. and Choudhury, A. (1994). *Mangroves of the Sundarbans, volume 1: India*. Bankok: IUCN.

Chen, F. H., Bloemendal, J., Wang, J. M., Li, J. J., and Oldfield, F. (1997). 'High-resolution multi-proxy records from Chinese loess: evidence for rapid climate change over the last 75hyr'. *Palaeogeography, Palaeoclimatology, Palaeoecology* 130: 323–35.

Cheung, W. W. L., Lam, V. W. Y., Sarmiento, J. L., Kearney, K., Watson, R., and Pauly, D. (2009). 'Projecting global marine biodiversity impacts under climate change scenarios'. *Fish and Fisheries* 10: 235–51.

Childe, V. G. (1936). *Man Makes Himself*. London: Watts.

—— (1958). 'Retrospect'. *Antiquity* 32: 69–74.

Chopra, K., Kapuria, P., and Kumar, P. (2009). *Biodiversity, Land-Use Change, and Human Well-Being: A Study of Aquaculture in the Indian Sundarbans*. New Delhi: Oxford University Press.

Church, J. A., Gregory, J. M., Huybrechts, P., Kuhn, M., Lambeck, K., Nhuan, M. T., Qin, D., and Woodworth, P. L. (2001). 'Changes in sea-level', in J. T. Houghton, Y. Ding, D. J. Griggs, M. Noguer, P. J. van der Linden, X. Dai, K. Maskell, and C. A. Johnson (eds.), *Climate Change 2001: The Scientific Basis. Contribution of Working Group I to the Third Assessment Report of the Intergovernmental Panel on Climate Change*. Cambridge: Cambridge University Press, 639–94.

Clark, J. G. D. (1936). *The Mesolithic Settlement of Northern Europe*. Cambridge: Cambridge University Press.

—— (1954). *Excavations at Star Carr*. Cambridge: Cambridge University Press.

—— (1961). *World Prehistory: An Outline*. Cambridge: Cambridge University Press.

Clark, P. U., Dyke, A. S., Shakun, J. D., Carlson, A. E., Clark, J., Wohlfarth, B., Mitrovica, J. X., Hostetler, S. W., and McCabe, A. M. (2009). 'The Last Glacial Maximum'. *Science* 325: 710–14.

Clarke, D. L. (1972). *Models in Archaeology*. London: Methuen.

Clemmensen, L. B., Pye, K., Murray, A., and Heinemeier, J. (2001). 'Sedimentology, stratigraphy and landscape evolution of a Holocene coastal dune system, Lodbjerg, NW Jutland, Denmark'. *Sedimentology* 48: 3–27.

Coast, E. (2002). 'Marsh Arab demography', in E. Nicholson and P. Clark (eds.) (2002), *The Iraqi Marshlands: A Human and Environmental Study*. London: Politico Publishing, 19–35.

Cochrane, K., De Young, C., Soto, D., and Bahri, T. (eds.) (2009). *Climate Change Implications for Fisheries and Aquaculture: Overview of Current Scientific Knowledge*. Rome: FAO Fisheries and Aquaculture Technical Paper 530.

Coen, L. D., Brumbaugh, R. D., Bushek, D., Grizzle, R., Luckenbach, M. W., Posey, M. H., Powers, S. P., and Tolley, S. G. (2007). 'Ecosystem services related to oyster restoration'. *Marine Ecology Progress Series* 341: 303–7.

Cohen, M. N. (1977). *The Food Crisis in Prehistory*. New Haven, CT: Yale University Press.

Cola, S., Sanavia, L., Simonini, P., and Schrefler, B. (2008). 'Coupled thermohydromechanical analysis of Venice lagoon salt marshes'. *Water Resources Research* 44: W00C05.

Coleman, J. M., Huh, O. K., and DeWitt, B. Jr (2008). 'Wetland loss in world deltas'. *Journal of Coastal Research* 1: 1–14.

Coles, B. J. (1998). 'Doggerland: a speculative survey'. *Proceedings of the Prehistoric Society* 64: 45–82.

—— (1999). 'Doggerland's loss and the Neolithic', in B. J. Coles, J. M. Coles, and M. Shou Jørgenson (eds.), *Bog Bodies, Sacred Sites and Wetland Archaeology*. Exeter: WARP, 51–8.

Collins, M. (2000). 'The El Niño–Southern Oscillation in the second Hadley Centre coupled model and its response to greenhouse warming'. *Journal of Climate* 13: 1,299–312.

Costanza, R., Graumlich, L. J, and Steffen, W. (2007). *Sustainability or Collapse? An Integrated History and Future of People on Earth*. Cambridge, MA: MIT Press.

Cooke, C. W. (1945). *The Geology of Florida*. Tallahassee, FL: The Florida Geological Survey.

Correljé, A., Broekhans, B., and Roos, W. (2010). 'Van Delatacommissie 1 tot Deltacommissie 2. Alleen preventie of ook kijken naar overstromingskansen?', in H. van der Most, S. de Wit, B. Broekhans, and W. Roos (eds.), *Kijk op Waterveiligheid; Perceptie and communicatie van risico's van overstromingen*. Delft: Eburon, 11–37.

Cosgrove, D. E. (1984). *Social Formation and Symbolic Landscape*. Madison, WI: University of Wisconsin.

Courtenay, W. R., Hartig, B. C., and Loisel, G. R. (1980). *Ecological Evaluation of a Beach Nourishment Project at Hallandale (Broward County), Florida; volume I: Evaluation of Fish Populations Adjacent to Borrow Areas of Beach Nourishment Project, Hallandale (Broward County), Florida*. Boca Raton, FL: Florida Atlantic University.

Cox, P. M., Betts, R. A., Jones, C. D., Spall, S. A., and Totterdell, I. J. (2000). 'Acceleration of global warming due to carbon-cycle feedbacks in a coupled climate model'. *Nature* 408: 184–7.

Cressey, G. B. (1958). 'The Shatt al-Arab basin'. *The Middle East Journal* 12: 448–60.

Croll, J. (1875). *Climate and Change in their Geological Relations: A Theory of Secular Changes of the Earth's Climate.* London: Dalby, Tsbister, and Co.

Crowson, A., Lane, T. Penn, K., and Trimble, D. (2005). *Anglo-Saxon Settlement on the Siltland of eastern England.* Sleaford: Heritage Trust for Lincolnshire (Lincolnshire Archaeology and Heritage Reports Series 7).

Crumley, C. L. (1993). 'Analyzing historical ecotonal shifts'. *Ecological Applications* 3: 377–84.

—— (2006). 'Historical Ecology: integrated thinking at multiple temporal and spatial scales', in A. Hornborg and C. Crumley (eds.), *The World System and the Earth System: Global Socioenvironmental Change and Sustainability since the Neolithic.* Walnut Creek, CA: Left Coast Press, 15–28.

—— (ed.) (1994). *Historical Ecology: Cultural Knowledge and Changing Landscapes.* Santa Fe, NM: School of American Research Press.

Cushing, F. H. (1897). 'Exploration of ancient Key-dweller remains on the Gulf Coast of Florida'. *Proceedings of the American Philosophical Society* 25: 329–448.

Dahl-Jensen, D., Gundestrup, N. S., Miller, H., Watanabe, O., Johnsen, S. J., Steffensen, J. P., Clausen, H. B., Svensson, A., and Larsen, L. B. (2002). 'The NorthGRIP deep drilling programme'. *Annals of Glaciology* 35: 1–4.

Dalley, S. (1989). *Myths from Mesopotamia: Creation, the Flood, Gilgamesh, and Others.* Oxford: Oxford University Press.

Davis, R. A. (1997). 'Geology of the Florida coast', in A. F. Randazzo and D. S. Jones (eds.), *The Geology of Florida.* Gainesville, FL: University Press of Florida, 155–68.

——, Yale, K. E., Pekala, J. M., and Hamilton, M. V. (2003). 'Barrier island stratigraphy and Holocene history of west-central Florida'. *Marine Geology* 200: 103–23.

Davis, S. M. and Ogden, J. C. (eds.) (1994). *Everglades: The Ecosystem and its Restoration.* Boca Raton, FL: St Lucie Press.

Daw, T. and Gray, T. (2005). 'Fisheries science and sustainability in international policy: a study of failure in the European Union's Common Fisheries Policy'. *Marine Policy* 29: 189–97.

Day, J. W., Boesch, D. F., Clairain, E. J., Kemp, G. P., Laska, S. B., Mitch, W. J., Orth, K., Mashriqui, H., Reed, D. J., Shabman, L., Simenstad, C. A., Streever, B. J., Twilley, R. R., Watson, C. C., Wells, J. T., and Whigham, D. F. (2007). 'Restoration of the Mississippi Delta: lessons from hurricanes Katrina and Rita'. *Science* 315: 1,679–84.

——, Christian, R. R., Boesch, D. M., Yáñez-Arancibia Morris, J., Twilley, R. R., Naylor, L., Schaffner, L., and Stevenson, C. (2008). 'Consequences of climate change on the ecomorphology of coastal wetlands'. *Estuaries and Coasts* 31: 477–91.

De Geer, G. (1912). 'A geochronology of the last 12000 years'. *Congresse Géologique Internationale Stockholm 1910*: 241–53

De la Vega-Leinert, A. C. and Nicholls, R. J. (2008). 'Potential implications of sea-level rise for Great Britain'. *Journal of Coastal Research* 24: 342–57.

De Ridder, T. (1999). 'De oudste deltawerken van West-Europa; tweeduizend jaar oude dammen en duikers te Vlaardingen'. *Tijdschrift voor Waterstaatsgeschiedenis* 8: 10–22.

De Silva, S. S. and Soto, D. (2009). 'Climate change and aquaculture: potential impacts, adaptation and mitigation', in K. Cochrane, C. De Young, D. Soto, and T. Bahri (eds.) (2009), *Climate Change Implications for Fisheries and Aquaculture: Overview of Current Scientific Knowledge*. Rome: FAO Fisheries and Aquaculture Technical Paper 530, 151–212.

Deagan, K. A. (1981). 'Downtown survey: the discovery of sixteenth-century St. Augustine in an urban area'. *American Antiquity* 46: 626–33.

—— (1983). *Spanish St. Augustine: The Archaeology of a Colonial Creole Community*. New York: Academic Press.

—— (1985). 'The archaeology of sixteenth-century St. Augustine'. *Florida Anthropologist* 38: 6–34.

Defeo, O., McLachlan, A., Schoeman, D. S., Schlacher, T. A., Dugan, J., Jones, A., Lastra, M., and Scapini, F. (2008). 'Threats to sandy beach ecosystems: a review'. *Estuarine, Coastal, and Shelf Science* 81: 1–12.

deMenocal, P. B. (2001). 'Cultural responses to climate change during the Late Holocene'. *Science* 292: 667–73.

Denton, G. H. and Karlén, W. (1973). 'Holocene climatic variations: their pattern and possible cause'. *Quaternary Research* 3: 155–205.

Devine-Wright, P. (2009). 'Rethinking NIMBYism: the role of place attachment and place identity in explaining place-protective action'. *Journal of Community & Applied Social Psychology* 19: 426–41.

Diamond, J. A. (2005). *Collapse: How Societies Choose to Fail or Succeed*. New York: Viking.

Dincauze, D. F. (2000). *Environmental Archaeology: Principles and Practice*. Cambridge: Cambridge University Press.

Donnelly, J. P. and Giosan, L. (2008). 'Tempestuous highs and lows in the Gulf of Mexico'. *Geology* 36: 751–2.

—— and Woodruff, J. D. (2007). 'Intense hurricane activity over the past 5000 years controlled by El Niño and the West African monsoon'. *Nature* 447: 465–8.

Duane, D. B. and Meisburger, E. P. (1969). *Geomorphology and Sediments of the Nearshore Continental Shelf, Miami to Palm Beach, Florida*. Washington, DC: Army Coastal Engineering Research Centre.

Dunbar, J. S. (1983). 'A model for the predictability of Clovis/Suwannee Paleo-Indian site clusters in Florida—a revival of W. T. Neill's Oasis Hypothesis', in R. Bonnichsen, and K. Turnmier (eds.), *Clovis: Origins and Adaptations*. Corvallis, OR: Centre for the First Americans, Oregon State University, 185–213.

—— (1997). 'Inundated Paleoindian and Archaic sites of the coastal margins and continental shelf in the Southeastern United States'. Paper presented at the 163rd Annual Meeting of the American Association for the Advancement of Science, Seattle.

Durnez, J. (2006). 'The provincial Museum of Walraversijde, from archaeological research to a cultural tourism project', in M. Pieters, G. Gevaert, J. Mees, and J. Seys (eds.), *Colloquium: To Sea or not to Sea: 2nd International Colloquium on Maritime and Fluvial Archaeology in the Southern North Sea Area, Brugge (B), 21–23 September 2006*. Oostende: Vlaams Institute voor de Zee (VLIZ Special Publication 32), 103–5.

Duxbury, A. (2000). *An Introduction to the World's Oceans*, 6th edn. New York: McGraw-Hill.

Edmonds, M. (1995). *Stone Tools and Society: Working Stone in Neolithic and Bronze Age Britain*. London: Routledge.

Edwards, A. M. C. and Winn, P. S. J. (2006). 'The Humber estuary, eastern England: strategic planning of flood defences and habitats'. *Marine Pollution Bulletin* 53: 165–74.

Edwards, R. (2008). 'Sea levels: science and society'. *Progress in Physical Geography* 32: 557–74.

Ellis, S., Fenwick, H., Lillie, M., and Van de Noort, R. (eds.) (2001). *Wetland Heritage of the Lincolnshire Marsh: An Archaeological Survey*. Hull: University of Hull.

Ellis, S. and Van de Noort, R. (1999). *The Humber Estuary: A Strategic Approach*. Hull: University of Hull (Research Institute of Environmental Studies and Management Working Paper 5).

Ellison, J. C. (2008). 'Long-term retrospection on mangrove development using sediment cores and pollen analysis: a review'. *Aquatic Botany* 89: 93–104.

Engelhart, S. E., Peltier, W. R., and Horton, B. P. (2012). 'Holocene relative sea-level changes and glacial isostatic adjustment of the U.S. Atlantic coast'. *Geology* 39: 751–4.

Environment Agency (2000). *Planning for the Rising Tide*. Almondsbury: Environment Agency.

Epstein, S. and Mayeda, T. K. (1953). 'Variation of ^{18}O content of waters from natural sources'. *Geochimica et Cosmochimica Acta* 4: 213–24.

Esmark, J. (1824). 'Bidrag til vor Jordklodes Historie'. *Magazin for Naturvidenskaberne* 2: 29–54.

Evans, J. G. (2003). *Environmental Archaeology and the Social Order*. London: Routledge.

Evans, M. I. (2002). 'The ecosystem', in E. Nicholson and P. Clark (eds.) (2002), *The Iraqi Marshlands: A Human and Environmental Study*. London: Politico Publishing, 201–19.

Fagan, B. (2008). *The Great Warming: Climate Change and the Rise and Fall of Civilizations*. New York: Bloomsbury Press.

—— (2009). 'Searching for clues', in B. Fagan (ed.), *The Complete Ice Age: How Climate Change Shaped the World*. London: Thames and Hudson, 30–47.

Fairbridge, R.W. (1992). 'Holocene marine coastal evolution of the United States: Quaternary coasts of the United States: marine and lacustrine systems', in *SEPM Special Publication* 48: 9–20.

Fan, H., Huang, H., and Zeng, T. (2006). 'Impacts of anthropogenic activity on the recent evolution of the Huanghe (Yellow) River Delta'. *Journal of Coastal Research* 22: 919–29.

Faught, M. K. (2004). 'The underwater archaeology of paleolandscapes, Apalachee Bay, Florida'. *American Antiquity* 69: 275–89.

Fenger, J., Buch, E., Roed Jakobsen, P., and Vestergaard, P. (2008). 'Danish attitudes and reactions to the threat of sea-level rise'. *Journal of Coastal Research* 24: 394–402.

Finney, B. P., Alheit, J., Emeis, K.-J., Field, D. B., Gutierrez, D., and Struck, U. (2010). 'Paleoecological studies on variability in marine fish populations: a long-term perspective on the impacts of climatic change on marine ecosystems'. *Journal of Marine Systems* 79: 316–26.

Fisher, A. (1995). 'An entrance to the Mesolithic world below the ocean: status of ten years' work on the Danish sea floor', in A. Fisher (ed.) (1995), *Man and Sea in the Mesolithic: Coastal Settlement above and below Present Sea-Level.* Oxford: Oxbow, 371–84.

—— (ed.) (1995). *Man and Sea in the Mesolithic: Coastal Settlement above and below Present Sea-Level.* Oxford: Oxbow.

Flannery, K. V. (1968). 'Archaeological systems theory and early Mesoamerica', in B. J. Meggers (ed.), *Anthropological Archaeology in the Americas.* Washington, DC: Anthropological Society of Washington, 67–87.

—— (1969). 'Origins and ecological effects of early domestication in Iran and the Near East', in P. J. Ucko and G. W. Dimbleby (eds.), *The Domestication and Exploitation of Plants and Animals.* Chicago, IL: Aldine, 73–100.

Flemming, B. W. (2002). 'Geographic distribution of muddy coasts', in T. Healy, Y. Wang, and J.-A. Healy (eds.), *Muddy Coasts of the World: Processes, Deposits and Function.* Amsterdam: Elsevier Science, 99–201.

Flemming, N. C. (ed.) (2004). *Submarine Prehistoric Archaeology of the North Sea: Research Priorities and Collaboration with Industry.* York: CBA (Research Report 141).

FOCC (2009). *The Effects of Climate Change on Florida's Ocean and Coastal Resources: A Special Report to the Florida Energy and Climate Commission and the People of Florida.* Tallahassee, FL: Florida Oceans and Coastal Council.

Fox, C. F. (1932). *The Personality of Britain: Its Influence on Inhabitants and Invaders in Prehistoric and Early Historic Times.* Cardiff: The National Museum of Wales.

Friedman, J. and Chase-Dunn, C. (eds.) (2005). *Hegemonic Declines: Past and Present.* Boulder, CO: Paradigm Publishers.

Fries, M. (1967). 'Lennart von Post's pollen diagram series of 1916'. *Review of Palaeobotany and Palynology* 4: 9–13.

Fulanain (1927). *Haji Rikkan: Marsh Arabs.* London: Chatto & Windus.

Fuller, D. Q. (2006). 'Agricultural origins and frontiers in South Asia: a working synthesis'. *Journal of World Prehistory* 20: 1–86.

Fyall, A. and Garrod, B. (1998). 'Heritage tourism: at what price?'. *Managing Leisure* 3: 213–28.

Geikie, J. (1874). *The Great Ice Age and its Relation to the Antiquity of Man.* London: Tsbister.

Geyer, B. and Sanlaville, P. (1996). 'Nouvelle contribution à l'étude géomorphologique de la région de Larsa-Oueili (Iraq)', in J.-L. Hunt (ed.), *Oueili: Travaux de 1987 et 1989.* Paris: Editions Recherche sur les Civilisations, 391–408.

Ghosh, A. (2004). *The Hungry Tide.* London: HarperCollins.

Gilliland, M. S. (1975). *The Material Culture of Key Marco, Florida.* Gainesville, FL: University Press of Florida.

—— (1989). *Key Marco's Buried Treasure: Archaeology and Adventure in the Nineteenth Century.* Gainesville, FL: University Press of Florida.

Gilman, E. L., Ellison, J., Duke, N. C., and Field, C. (2008). 'Threats to mangroves from climate change and adaptation options: a review'. *Aquatic Botany* 89: 237–50.

Gleason, P. J., Cohen, A. D., Smith, W. G., Brooks, H. K., Stone, P. A., Goodrick, R. L., and Spackman, W. (1984). 'The environmental significance of Holocene sediments from the Everglades and Saline Tidal Plain', in P. J. Gleason (ed.), *Environments of South Florida: Present and Past II*. Miami, FL: Geological Society, 297–351.

Glimmerveen, J., Mol, D., Post, K., Reumer, J. W. F., Van der Plicht, H., De Vos, J., Van Geel, B., Van Reenen, G., and Pals, J. P. (2004). 'The North Sea project: the first palaeontological, palynological and archaeological results', in N. C. Flemming (ed.) (2004), *Submarine Prehistoric Archaeology of the North Sea: Research Priorities and Collaboration with Industry*. York: CBA (Research Report 141), 43–52.

——, ——, and Van der Plicht, H. (2006). 'The Pleistocene reindeer of the North Sea: initial palaeontological data and archaeological remarks'. *Quaternary International* 142–3: 242–6.

Godwin, H. (1940). 'Pollen analysis and forest history of England and Wales'. *New Phytologist* 39: 370–400.

—— (1943). 'Coastal peat beds of the British Isles and North Sea: presidential address to the British Ecological Society 1943'. *Journal of Ecology* 31: 199–247.

—— (1956). *History of the British Flora*. Cambridge: Cambridge University Press.

Goodbred Jr, S. L. (2003). 'Response of the Ganges dispersal system to climate change: a source-to-sink view since the last interstade'. *Sedimentary Geology* 162: 83–104.

—— and Kuehl, S. A. (2000). 'Enormous Ganges—Brahmaputra sediment discharge during strengthened early Holocene monsoon'. *Geology* 28: 1,083–6.

Goodland, R. (1995). 'The concept of environmental sustainability'. *Annual Review of Ecological Systems* 26: 1–24.

Goodyear, A. C. and Warren, L. O. (1972). 'Further observations on the submarine oyster shell deposits of Tampa Bay'. *Florida Anthropologist* 25: 52–66.

——, Upchurch, S. B., Brooks, M. J., and Goodyear, N. N. (1983). 'Palaeo-indian manifestations in the Tampa Bay region, Florida'. *Florida Anthropologist* 36: 40–66.

Gosden, C. and Lock, G. (1998). 'Prehistoric histories'. *World Archaeology* 30: 2–12.

Goudie, A. (1981). *The Human Impact on the Natural Environment: Past, Present, and Future*. London: Blackwell.

Gradmann, R. (1906). 'Beziehung zwischen Pflanzengeographie und Siedlungsgeschichte'. *Geographische Zeitschrift* 12: 305–25.

Graham, R. W. and Grimm, E. C. (1990). 'Effects of global climate change on the patterns of terrestrial biological communities'. *Tree* 5: 289–92.

Gray, J. M. (1974). 'Late glacial and postglacial shorelines in western Scotland'. *Boreas* 3: 129–38.

Griffin, J. W. (2002). *Archaeology of the Everglades*. Gainesville, FL: University Press of Florida.

—— and Smith, H. G. (1948). *The Goodnow Mound, Highlands County, Florida*. Tallahassee, FL: Florida Park Service.

Groucutt, H. S. and Petraglia, M. D. (2012). 'The prehistory of the Arabian Peninsula: deserts, dispersals and demography'. *Evolutionary Anthropology* 21: 113–25.

Gunderson, L. and Holling, C. S. (eds.) (2002). *Panarchy: Understanding Transformations in Human and Natural Systems*. Washington, DC: Island Press.

——, ——, and Light, S. S. (1995). *Barriers and Bridges to the Renewal of Ecosystems and Institutions*. New York: Columbia University Press.

Haarnagel, W. (1979). *Die Grabung Feddersen Wierde. Methode, Hausbau, Siedlungs- und Wirtshaftsformen sowie Sozialstruktur.* Wiesbaden: Steiner.

Hall, D. and Coles, J. M. (1994). *Fenland Survey: An Essay in Landscape and Persistence.* London: English Heritage.

Hall, H. R. (1919). 'Ur and Eridu: the British Museum excavations of 1919'. *Journal of Egyptian Archaeology* 9: 177–95.

—— (1929). 'An early Sumerian sculptured trough'. *British Museum Quarterly* 3: 40–1.

—— and Woolley, L. (1927). *Ur Excavations, Volume I: Al-'Ubaid.* Oxford: Oxford University Press.

Hall, S. J., Delaporte, A., Phillips, M. J., Beveridge, M., and O'Keefe, M. (2011). *Blue Frontiers: Managing the Environmental Costs of Aquaculture.* Penang: The World-Fish Center.

Hanna, S. S., Folke, C., and Mäler, K. G. (1996). *Rights to Nature.* Washington, DC: Island Press.

Hansen, J. E. and Lacis, A. A. (1990). 'Sun and dust versus greenhouse gases: an assessment of their relative roles in global climate change'. *Nature* 346: 713–19.

Hansman, J. F. (1978). 'The Mesopotamian delta in the first millennium'. *Geographical Journal* 144: 49–61.

Harff, J., Lampe, R., Lemke, W., Lübke, H., Lüth, F., Meyer, M., and Tauber, F. (2005). 'The Baltic Sea: a model ocean to study interrelations of geosphere, ecosphere, and anthroposphere in the coastal zone'. *Journal of Coastal Research* 21: 441–6.

Hartig, E. K., Gormitz, V., Kolker, A., Mushacke, F., and Fallon, D. (2002). 'Anthropogenic and climate-change impacts on salt marshes of Jamaica Bay, New York City'. *Wetlands* 22: 71–89.

Hayden, B. (1995). 'A new overview of domestication', in A. B. Gebauer and T. D. Price (eds.), *Transitions to Agriculture in Prehistory.* Madison, WI: Prehistory Press, 11–19.

Haykal, H. T. (2011). 'Arts and crafts', in S. A. A. Kubba (ed.) (2011), *The Iraqi Marshlands and the Marsh Arabs: The Ma'dan, their Culture and the Environment.* Reading: Ithaca Press, 44–71.

Higgs, E. S. (ed.) (1972). *Papers in Economic Prehistory.* Cambridge: Cambridge University Press.

Hodder, I. (1986). *Reading the Past; Current approaches to interpretation in archaeology.* Cambridge: Cambridge University Press.

—— (1990). *The Domestication of Europe: Structure and Contingency in Neolithic Societies.* Oxford: Blackwell.

—— and Orton, C. (1976). *Spatial Analysis in Archaeology.* Cambridge: Cambridge Archaeology Press.

Hodell, D. A., Curtis, J. A., and Brenner, M. (1995). 'Possible role of climate in the collapse of Classic Maya civilization'. *Nature* 375: 394.

Hoffmann, T., Erkens, G., Cohen, K. M., Houben, P, Seidel, J., and Dikau, R. (2007). 'Holocene floodplain sediment storage and hillslope erosion within the Rhine catchment'. *Holocene* 17: 105–18.

Holling, C. S. (1973). 'Resilience and stability of ecological systems'. *Annual Review of Ecology and Systematics* 4: 1–23.

——, Gunderson, L. H., and Peterson, G. D. (2002). 'Sustainability and panarchies', in L. Gunderson and C. S. Holling (eds.) (2002), *Panarchy: Understanding Transformations in Human and Natural Systems.* Washington, DC: Island Press, 63–102.

Housley, R. (1991). 'AMS dates from the late glacial and early postglacial in north-west Europe: a review', in N. Barton, A. J. Roberts, and D. A. Roe (eds.) (1991), *The Late Glacial in North-West Europe.* London: CBA (Research Report 77), 25–39.

Houston, J. R. and Dean, R. G. (2011). 'Sea-level acceleration based on U.S. tide gauges and extensions of previous global-gauge analysis'. *Journal of Coastal Research* 27: 409–17.

Hovland, M. and Dukefoss, K. M. (1981). 'A submerged beach between Norway and Ekofisk in the North Sea'. *Marine Geology* 43: M19–28.

Hritz, C., Pournelle, J., Smith, J., Albadran, B., Majeed Issa, B., and Al-Handal, A. (2012). 'Mid-Holocene dates from organic-rich sediments, palustrine shell, and charcoal from southern Iraq'. *Radiocarbon* 54: 65–79.

Hueting R. (1990). 'The Brundtland Report: a matter of conflicting goals'. *Ecological Economy* 2: 109–1.

Hughes, P. D. M., Blundell, A., Charman, D. J., Bartlett, S., Daniell, J. R. G., Wojatschke, A., and Chambers, A. F. (2006). 'An 8500 cal. year multi-proxy climate record from a bog in eastern Newfoundland; contributions of meltwater discharge and solar forcing'. *Quaternary Science Reviews* 25: 1,208–27.

Hunt, T. L. and Lipo, C. P. (2010). 'Ecological catastrophe, collapse, and the myth of "ecocide" on Rapa Nui (Easter Island)', in P. A. McAnany and N. Yoffee (eds.) (2010), *Questioning Collapse: Human Resilience, Ecological Vulnerability and the Aftermath of Empire.* Cambridge: Cambridge University Press, 21–44.

Hunter, W. W. (1875). *A Statistical Account of Bengal. Volume 1, District of the 24 Parganas and Sundarbans.* London: D. K. Publishing.

Huntley, B. (1991). 'How plants respond to climate change: migration rates, individualism and the consequences for plant communities'. *Annals of Botany* 67 (supplement 1): 15–22.

—— and Webb III, T. (1989). 'Migration: species' response to climatic variations caused by changes in the earth's orbit'. *Journal of Biogeography* 16: 5–19.

Immerzeel, W. W., Van Beek, L. P. H., and Bierkens, M. F. P. (2010). 'Climate change will affect the Asian water towers'. *Science* 328: 1,382–5.

IPCC (2000). *Special Report Emissions Scenarios.* (http://www.ipcc.ch/pdf/special-reports/spm/sre.)

IPCC (2007). 'Summary for policymakers', in S. Solomon, D. Qin, M. Manning, Z. Chen, M. Marquis, K. B. Averyt, M. Tignor, and H. T. Miller (eds.) (2007), *Climate Change 2007: The Physical Science Base. Contribution of Working Group 1 to the Fourth Assessment Report of the Intergovernmental Panel on Climate Change.* Cambridge: Cambridge University Press, 1–18.

Islam, W., Alam, S., and Islam, M. (2007). 'Study of human casualties by Bengal tigers (Panthera tigris tigris L.) in the Sunderbans forest of Bangladesh'. *Tiger Paper* 34: 1–6.

Jackson, J. B. C., Kirby, M. X., Berger, W. H., Bjomdal, K. A., Botsford, L. W., Bourque, B. J., Bradbury, R. H., Cooke, R., Erlandson, J., Estes, J. A., Hugh, T. P., Kidwell, S., Lange, C. B., Lenihan, H. S., Pandolfi, J. M., Peterson, C. H., Steneck, R. S., Tegner,

M. J., and Wamer, R. R. (2001). 'Historical overfishing and the recent collapse of coastal ecosystems'. *Science* 293: 629–37.

Jackson, R. P. J. and Potter, T. W. (1996). *Excavations at Stonea, Cambridgeshire, 1980–85*. London: British Museum.

Jacobsen, T. and Adams, R. M. (1958). 'Salt and silt in ancient Mesopotamian agriculture'. *Science* 128: 1,251–8.

Jahan, S. (2010). 'Archaeology of Wari-Bateshwar'. *Ancient Asia* 2: 135–46.

Jalais, A. (2010). *Forest of Tigers: People, Politics and Environment in the Sundarbans*. London: Routledge.

Jansen, E., Overpeck, J., Briffa, K. R., Duplessy, J.-C., Joos, F., Masson-Delmotte, V., Olago, D., Otto-Bliesner, B., Peltier, W. R., Rahmstorf, S., Ramesh, R., Raynaud, D., Rind, D., Solomina, O., Villalba, R., and Zhang, D. (2007). 'Palaeoclimate', in S. Solomon, D. Qin, M. Manning, Z. Chen, M. Marquis, K. B. Averyt, M. Tignor, and H. T. Miller (eds.) (2007), *Climate Change 2007: The Physical Science Base. Contribution of Working Group 1 to the Fourth Assessment Report of the Intergovernmental Panel on Climate Change*. Cambridge: Cambridge University Press, 433–97.

Jessen, K. (1935). 'Archaeological dating in the history of north Jutland's vegetation'. *Acta Archaeologica* 5: 185–98.

Jiongxin, X. (2003). 'Sediment flux to the sea as influenced by changing human activities and precipitation: example of the Yellow River, China'. *Environmental Management* 31: 328–41.

Johnston, S. A. (1981). 'Estuarine dredge-and-fill activities: a review of impacts'. *Environmental Management* 5: 427–40.

Kanjilal, T. (2000). *Who Killed the Sundarbans?* Kolkata: Tagore Society for Rural Development.

Kates, R. W., Clark, W. C., Corell, R., Hall, J. M., Jaeger, C. C., Lowe, I., McCarthy, J. J., Schellnhuber, H. J., Bolin, B., Dickson, N. M., Faucheux, S., Gallopin, G. C., Grübler, A., Huntley, B., Jäger, J., Jodha, N. S., Kasperson, R. E., Mabogunje, A., Matson, P., Mooney, H., Moore III, B., O'Riordan, T., and Svedin, U. (2001). 'Sustainability science'. *Science* 27: 641–2.

Kathiresan, K. and Rajendra, N. (2005). 'Coastal mangrove forests mitigated tsunami'. *Estuarine, Coastal, and Shelf Science* 65: 601–6.

Keeling, C.D. (1961). 'The concentration and isotopic abundances of carbon dioxide in rural and marine air'. *Geochimica et Cosmochimica Acta* 24: 277–98.

Keiner, M. (2005). *History, Definition(s) and Models of Sustainable Development*. Zurich: Eidgenössische Technische Hochschule.

Kennedy, A. J., Griffin, M. L., Morey, S. L., Smith, S. R., and O'Brein, J. J. (2007). 'Effects of El Niño-Southern Oscillation on sea level anomalies along the Gulf of Mexico coast'. *Journal of Geophysical Research* 12: C05047.

Kennedy, J., Morice, C., and Parker, D. (2011). 'Global and regional climate in 2010'. *Weather* 66: 188–94.

Kennett, D. J. and Kennett, J. P. (2007). 'Influence of Holocene marine transgression and climate change on cultural evolution in southern Mesopotamia', in D. G. Anderson, K. A. Maasch, and D. H. Sandweiss (eds.), *Climate Change and Cultural Dynamics: A Global Perspective on Mid-Holocene Transitions*. Amsterdam: Elsevier, 229–64.

Kirwan, M. L., Guntenspergen, G. R., D'Alpaos, A., Morris, J. T., Mudd, S.M., and Temmerman, S. (2010). 'Limits on the adaptability of coastal marshes to rising sea level'. *Geophysical Research Letters* 37: L23401.

Klein, R. G., Avery, G., Cruz-Uribe, K., Halkett, D., Parkington, J. E., Steele, T., Volman, T. P., and Yates, R. (2004). 'The Ysterfontein 1 Middle Stone Age site, South Africa, and early human exploitation of coastal resources'. *Proceedings of the National Academy of Sciences* 101: 5,708–15.

Klemperer, S. L. (1988). 'Crustal thinning and nature of extension in the northern North Sea from deep seismic reflection profiling'. *Tectonics* 7: 803–21.

Koch, E. (1998). *Neolithic Bog Pots from Zealand, Møn, Lolland and Falster*. Copenhagen: Kongelige Nordiske Oldskriftselskab.

Köppen, W. (1873). 'Über mehrjährige Perioden der Witterung, insbesondere über die 11-jährige Periode der Temperatur'. *Zeitschrift der Österreichischen Gesellschaft für Meteorologie* VIII: 241–8; 257–67.

—— and Wegener, A. (1924). *Klimate der geologischen Vorzeit*. Berlin: Gebrüder Borntraeger.

Kubba, S. A. A. (ed.) (2011). *The Iraqi Marshlands and the Marsh Arabs: The Ma'dan, their Culture and the Environment*. Reading: Ithaca Press.

—— (2011). 'People and culture', in S. A. A. Kubba (ed.) (2011), *The Iraqi Marshlands and the Marsh Arabs: The Ma'dan, their Culture and the Environment*. Reading: Ithaca Press, 25–43.

—— and Jamali, A. F. (2011). 'History of the Iraqi Marshes', in S. A. A. Kubba (ed.) (2011), *The Iraqi Marshlands and the Marsh Arabs: The Ma'dan, their Culture and the Environment*. Reading: Ithaca Press, 4–24.

—— and Salim, M. (2011). 'The wetlands wildlife and ecosystem', in S. A. A. Kubba (ed.) (2011), *The Iraqi Marshlands and the Marsh Arabs: The Ma'dan, their Culture and the Environment*. Reading: Ithaca Press, 115–42.

Kuehl, S. A., Allison, M. A., Goodbred, S. L., and Kudrass, H. (2005). 'Sedimentary processes and products of the modern Ganges–Brahmaputra delta', in L. Giosan and J. Bhattacharya (eds.), *River Deltas: Concepts, Models and Examples*. Tulsa, OK: Society for Sedimentary Geology, 413—34.

Lal, B. B. (1958). 'A microlithic site in the Damodal Valley, West Bengal'. *Ancient India* 14: 4–48.

Langley, J. A., McKee, K. L., Cahoon, D. R., Cherry, J. A., Megonigal, J. P. (2009). 'Elevated CO_2 stimulates marsh elevation gain, counterbalancing sea-level rise'. *Proceedings of the National Academy of Sciences* 106: 6,182–6.

Larocque, I., Hall, R. I., and Grahn, E. (2001). 'Chironomids as indicators of climate change: a 100–lake training set from a subarctic region of northern Sweden (Lapland)'. *Journal of Paleolimnology* 26: 307–22.

Latour, B. (1993). *We Have Never Been Modern*. Cambridge, MA: Harvard University Press.

Le Pichon, X., Fournier, M., and Jolivet, L. (1992). 'Kinematics, topography, shortening, and extrusion in the India-Eurasia collision'. *Tectonics* 11: 1,085–98.

Le Treut, H., Somerville, R., Cubasch, U., Ding, Y., Mauritzen, C., Mokssit, A., Peterson, T., and Prather, M. (2007). 'Historical overview of climate change', in S. Solomon, D. Qin, M. Manning, Z. Chen, M. Marquis, K. B. Averyt, M. Tignor, and H. T. Miller (eds.) (2007). *Climate Change 2007: The Physical Science Base. Contribution of Working Group 1 to the Fourth Assessment Report of the Intergovernmental Panel on Climate Change.* Cambridge: Cambridge University Press, 93–127.

Leatherman, S. P., Zhang, K., and Douglas, B. C. (2000). 'Sea-level rise shown to drive coastal erosion'. *Eos* 81: 55–7.

Ling, S. D., Johnson, C. R., Frusher, S. D., and Ridgway, K. R. (2009). 'Overfishing reduces resilience of kelp beds to climate-driven catastrophic phase shift'. *Proceedings of the National Academy of Sciences* 106: 22,341–5.

Liu K.-B. (2004). 'Paleotempestology: principles, methods, and examples from Gulf Coast lake-sediments', in R. Murnane and K.-B. Liu (eds.), *Hurricanes and Typhoons: Past, Present, and Future.* New York: Columbia University Press: 13–57.

Lloyd, S., and Safar, F. (1948). 'Eridu: Preliminary report on the second season's excavations'. *Sumer* 4: 115–25.

Long, D., Wickham-Jones, C. R., and Ruckley, N. A. (1986). 'A flint artefact from the North Sea', in D. Roe (ed.), *Studies in the Upper Paleolithic of Britain and Northwest Europe.* Oxford: BAR (British Series 296), 55–62.

Longshore, D. (2000). *Encyclopedia of Hurricanes, Typhoons, and Cyclones.* New York: Facts on File.

Lotze, H. K., and Reise, K., Worm, B., Van Beusekom, J., Busch, M., Ehlers, A., Heinrich, D., Hoffmann, R. C., Holm, P., Jensen, C., Knottnerus, O. S., Langhanki, N., Prummel, W., Vollmer, M., and Wolff, W. J. (2005). 'Human transformations of the Wadden Sea ecosystem through time: a synthesis', in H. K. Lotze and K. Reise (eds.), *Ecological History of the Wadden Sea.* Helgoland: Helgoland Marine Research, 84–95.

Loucks, C., Barber-Meyer, S., Hossain, Md., Barlow, A., and Chowdhury, R. (2010). 'Sea level rise and tigers: predicted impacts to Bangladesh's Sundarbans mangroves'. *Climatic Change*: 98, 291–8.

Lough, J. M. (2008). '10th Anniversary review: a changing climate for coral reefs'. *Journal of Environmental Monitoring* 10: 21–9.

Louwe Kooijmans, L. P. (1971). 'Mesolithic bone and antler implements from the North Sea and from the Netherlands'. *Berichten van de Rijksdienst voor het Oudheidkundig Bodemonderzoek* 20–1: 27–73.

—— (1985). *Sporen in het Land: de Nederlandse delta in de prehistorie.* Amsterdam: Meulenhoff Informatief.

Ludwig, D., Walker, B., and Holling, C. S. (1997). 'Sustainability, stability and resilience'. *Conservation Ecology* 1 (URL: http://www.consecol.org/vol1/iss1/art7).

Luer, G. M. (1989). 'Calusa canals in southwestern Florida: routes of tribute and exchange'. *The Florida Anthropologist* 42: 89–130.

Maarleveld, T. (1984). *Archeologie in Troebel Water; Archeologie onder water. 1e onderzoeksrapport.* Amsterdam: Ministerie van Welzijn, Volksgezondheid en Cultuur.

Mack, J. (2011). *The Sea: A Cultural History.* London: Reaktion Books.

Maltby, E. (1994). *An Environmental and Ecological Study of the Marshlands of Mesopotamia.* London: Amar Appeal Trust, and Exeter: Wetland Ecosystems Research Group.

Mandal, A. (2004). *Sundarbans: An Ecological History.* Kolkata: Readers Service.

Manten, A. A. (1967). 'Lennart von Post and the foundation of modern palynology'. *Review of Palaeobotany and Palynology* 1: 11–22.

Marcus, N. H. (1991). 'Planktonic copepods in a sub-tropical estuary: seasonal patterns in the abundance of adults, copepodites, nauplii, and eggs in the sea bed'. *The Biological Bulletin* 181: 269–74.

Marquardt, W. H. (1999). 'The archaeology of Useppa Island: a summary', in W. H. Marquardt (ed.), *The Archaeology of Useppa Island*. Gainesville, FL: University of Florida, 241–53.

—— and Walker, K. J. (2001). 'Pineland: a coastal wet site in southwest Florida', in B. Purdy (ed.), *Enduring Records: The Environmental and Cultural Heritage of Wetlands*. Oxford: Oxbow Press, 48–60.

——, —— (2012). 'Southwest Florida during the Mississippi Period', in K. Ashley and N. M. White (eds.), *Late Prehistoric Florida: Archaeology at the Edge of the Mississippian World*. Gainesville, FL: University Press of Florida, 29–61.

Mars, J. C. and Houseknecht, D. W. (2007). 'Quantitative remote sensing study indicates doubling of coastal erosion rate in past 50 yrs along a segment of the Arctic coast of Alaska'. *Geology* 35: 583–6.

Marshall, J. D., Lang, B., Crowley, S. F., Weedon, G. P., Van Calsteren, P., Fisher, E. H., Holme, R., Holmes, J. A., Jones, R. T., Bedford, A., Brooks, S. J., Bloemendal, J., Kiriakoulakis, K., and Ball, J. D. (2007). 'Terrestrial impact of abrupt changes in the North Atlantic thermohaline circulation: Early Holocene, UK'. *Geology* 35: 639–42.

Masselink, G. and Hughes, M. G. (2003). *Introduction to Coastal Processes & Geomorphology*. London: Hodder Education.

Matthews, R. (2003). *The Archaeology of Mesopotamia: Theories and Approaches*. London: Routledge.

Maxwell, G. (1957). *People of the Reeds*. New York: Harper & Brothers.

McAnany, P. A. and Yoffee, N. (eds.) (2010). *Questioning Collapse: Human Resilience, Ecological Vulnerability and the Aftermath of Empire*. Cambridge: Cambridge University Press.

McCally, D. (2005). 'The Everglades and the Florida dream', in J. E. Davis and R. Arsenault (eds.), *Paradise Lost? The Environmental History of Florida*. Gainesville, FL: The University Press of Florida, 141–59.

McClenachan, L. (2008). 'Social conflict, over-fishing and disease in the Florida sponge fishery, 1849–1939', in D. J. Starkey, P. Holm, and M. Barnard (eds.), *Oceans Past: Management Insight from the History of Marine Animal Populations*. London: Earthscan, 25–46.

McEwan, I. and Gormley, A. (2006). 'A conversation about art and nature'. *The Kenyon Review, New Series* 28: 104–12.

McFadden, L., Spencer, T., and Nicholls, R. J. (2007). 'Broad-scale modelling of coastal wetlands: what is required?'. *Hydrobiologia* 577: 5–15.

McIntosh, R. J., Tainter, J. A., and McIntosh, S. K. (2000). 'Climate, history and human action', in R. J. McIntosh, J. A. Tainter, and S. K. McIntosh (eds.), *The Way the Wind Blows: Climate, History and Human Action*. New York: Columbia University Press, 1–42.

McMichael, A. J., Woodruff, R. E., and Hales, S. (2006). 'Climate change and human health: present and future risks'. *The Lancet* 367: 859–69.

McNeil, J. R. (2010). 'Sustainable survival', in P. A. McAnany and N. Yoffee (eds.) (2010). *Questioning Collapse: Human Resilience, Ecological Vulnerability and the Aftermath of Empire.* Cambridge: Cambridge University Press, 355–66.

Meddens, F. M. (1996). 'Sites from the Thames estuary, England, and their Bronze Age use'. *Antiquity* 70: 325–34.

Meehl, G., Stocker, T., Collins, W., Friedlingstein, P., Gaye, A., Gregory, J., Kitoh, A., Knutti, R., Murphy, J. M., Noda, A., Raper, S. C. B., Watterson, I. G., Weaver, A. J., and Zhao, Z.-C. (2007). 'Global climate projections', in S. Solomon, D. Qin, M. Manning, Z. Chen, M. Marquis, K. B. Averyt, M. Tignor, and H. T. Miller (eds.) (2007), *Climate Change 2007: The Physical Science Base. Contribution of Working Group 1 to the Fourth Assessment Report of the Intergovernmental Panel on Climate Change.* Cambridge: Cambridge University Press, 747–843.

Meese, D. A., Gow, A. J., Alley, R. B., Zielinski, G. A., Grootes, P. M., Ram, M., Taylor, K. C., Mayewski, P. A., and Bolzan, J. F. (1997). 'The Greenland Ice Sheet Project 2 depth-age scale: methods and results'. *Journal of Geophysical Research* 102: 411–26.

Meier, D. (2004). 'Man and environment in the marsh area of Schleswig-Holstein from Roman until late medieval times'. *Quaternary International* 112: 55–69.

Metz, B., Davidson, O. R., Bosch, P. R., Dave, R., and Meyer, L. A. (eds.) (2007). *Climate Change 2007: Mitigation of Climate Change. Contribution of Working Group 3 to the Fourth Assessment Report of the Intergovernmental Panel on Climate Change.* Cambridge: Cambridge University Press.

Micheli, F., Bishop, M. J., Peterson, C. H., and Rivera, J. (2008). 'Alteration of seagrass species composition and function over two decades'. *Ecological monographs* 78: 225–44.

Milliken, K. T., Anderson, J. B., and Rodriguez, A. B. (2008). 'A new composite Holocene sea-level curve for the northern Gulf of Mexico', in J. B. Anderson and A. B. Rodriguez (eds.), *Response of Upper Gulf Coast Estuaries to Holocene Climate Change and Sea Level Change.* Washington, DC: The Geological Society of America (Special Paper 443), 1–11.

Milliman, J. D. and Meade, R. H. (1983). 'World-wide delivery of river sediments to the oceans'. *Journal of Geology* 91: 1–21.

Mitchell, R. J., Auld, M. H. D., Le Duc, M. G., and Robert, M. A. (2000). 'Ecosystem stability and resilience: a review of their relevance for the conservation management of lowland heaths'. *Perspectives in Plant Ecology, Evolution and Systematics* 3: 142–160.

Mook, W. G. (1986). 'Business meeting: recommendations/resolutions adopted by the Twelfth International Radiocarbon Conference'. *Radiocarbon* 28: 799.

Moore, C. B. (1907). 'Notes on the Ten Thousand Islands'. *Journal of the Academy of Natural Sciences of Philadelphia* 17: 458–70.

Morlot, A. (1861). 'General views on archaeology'. Washington, DC: *Annual Report of the Smithsonian Institution for 1860*: 284–343.

Morony, M. (2004). 'Economic boundaries? Late Antiquity and Early Islam'. *Journal of the Economic and Social History of the Orient* 47: 166–94.

Muhs, D. R., Simmons, K. R., Schumann, R. R., and Halley, R. B. (2011). 'Sea-level history of the past two interglacial periods: new evidence from U-series dating of reef corals from south Florida'. *Quaternary Science Reviews* 30: 570–90.

Müller, U. C., Peoss, J., Tzedakis, P. C., Gamble, C., Kotthoff, U., Schmiedl, G., Wulf, S., and Chistanis, K. (2011). 'The role of climate in the spread of modern humans in Europe'. *Quaternary Science Reviews* 30: 273–9.

Myatt, L. B., Scrimshaw, M. D., and Lester, J. N. (2003). 'Public perceptions and attitudes towards a current managed realignment scheme: Brancaster West Marsh, North Norfolk, U.K'. *Journal of Coastal Research* 19: 278–86.

Naff, T. and Hanna, G. (2002). 'The marshes of southern Iraq: a hydro-engineering and political profile', in E. Nicholson and P. Clark (eds.) (2002), *The Iraqi Marshlands: A Human and Environmental Study*. London: Politico Publishing, 169–200.

Nicholls, R. J. and Branson, J. (1998). 'Coastal resilience and planning for an uncertain future: an introduction'. *Geographical Journal* 164: 255–8.

——, Hoozemans, F. M. J., and Marchand, M. (1999). 'Increasing flood risk and wetland losses due to global sea-level rise: regional and global analyses'. *Global Environmental Change* 9: 69–87.

——, Wong, P. P., Burkett, V. R., Codignotto, J. O., Hay, J. E., McLean, R. F., Ragoonaden, S., and Woodroffe, C. D. (2007). 'Coastal systems and low-lying areas', in M. L. Parry, O. F. Canziani, J. P. Palutikof, P. J. Van der Linden, and C. E. Hanson (eds.) (2007), *Climate Change 2007: Impacts, Adaptation and Vulnerability. Contribution of Working Group II to the Fourth Assessment. Report of the Intergovernmental Panel on Climate Change.* Cambridge: Cambridge University Press, 315–56.

Nicholson, E. and Clark, P. (eds.) (2002). *The Iraqi Marshlands: A Human and Environmental Study.* London: Politico Publishing.

Nicholson, G., Jenkins, G. P., Sherwood, J., and Longmore, A. (2008). 'Physical environmental conditions, spawning and early-life stages of estuarine fish: climate change implications for recruitment in intermittently open estuaries'. *Marine and Freshwater Research* 59: 735–49.

Oates, J. L. (1960). 'Ur and Eridu: the prehistory'. *Iraq* 22: 32–50.

Ochsenschlager, E. L. (2004). *Iraq's Marsh Arabs in the Garden of Eden.* Philadelphia, PA: University of Pennsylvania Museum of Archaeology and Anthropology.

Orford, J. D., Wilson, P., Wintle, A. G., Knight, J., and Braley, S. (2000). 'Holocene coastal dune initiation in Northumberland and Norfolk, eastern UK: climate and sea level changes as possible forcing agents for dune initiation', in I. Shennan and J. E. and Andrews (eds.) (2000), *Holocene Land-Ocean Interaction and Environmental Change around the Western North Sea.* London: Geological Society (Special Publication 166), 197–217.

Otvos, E. G. (1995). 'Multiple Pliocene-Quaternary marine highstands, northeast Gulf Coastal plain: fallacies and facts'. *Journal of Coastal Research.* 11: 984–1,002.

—— (2001). 'Assumed Holocene highstands, Gulf of Mexico: basic issues of sedimentary and landform criteria: discussion'. *Journal of Sedimentary Research* 71: 645–7.

—— (2004). 'Holocene Gulf levels: recognition issues and an updated sea-level curve'. *Journal of Coastal Research* 20: 680–99.

—— (2011). 'Ground-penetrating radar study of North Padre Island: implications for barrier island internal architecture, model for growth of progradational microtidal barrier inlands, and Gulf of Mexico sea-level cyclicity—discussion'. *Journal of Sedimentary Research* 81: 392–3.

Pandolfi, J. M., Connolly, S. R., Marshall, D. J., and Cohen, A. L. (2011). 'Projecting coral reef futures under global warming and ocean acidification'. *Science* 333: 418–22.

Parry, M. L., Canziani, O. F., Palutikof, J. P., Van der Linden, P. J., and Hanson, C. E. (eds.) (2007). *Climate Change 2007: Impacts, Adaptation and Vulnerability. Contribution of Working Group II to the Fourth Assessment. Report of the Intergovernmental Panel on Climate Change.* Cambridge: Cambridge University Press.

Peel, D. A. (1992). 'Merely the tip of the ice core'. *Nature* 359: 274–5.

Peltier, W. R. (2002). 'On eustatic sea level history: Late Glacial Maximum to Holocene'. *Quaternary Science Reviews* 21: 377–96.

Peterson, T. C. and Vose, R. S. (1997). 'An overview of the global historical climatology network temperature database'. *Bulletin of the American Meteorological Society* 78: 2,937–49.

Pethick, J. (1993). 'Shoreline adjustments and coastal management: physical and biological processes under accelerated sea-level rise'. *Geographical Journal* 159: 162–8.

Petit, J. R., Jouzel, J., Raynaud, D., Barkov, N. I., Barnola, J.-M., Basile, I., Bender, M., Chappellaz, J., Davis, M., Delaygue, G., Delmotte, M., Kotlyakov, V. M., Legrand, M., Lipenkov, V. Y., Lorius, C., Pépin, L., Ritz, C., Saltzman, E., and Stievenard, M. (1999). 'Climate and atmospheric history of the past 420,000 years from the Vostok ice core, Antarctica'. *Nature* 399: 429–36.

Piatek, B. J. (1994). 'The Tomoka mound complex in northeast Florida'. *Southeastern Archaeology* 13: 109–17.

Pilarczyk, K., Van Overeem, J., and Bakker, W. (1986). 'Design of beach nourishment scheme'. *Proceedings of the International Conference on Coastal Engineering* 20: 1,456–70.

Pittman, C. and Waite, M. (2009). *Paving Paradise: Florida's Vanishing Wetlands and the Failure of No Net Loss.* Tallahassee, FL: University Press of Florida.

Pope, V. D., Gallani, M. L., Rowntree, P. R., and Stratton, R. A. (2000). 'The impact of new physical parametrizations in the Hadley Centre climate model: HadAM3'. *Climate Dynamics* 16: 123–46.

Post, K. (2000). 'Everzwijnjagers?'. *Cranium* 17: 13–14.

Postel, S. L., Daily, G. C., and Ehrlich, P. R. (1996). 'Human appropriation of renewable fresh water'. *Science* 271: 785–8.

Postgate J. N. (1980). 'Palm-trees, reeds and rushes in Iraq ancient and modern', in M. T. Barrelet (ed.), *L'archéologie de l'Iraq.* Paris: CNRS, 99–110.

Pournelle, J. R. (2003). *Marshland of Cities: Deltaic Landscapes and the Evolution of Early Mesopotamian Civilization.* PhD thesis, University of California, San Diego.

——, Hritz, C. A., and Smith, J. R. (2010). *High Risk: Deltaic Resilience and the Genesis of Mesopotamian Cities (Iraq). Final Report.* On-line at: sc.academia.edu.

Powell, M. A. (1985). 'Salt, seed and yields in Sumerian agriculture: a critique of the theory of progressive salinization'. *Zeitschrift der Assyrologie* 75: 7–38.

Powers, L. A., Johnson, T. C., Werne, J. P., Castañeda, I. S., Hopmans, E., Sinninghe Damsté, J. S., and Schouten, S. (2005). 'Large temperature variability in the southern African tropics since the Last Glacial Maximum'. *Geophysical Research Letters* 32: L08706.

Preusser, F. (2009). 'Chronology of the impact of Quaternary climate change on continental environments in the Arabian Peninsula'. *Comptes Rendus Geoscience* 341: 621–32.

Pryor, F. (ed.) (2001). *The Flag Fen Basin: Archaeology and Environment of a Fenland Landscape.* Swindon: English Heritage.

Pumpelly, R. (ed.) (1908). *Explorations in Turkestan.* Washington, DC: Carnegie Institution.

Purdy, B. A. (1991). *The Art and Archaeology of Florida's Wetlands.* Boca Raton, FL: CRC Press.

Purser, B. H., Al-Azzawi, M., Al-Hassani, H. H., Baltzer, F., Hassan, K. M., Orszag-Sperber, F., Plaziat, J.-C., Yacoub, S. Y., and Younis, W. R. (1982). 'Caractères et évolution du complexe deltaïque Tigre-Euphrate'. *Mémoires de la Société Géologique de France* 144: 207–16.

Rahman, A. F., Dragoni, D., and El-Masri, B. (2011). 'Response of the Sundarbans coastline to sea level rise and decreased sediment flow: a remote sensing assessment'. *Remote Sensing of Environment* 115: 3,121–8.

Rahmstorf, S. (2010). 'A new view on sea level rise'. *Nature Reports on Climate Change* 4: 44–5.

Ranger, N., Hallegatte, S., Bhattacharya, S., Bachu, M., Priya, S., Dhore, K., Rafique, F., Mathur, P., Naville, N., Henriet, F., Herweijer, C., Pohit, S., and Corfee-Morlot, J. (2011). 'An assessment of the potential impact of climate change on flood risk in Mumbai'. *Climatic Change* 104: 139–67.

Rao, D. G., Krishna, K. S., and Sar, D. (1997). 'Crustal evolution and sedimentation history of the Bay of Bengal since the Cretaceous'. *Journal of Geophysical Research* 102: 17,747–68.

Ravelo, A. C., Andreasen, D. H., Lyle, M., Olivarez Lyle, A., and Ware, M. W. (2004). 'Regional climate shifts caused by gradual cooling in the Pliocene epoch'. *Nature* 429: 263–7.

Ray, B. (2010). *Climate Change: IPCC, Water Crisis, and Policy Riddles with Reference to India and her Surroundings.* Lanham: Lexington Books.

Redman, C. L. (2005). 'Resilience theory in archaeology'. *American Anthropologist* 107: 70–7.

Reid, C. (1913). *Submerged Forests.* Oxford: Cambridge Series of Manuals of Literature and Science.

Renewable UK (2010). *UKWED: Operational Wind Farms.* (http://www.bwea.com//.asp; accessed 3 December 2010).

Reuswigg, F. (2007). Review of: B. H. Walker, J. M. Andries, A. P. Kinzig, and P. Ryan (eds.), 'Exploring resilience in social-ecological systems: comparative studies and theory development'. *Regional Environmental Change* 7: 121–2.

Richards, J. F. and Flint, E. P. (1990). 'Long-term transformations in the Sundarbans wetlands forests of Bengal'. *Agriculture and Human Values* 7: 17–33.

Richardson, C. J. and Hussain, N. A. (2006). 'Restoring the Garden of Eden: an ecological assessment of the marshes of Iraq.' *BioScience* 56: 477–89.

Richerson, P. J., Boyd, R., and Bettinger, R. L. (2001). 'Was agriculture impossible during the Pleistocene but mandatory during the Holocene? A climate change hypothesis'. *American Antiquity* 66: 387–411.

Rick, T. C., Erlandson, J. M., and Vellanoweth, R. L. (2001). 'Paleocoastal marine fishing on the pacific coast of the Americas: perspectives from Daisy Cave, California'. *American Antiquity* 66: 595–613.

Rijnsdorp, A. D., Peck, M. A., Engelhard, G. H., Möllmann, C., and Pinnegar, J. K. (2009). 'Resolving the effect of climate change on fish populations'. *ICES Journal of Marine Science* 66: 1570–83.

Rippon, S. (2000). *The Transformation of Coastal Wetlands: Exploitation and Management of Marshland Landscapes in North West Europe during the Roman and Medieval Periods*. London: British Academy.

Robb, J. E. (1998). 'The archaeology of symbols'. *Annual Review of Anthropology* 27: 329–46.

Robinson, R. A., Crick, H. Q. P., Learmonth, J. A., Maclean, I. M. D., Thomas, C. D., Bairlein, F., Forchhammer, M. C., Francis, C. M., Gill, J. A., Godley, B. J., Harwood, J., Hays, G. C., Huntley, B., Hutson, A. M., Pierce, G. J., Rehfisch, M. M., Santos, M. B., Sims, D. W., Sparks, T. H., Stroud, D. A., and Visser, M. E. (2009). 'Travelling through a warming world: climate change and migratory species'. *Endangered Species Research* 7: 87–99.

Rosenzweig, C., Casassa, G., Imeson, A., Karoly, D. J., Liu Chunzhen, Menzel, A., Rawlins, S., Root T. L., Seguin, B., and Tryjanowski, P. (2007). 'Assessment of observed changes and responses in natural and managed systems', in M. L. Parry, O. F. Canziani, J. P. Palutikof, P. J. Van der Linden, and C. E. Hanson (eds.) (2007), *Climate Change 2007: Impacts, Adaptation and Vulnerability. Contribution of Working Group II to the Fourth Assessment. Report of the Intergovernmental Panel on Climate Change*. Cambridge: Cambridge University Press, 79–131.

Rowley-Conwy, P. (2007). *From Genesis to Prehistory: The Archaeological Three Age System and its Contested Reception in Denmark, Britain and Ireland*. Oxford: Oxford University Press.

Ruddiman, W. F. (2003). 'The anthropogenic greenhouse era began thousands of years ago'. *Climatic Change* 61: 261–93.

—— and Ellis, E. C. (2009). 'Effect of per-capita land use changes on Holocene forest clearance and CO_2 emissions'. *Quaternary Science Reviews* 28: 3,011–15.

Russo, M. (1991). *Archaic Sedentism on the Florida Coast: A Case Study from Horr's Island*. PhD theses, University of Florida, Gainesville.

—— (1994). 'Why we don't believe in Archaic ceremonial mounds and why we should: the case from Florida'. *Southeastern Archaeology* 13: 93–108.

—— (1998). 'Measuring sedentism with fauna: archaic cultures along the southwest Florida coast', in T. R. Rocek and O. Bar-Yosef (eds.), *Seasonality and Sedentism: Archaeological Perspectives from Old and New World Sites*. Cambridge, MA: Peabody Museum, 143–64.

SAARC Meteorological Research Center (2003). *The Vulnerability Assessment of the SAARC Coastal Region due to Sea Level Rise: Bangladesh Case Study*. Dhaka: SAARC.

Sabloff, J. A. (2008). *Archaeology Matters: Action Archaeology in the Modern World*. Walnut Creek, CA: Left Coast Press.

Safar, F., Mustafa, M. A., and Lloyd, S. (1981). *Eridu*. Baghdad: Ministry of Culture and Information, State Organization of Antiquities and Heritage.

Sahlins, M. D. (1968). *Tribesmen*. Englewood Cliffs, NJ: Prentice-Hall.

Salim, S. M. (1962). *Marsh Dwellers of the Euphrates Delta*. London: Athlone.

Sanlaville, P. (2002). 'The deltaic complex of the Lower Mesopotamian Plain and its evolution through the millennia', in E. Nicholson and P. Clark (eds.) (2002), *The*

Iraqi Marshlands: A Human and Environmental Study. London: Politico Publishing, 133–50.

Sarkar, R. M. (2009). *Sundarban Fishermen in the World Heritage Setting: A Community Striving in the Mystic Mangrove Ecosystem*. New Delhi: Serials.

Sarntheim, M. (1972). 'Sediments and history of the postglacial transgression in the Persian Gulf and the northwest Gulf of Oman'. *Marine Geology* 12: 245–66.

Sawyer, R. K. (2005). *Social Emergence: Societies as Complex Systems*. Cambridge: Cambridge University Press.

Schimper, F. (1837). 'Auszug aus dem briefe des herrn Dr Schimper ueber die eiszeit, an Pr. Agassiz, President der Gesellschaft'. *Actes de la Société helvétique des sciences naturelles, réunie à Neuchâtel* 38: 5.

Schwartz, M. L. (1976). 'The Bruun theory of sea-level rise as a cause of shore erosion'. *Journal of Geology* 75: 76–92.

Service, E. R. (1962). *Primitive Social Organization*. New York: Random House.

—— (1975). *Origins of the State and Civilization*. New York: Norton.

Severinghaus, J. P. and Brook, E. J. (1999). 'Abrupt climate change at the end of the Last Glacial inferred from trapped air in polar ice'. *Science* 286: 930–4.

Shackleton, N. J. and Opdyke, N. D. (1973). 'Oxygen isotope and paleomagnetic stratigraphy of equatorial Pacific Core V28–238: oxygen isotope temperatures and ice volume on a 10^5 year and 10^6 year scale'. *Quaternary Research* 3: 39–55.

Shennan, I. and Andrews, J. E. (eds.) (2000). *Holocene Land-Ocean Interaction and Environmental Change around the Western North Sea*. London: Geological Society (Special Publication 166).

—— and Horton, B. (2002). 'Holocene land- and sea-level changes in Great Britain'. *Journal of Quaternary Science* 17: 511–52.

——, Lambeck, K., Horton, B. P., Innes, J., Lloyd, J., McArthur, J., and Rutherford, M. (2000). 'Holocene isostacy and relative sea-level on the east coast of England', in I. Shennan and J. E. and Andrews (eds.) (2000), *Holocene Land-Ocean Interaction and Environmental Change around the Western North Sea*. London: Geological Society (Special Publication 166), 275–98.

Shennan, S. J. (2002). *Genes, Memes and Human History: Darwinian Archaeology and Cultural Evolution*, London: Thames and Hudson.

Sherratt, A. (1980). 'Water, soil and seasonality in early cereal cultivation'. *World Archaeology* 11: 313–30.

Siegenthaler, U., Stocker, T. F., Monnin, E., Lüth, D., Schwander, J., Stauffer, B., Raynaud, D., Bernola, J.-M., Fischer, H., Masson-Delmotte, V., and Jouzel, J. (2005). 'Stable carbon cycle: climate relationship during the Late Pleistocene'. *Science* 310: 1,313–17.

Simenstad, C., Reed, D., and Ford, M. (2006). 'When is restoration not? Incorporating landscape-scale processes to restore self-sustaining ecosystems in coastal wetland restoration'. *Ecological Engineering* 26: 27–39.

Smith, D. L. and Lord, K. M. (1997). 'Tectonic evolution and geophysics of the Florida Basement', in A. F. Randazzo and D. S. Jones (eds.), *The Geology of Florida*. Gainesville, FL: University Press of Florida, 13–26.

Snoussi, M., Ouchani, T., Khouakhi, A., and Niang-Diop, I. (2009). 'Impacts of sea-level rise on the Moroccan coastal zone: quantifying coastal erosion and flooding in the Tangier Bay'. *Geomorphology* 107: 32–40.

Solomon, S., Qin, D., Manning, M., Chen, Z., Marquis, M., Averyt, K. B., Tignor, M., and Miller, H. T. (eds.) (2007). *Climate Change 2007: The Physical Science Base. Contribution of Working Group 1 to the Fourth Assessment Report of the Intergovernmental Panel on Climate Change.* Cambridge: Cambridge University Press.

Stanley, D. J. and Hait, A. H. (2000). 'Holocene depositional patterns, neotectonics and Sundarban mangroves in the western Ganges–Brahmaputra Delta'. *Journal of Coastal Research* 16: 26–39.

Stapor, F. W., Mathews, T. D., and Lindfors-Kearns, F. E. (1991). 'Barrier island progradation and Holocene sea-level history in southeast Florida'. *Journal of Coastal Research* 7: 815–38.

—— and Tanner, W. F. (1977). 'Late Holocene mean sea level data from St. Vincent Island and the shape of the Late Holocene mean sea level curve', in W. F. Tanner (ed.) *Coastal Sedimentology.* Tallahassee, FL: Florida State University, 35–68.

Steffen, H. and Wu, P. (2011). 'Glacial isostatic adjustment in Fennoscandia: a review of data and modeling'. *Journal of Geodynamics* 52: 169–204.

Stern, N. (2006). *Stern Review on the Economics of Climate Change.* London: HM Treasury and Cambridge: Cambridge University Press.

Sterr, H. (2008). 'Assessment of vulnerability and adaptation to sea-level rise for the coastal zone of Germany'. *Journal of Coastal Research* 24: 380–93.

Stevenson, J. C. and Kearney, M. S. (2009). 'Impacts of global climate change and sea-level rise on tidal wetlands', in B. R. Silliman, E. Grosholz, and M. D. Bertness (eds.), *Human Impacts on Salt Marshes: A Global Perspective.* Berkeley, CA: University of California Press, 171–206.

Steward, J. H. (1937). 'Ecological aspects of southwestern society'. *Anthropos* 32: 87–107.

—— and Setzler, F. M. (1938). 'Function and configuration in archaeology'. *American Antiquity* 4: 4–10.

Stige, L. C., Ottersen, G., Dalpadado, P., Chan, K.-S., Hjermann, D. Ø., Lajus, D. L., Yaragina, N. A., and Stenseth, N. C. (2010). 'Direct and indirect climate forcing in a multi-species marine system'. *Proceedings of the Royal Society B* 277: 3,411–20.

Street-Perrott, F. A., Huang, Y., Perrott, R. A., Eglington, G., Barker, P., Khelifa, L. B., Harkness, D. D., and Olago, D. O. (1997). 'Impact of lower atmospheric carbon dioxide on tropical mountain ecosystems'. *Science* 278: 1,422–6.

Stuiver, M. and Reimer, P. J. (1986). 'A computer program for radiocarbon age calculations'. *Radiocarbon* 28: 1,022–30.

Swetnam, T. W., Allen, C. D., and Betancourt, J. L. (1999). 'Applied historical ecology: using the past to manage for the future'. *Ecological Applications* 9: 1,189–206.

Tainter, J. A. (1988). *The Collapse of Complex Societies.* Cambridge: Cambridge University Press.

Tanner, W. F. (1992). 'An 8000-year record of sea-level change for grain-size parameters: data from beach ridges in Denmark'. *Holocene* 3: 220–31.

Tanzeema, S. and Faisal, I. M. (2001). 'Sharing the Ganges: a critical analysis of the water-sharing treaties'. *Water Policy* 3: 13–28.

Taylor, J. A., Murdock, A. P., and Pontee, N. I. (2004). 'A macroscale analysis of coastal steepening around the coast of England and Wales'. *Geographical Journal* 170: 179–88.

Taylor, J. E. (1855). 'Notes on Abu Shahrein and Tell el Lahm'. *Journal of the Royal Asiatic Society* 15: 404–15.

Teasdale, P. A., Collins, P. E. F., Firth, C. R., and Cundy, A. B. (2011). 'Recent estuarine sedimentation rates from shallow inter-tidal environments in western Scotland: implications for future sea-level trends and coastal wetland development'. *Quaternary Science Reviews* 30: 109–29.

Teller, J., Glennie, K. W., Lancaster, N., and Singhvi, A. K. (2000). 'Calcareous dunes of the United Arab Emirates and Noah's flood; the postglacial reflooding of the Persian (Arabian) Gulg'. *Quaternary International* 68–71: 297–308.

Thesiger, W. (1964). *The Marsh Arabs*. London: Longman, Green & Co.; New York: Dutton.

Thomas, J. (1991). *Rethinking the Neolithic*. Cambridge: Cambridge University Press.

Thompson, R. S. and Fleming, R. F. (1996). 'Middle Pliocene vegetation: reconstructions, palaeoclimatic inferences, and boundary conditions for climate modeling'. *Marine Micropalaeontology* 27: 27–49.

Thompson, V. D. and Pluckham, T. J. (2010). 'History, complex hunter-gatherers, and the mounds and monuments of Crystal River, Florida, USA; a geophysical perspective'. *Journal of Island and Coastal Archaeology* 5: 33–51.

—— and Turck, J. A. (2009). 'Adaptive cycles of coastal hunter-gatherers'. *American Antiquity* 74: 255–78.

—— and Worth, J. (2011). 'Dwellers by the sea: native American adaptations along the southern coasts of eastern North America'. *Journal of Archaeological Research* 19: 51–101.

Tilley, C. (1994). *A Phenomenology of Landscape*. Oxford: Berg.

Tol, R. S. J. and Langen, A. (2000). 'A concise history of Dutch river floods'. *Climate Change* 46: 357–69.

Törnqvist, T. E., González, J. L., Newsom, L. E., Van der Borg, K., De Jong, A. F. M., and Kumik, C. W. (2004). 'Deciphering Holocene sea-level history on the U.S. Gulf Coast: a high-resolution record from the Mississippi Delta'. *Geological Society of America Bulletin* 116: 1,026–39.

Townend, I. and Pethick, J. (2002). 'Estaurine flooding and managed retreat'. *Philosophical Transactions of the Royal Society A* 360: 1477–95.

Trigger, B. G. (2006). *A History of Archaeological Thought*, 2nd edn. Cambridge: Cambridge University Press.

Tunnicliffe, V., O'Connell, J. M., and McQuoid, M. R. (2001). 'A Holocene record of marine fish remains from the Northeastern Pacific'. *Marine Geology* 174: 197–210.

Turner, R. K., Burgess, D., Hadley, D., Coombes, E., and Jackson, N. (2007). 'A cost–benefit appraisal of coastal managed realignment policy'. *Global Environmental Change* 17: 397–407.

Turney, C. S. M., Baillie, M., Palmer, J., and Brown, D. (2006). 'Holocene climatic change and past Irish societal response'. *Journal of Archaeological Science* 33: 34–8.

Tweed, C. and Sutherland, M. (2007). 'Built cultural heritage and sustainable urban development'. *Landscape and Urban Planning* 83: 62–9.

UNEP (2003). *Environments in Iraq: UNEP Progress Report*. Geneva: United Nations Environment Programme.

—— (2009). *Support for Environmental Management of the Iraqi Marshlands 2004–2009*. Geneva: United Nations Environment Programme.

Valiela, I., Bowen, J. L., and York, J. K. (2001). 'Mangrove forests: one of the world's threatened major tropical environments'. *BioScience* 51: 807–15.

Van de Noort, R. (2004). *The Humber Wetlands: The Archaeology of a Dynamic Landscape.* Bollington: Windgather Press.

—— (2011a). 'Conceptualising climate change archaeology'. *Antiquity* 85: 1,039–48.

—— (2011b). *North Sea Archaeologies: A Maritime Biography 10,000 BC–AD 1500.* Oxford: Oxford University Press.

—— and Davies, P. (1993). *Wetland Heritage: An Archaeological Assessment of the Humber Wetlands.* Hull: University of Hull.

—— and Ellis, S. (eds.) (1995). *Wetland Heritage of Holderness: An Archaeological Survey.* Hull: University of Hull.

——, —— (eds.) (1997). *Wetland Heritage of the Humberhead Levels: An Archaeological Survey.* Hull: University of Hull.

——, —— (eds.) (1998). *Wetland Heritage of the Ancholme and Lower Trent Valleys: An Archaeological Survey.* Hull: University of Hull.

——, —— (eds.) (1999). *Wetland Heritage of the Vale of York: An Archaeological Survey.* Hull: University of Hull.

——, —— (eds.) (2000). *Wetland Heritage of the Hull Valley: An Archaeological Survey.* Hull: University of Hull.

Van den Dool, H. M., Krijnen, H. J., and Schuurmans, C. J. E. (1978). 'Average winter temperatures at de Bilt (the Netherlands): 1634–1977'. *Climate Change* 1: 319–30.

Van der Molen, J. and Van Dijck, B. (2000). 'The evolution of the Dutch and Belgian coasts and the role of sand supply from the North Sea'. *Global and Planetary Change* 27: 223–44.

Van der Ven, G. P. (2003). *Geschiedenis van de Waterbeheersing en Landaanwinning in Nederland,* 5th edn. Utrecht: Matrijs.

Van der Vleuten, E. and Disco, C. (2004). 'Water wizards: reshaping wet nature and society'. *History and Technology* 20: 291–309.

Van Geel, B., Buurman, J., and Waterbolk, H. T. (1996). 'Archaeological and palaeoecological indications of an abrupt climate change in The Netherlands, and evidence for climatological teleconnections around 2650 BP'. *Journal of Quaternary Science* 11: 451–60.

Van Kolfschoten, T. and Van Essen, H. (2004). 'Palaeozoological heritage from the bottom of the North Sea', in N. C. Flemming (ed.) (2004), *Submarine Prehistoric Archaeology of the North Sea: Research Priorities and Collaboration with Industry.* York: CBA (Research Report 141), 70–80.

Van Koningsveld, M., Muler, J. P. M., Stive, M. J. F., Van der Valk, L., and Van der Weck, A. W. (2008). 'Living with sea-level rise and climate change: a case study of the Netherlands'. *Journal of Coastal Research* 24: 367–79.

Van Wijk, M. N., Sanders, M. E., De Jong, J. J., and Van Veen, M. P. (2006). *Natuurbeheer in de Duinen; Achtergronden bij Natuurbalans 2005.* Wageningen: MNP/Expertise Groep Wageningen.

Van Zeist, W., Van Hoorn, T. C., Bottema, S., and Woldring, H. (1976). 'An agricultural experiment in the unprotected saltmarsh'. *Palaeohistoria* 18: 111–53.

Veenstra, H. J. (1965). 'Geology of the Dogger Bank area, North Sea'. *Marine Geology* 3: 245–62.

Verhart, L. B. M. (1995). 'Fishing for the Mesolithic; the North Sea: a submerged Mesolithic landscape', in Fisher (ed.) (1995), *Man and Sea in the Mesolithic: Coastal Settlement above and below Present Sea-Level.* Oxford: Oxbow, 291–302.

Vishnu, M. (1989). 'Forty years of archaeobotanical research in South Asia'. *Man and Environment* 14: 1–16.

Vita-Finzi, C. and Higgs, E. S. (1970). 'Prehistoric economy in the Mount Carmel area of Palestine: Site Catchment Analysis'. *Proceedings of the Prehistoric Society* 36: 1–37.

von Post, L. (1916). 'Om skogsträdpollen i sydsvenska torfmosselagerföljder', *Geologiska föreningens i Stockholm Förhandlingar* 38: 384–90.

—— (1918). 'Skogsträdpollen i sydsvenska torfmosselagerföljder'. *Förhandlingar ved de Skandinaviske Naturforskeres ottende Möde i Kjöbenhavn* 16: 432–65.

—— (1946). 'The prospect for pollen analysis in the study of the earth's climatic history'. *New Phytologist* 45: 193–217.

Vos, P. C. (2006). 'Toelichting bij de nieuwe paleogeografische kaarten van Nederland'. *Nationale Onderzoeksagenda Archeology.* Amersfoort: RACM.

—— and Van Kesteren, W. (2000). 'The long-term evolution of intertidal mudflats in the northern Netherlands during the Holocene: natural and anthropogenic processes'. *Continental Shelf Research* 20: 1,687–710.

Walcott, R. I. (1972). 'Late Quaternary vertical movements in eastern North America: quantitative evidence of glacio-isostatic rebound'. *Review of Geophysics* 10: 849–84.

Walkden, M. and Dickson, M. (2008). 'Equilibrium erosion of soft rock shores with a shallow or absent beach under increased sea-level rise'. *Marine Geology* 251: 75–84.

Wallis, I. and Beale, I. (1669). 'Some observations concerning the baroscope and thermoscope, made and communicated by Doctor I. Wallis at Oxford, and Dr. I. Beale at Yeovil in Somerset, deliver'd here according to the several dates, when they were imparted. Dr. Beale in those letters of his dated Decemb.18. Decemb. 29. 1669. and Januar. 3. 1670'. *Philosophical Transactions* 4: 1,113–20.

Wang, H., Yang, Z., Saito, Y., Paul Liu, J., Sun, X., and Wang, Y. (2007). 'Stepwise decreases of the Huanghe (Yellow River) sediment load (1950–2005): Impacts of climate change and human activities'. *Global and Planetary Change* 57: 331–54.

Wang, N., Yao, T., Thompson, L., Henderson, K., and Davis, M. (2002). 'Evidence for cold events in the early Holocene from the Guliya ice core, Tibetan Plateau, China'. *Chinese Science Bulletin* 47: 1,422–7.

Wang, Y. and Frei, B. (2011). 'Stressed food—the impact of abiotic environmental stresses on crop quality'. *Agriculture, ecosystems and environment* 141: 271–86.

Warner, B. G., Douabul, A., and Abaychi, J. (2011). 'Restoration of the marshes of southern Iraq: prospects and challenges', in S. A. A. Kubba (ed.) (2011), *The Iraqi Marshlands and the Marsh Arabs: The Ma'dan, their Culture and the Environment.* Reading: Ithaca Press, 210–29.

Wassmann, R., Hien, N. X., Hoanh, C. T., and Tuong, T. P. (2004). 'Sea level rise affecting the Vietnamese Mekong Delta: water elevation in the flood season and implications for rice production'. *Climatic Change* 66: 89–107.

——, Jagadish, S. V. K., Heuer, S., Ismail, A., Redona, E., Serraj, R., Singh, R. K., Howell, G., Pathak, H., and Sumfleth, K. (2009). 'Climate change affecting rice production: the physiological and agronomic basis for possible adaptation strategies'. *Advances in Agronomy* 101: 59–122.

Webb, T. and Clark, D. R. (1977). 'Calibrating micropaleontological data in climatic terms: a critical review'. *Annals of the New York Academy of Sciences* 288: 93–118.

Weerts, H. J. T., Westerhoff, W. E., Cleveringa, P., Bierkens, M. F. P., Veldkamp, J. G., and Rijsdijk, K. F. (2005). 'Quaternary geological mapping of the lowlands of The Netherlands, a 21st-century perspective'. *Quaternary International* 133–4: 159–78.

Weisman, B. R. (1995). *The Roberts Island Shell Mound Complex and its archaeological significance.* Tallahassee: CARL Archaeological Survey Report.

—— (2003). 'Why Florida archaeology matters'. *Southeastern Archaeology* 22: 210–26.

Weiss, J. L., Overpeck, J. T., and Strauss, B. (2011). 'Implications of recent sea- level rise science for low-elevation areas in coastal cities of the conterminous U.S.A.'. *Climate Change* 105: 635–45.

Weninger, B., Schulting, R., Bradtmöller, M., Clare, L., Collard, M., Edinborough, K., Hilpert, J., Jöris, O., Niekus, M., Rohling, E. J., and Wagner, B. (2008). 'The catastrophic final flooding of Doggerland by the Storegga Slide tsunami'. *Documenta Praehistorica* 35: 1–24.

Wernberg, T., Thomsen, M. S., Tuya, F., Kendrick, G. A., Staehr, P. A., and Toohey, B. D. (2010). 'Decreasing resilience of kelp beds along latitudinal temperature gradient: potential implications for a warmer future'. *Ecology Letters* 13: 685–94.

Wernstedt, F. L. (1972). *World Climatic Data.* Lemont: Climatic Data Press.

West, R. G. (1970). 'Pollen zones in the Pleistocene of Great Britain and their correlation'. *New Phytologist* 69: 1,179–83.

Widmer, R. J. (1988). *The Evolution of the Calusa: A Nonagricultural Chiefdom on the Southwest Florida Coast.* Tuscaloosa, AL: University of Alabama Press.

Wilcox, M. (2010). 'Marketing conquest and the vanishing Indian: an indigenous response to Jared Diamond's archaeology of the American Southwest', in P. A. McAnany and N. Yoffee (eds.) (2010), *Questioning Collapse: Human Resilience, Ecological Vulnerability and the Aftermath of Empire.* Cambridge: Cambridge University Press, 113–41.

Willcocks, W. (1930). *Ancient Systems of Irrigation in Bengal and its Application to Modern Problems.* Delhi: B. R. Publishing.

Willett, M. A. (2005). *Effect of Dissolution of the Florida Carbonate Plateau on Isostatic Uplift and Relative Sea-Level Change.* MSc thesis, University of South Florida, Tampa.

Willey, G. R. (1953). *Prehistoric Settlement Patterns in the Virú Valley, Peru.* Washington, DC: Bureau of American Ethnology Bulletin 155.

Wolsink, M. (2007). 'Wind power implementation; the nature of public attitudes: equity and fairness instead of "backyard motives"'. *Renewable and Sustainable Energy Reviews* 11: 1,188–207.

Woolley, L. (1938). *Ur of the Chaldees.* London: Herbert Press.

Wright, E. E., Hine, A. C., Goodbred, S. L., and Locker, S. D. (2005). 'The effect of sea-level and climate change on the development of a mixed siliciclastic carbonate, deltaic coastline: Suwannee River, Florida, USA'. *Journal of Sedimentary Research* 75: 621–35.

WWF-India (2010). *Sundarbans: Future Imperfect: Climate Adaptation Report.* New Delhi: WWF-India.

—— (2011). *Indian Sundarbans Delta: A Vision*. New Delhi: WWF-India.

Yizhaq, H., Ashkenazy, Y., and Tsoar, H. (2009). 'Sand dune dynamics and climate change: a modeling approach'. *Journal of Geophysical Research* 114: F01023.

Yoffee, N. (2010). 'Collapse in ancient Mesopotamia: what happened, what didn't', in P. A. McAnany and N. Yoffee (eds.) (2010), *Questioning Collapse: Human Resilience, Ecological Vulnerability and the Aftermath of Empire*. Cambridge: Cambridge University Press, 176–203.

Yu, S., Zhu, C., Song, J., and Qu, W. (2000). 'Role of climate in the rise and fall of Neolithic cultures on the Yangtze Delta'. *Boreas* 29: 157–65.

Zhang, Z., Yao, Q., Li, J., Nie, L., and Yuan, Y. (2009). 'The damage and mechanism of sediment by dredging from irrigation districts to environment in the Lower Yellow River'. *Advances in Water Resources and Hydraulic Engineering*. Proceedings of the 16th IAHR-APD Congress and 3rd Symposium of IAHR-ISHS: 1,084–9.

Zwart, H. (2003). 'Aquaphobia, tulipmania, biophilia: a moral geography of the Dutch landscape'. *Environmental Values* 12: 107–28.

Index

Italic page numbers refer to illustrations